Probability and Mathematical Statistics (Continued)

SEN • Sequential Nonparametrics: Invariance Principles and Statistical Inference
SERFLING • Approximation Theorems of Mathematical Statistics
TJUR • Probability Based on Radon Measures
WILLIAMS • Diffusions, Markov Processes, and Martingales, Volume I: Foundations
ZACKS • Theory of Statistical Inference

Applied Probability and Statistics

ANDERSON, AUQUIER, HAUCK, OAKES, VANDAELE, and WEISBERG • Statistical Methods for Comparative Studies
ARTHANARI and DODGE • Mathematical Programming in Statistics
BAILEY • The Elements of Stochastic Processes with Applications to the Natural Sciences
BAILEY • Mathematics, Statistics and Systems for Health
BARNETT • Interpreting Multivariate Data
BARNETT and LEWIS • Outliers in Statistical Data
BARTHOLOMEW • Stochastic Models for Social Processes, *Third Edition*
BARTHOLOMEW and FORBES • Statistical Techniques for Manpower Planning
BECK and ARNOLD • Parameter Estimation in Engineering and Science
BELSLEY, KUH, and WELSCH • Regression Diagnostics: Identifying Influential Data and Sources of Collinearity
BENNETT and FRANKLIN • Statistical Analysis in Chemistry and the Chemical Industry
BHAT • Elements of Applied Stochastic Processes
BLOOMFIELD • Fourier Analysis of Time Series: An Introduction
BOX • R. A. Fisher, The Life of a Scientist
BOX and DRAPER • Evolutionary Operation: A Statistical Method for Process Improvement
BOX, HUNTER, and HUNTER • Statistics for Experimenters: An Introduction to Design, Data Analysis, and Model Building
BROWN and HOLLANDER • Statistics: A Biomedical Introduction
BROWNLEE • Statistical Theory and Methodology in Science and Engineering, *Second Edition*
BURY • Statistical Models in Applied Science
CHAMBERS • Computational Methods for Data Analysis
CHATTERJEE and PRICE • Regression Analysis by Example
CHOW • Analysis and Control of Dynamic Economic Systems
CHOW • Econometric Analysis by Control Methods
CLELLAND, BROWN, and deCANI • Basic Statistics with Business Applications, *Second Edition*
COCHRAN • Sampling Techniques, *Third Edition*
COCHRAN and COX • Experimental Designs, *Second Edition*
CONOVER • Practical Nonparametric Statistics, *Second Edition*
CONOVER and IMAN • Introduction to Modern Business Statistics
CORNELL • Experiments with Mixtures: Designs, Models and The Analysis of Mixture Data
COX • Planning of Experiments
DANIEL • Biostatistics: A Foundation for Analysis in the Health Sciences, *Third Edition*
DANIEL • Applications of Statistics to Industrial Experimentation
DANIEL and WOOD • Fitting Equations to Data: Computer Analysis of Multifactor Data, *Second Edition*
DAVID • Order Statistics, *Second Edition*
DAVISON • Multidimensional Scaling
DEMING • Sample Design in Business Research
DODGE and ROMIG • Sampling Inspection Tables, *Second Edition*
DOWDY and WEARDEN • Statistics for Research

continued on back

Comparison Methods for Queues and Other Stochastic Models

Comparison Methods for Queues and Other Stochastic Models

Dietrich Stoyan

Bergakademie Freiberg
German Democratic Republic

Edited with revisions by

Daryl J. Daley

Australian National University

JOHN WILEY & SONS

Chichester · New York · Brisbane · Toronto · Singapore

Library of Congress Cataloging in Publication Data:

Stoyan, Dietrich.
Comparison methods for queues and other stochastic models.
(Wiley series in probability and mathematical statistics. Applied section)
Bibliography: p.
Includes indexes.
1. Stochastic processes. 2. Queuing theory.
I. Daley, Daryl J. II. Title. III. Series.
QA 274. S 838 519.2 81-16365
ISBN 0-471-10122-2 AACR 2

British Library Cataloguing in Publication Data:

Stoyan, Dietrich
Comparison methods for queues and other stochastic models. — (Wiley series in probability and mathematical statistics — applied section)
1. Mathematical models
2. Stochastic processes
I. Title II. Daley, Daryl J.
III. Qualitative Eigenschaften und Abschätzungen stochastischer Modelle. *English*
519.2 QA 402
ISBN 0-471-10122-2

Printed in GDR

Foreword to the German Edition[1]

Most of the specific *stochastic models* in this monograph are the mathematical models of queueing, reliability, inventory and sequencing etc., in which random influences are considered. Many of these models are sufficiently complex that bounds and approximations for their characteristics are of practical importance. Qualitative properties of stochastic models constitute an important theoretical basis for approximation methods, for these characterize the influence of quantities describing the behaviour of the constituent elements or components of the system, on parameters describing the system as a whole.

In textbooks on stochastic models, it is mostly exact methods that are presented: bounds and approximations are often treated like a stepmother, although for applications they are of great importance and have always been studied (recall the work of such pioneers of queueing theory as ERLANG, CROMMELIN, MOLINA and POLLACZEK). The present book deals with two especially important types of qualitative properties and approximation methods based on them. These monotonicity and robustness (or, continuity) properties of stochastic models underlie problems of the type formulated in the introductory section of GNEDENKO and KOVALENKO's (1966) "Introduction to Queueing Theory". Our interest in models centres on queueing theory and, to a lesser extent, reliability; other stochastic models are considered only briefly. However, the intention is that the methods be presented in such a way that their application to other models should be clear.

Monotonicity properties, which form the basis of many bounds, are described in detail, both theoretically, including a description of methods for proving them, and for practical applications, including exemplification of methods for their use on particular models. Chapter 1 presents *partial orderings* of distributions and Chapter 2 exhibits methods for proving monotonicity properties. These are used to study problems of *experimental design* in Chapter 3, and to establish monotonicity and camparability properties of *random processes*, especially of Markov chains, in Chapter 4. Chapters 5 and 6 deal with the *queueing models* GI/GI/1 and GI/GI/s, while in Chapter 7 some other models are studied, in particular *reliability models*. Where possible, useful approximation formulae are presented for the models considered.

---

[1] German Edition: Qualitative Eigenschaften und Abschätzungen stochastischer Modelle. © Akademie-Verlag Berlin 1977.

Methods for the proof of *robustness properties* are described in Chapter 8. The examples studied come mainly from queueing theory, and illustrate specific difficulties. It should be noted that it is by proving robustness properties that we can give theoretical justification for the use of such approximation methods as the Erlang phase method or simulation.

This monograph deals with a relatively young field of research that is still evolving. Consequently, it contains both new and recent results, some of which may be superseded shortly.

At the end of each chapter there are notes and bibliographic comments. There are stated some (open) problems whose solution may help to advance the theory. Their level is not uniform: some are so simple that their solution may well be known, while others may be too complicated to be solved in the form as given. I should be glad if their presentation stimulates further research.[1]

It is envisaged that the book may cater for a variety of readers: it should be of use to theoreticians studying qualitative properties and approximations for stochastic models and random processes, to operations research workers determining bounds or approximations for special models, and also to readers simply looking for formulae for specific stochastic models . . . Their application may often be much more efficient than simulation techniques when it is sought to compare some variants and no great accuracy is needed.

I wish to thank Professor D. König for many valuable suggestions, for reading the manuscript critically, and for his generous support of my work. I am also indebted to Professor P. Franken and Professor H.-J. Rossberg for many useful hints and suggestions for my work and the manuscript of this book; they were referees of my Ph. D. dissertation (Dissertation B, Bergakademie Freiberg, GDR, 1975) which formed the basis of this monograph. Furthermore, Dr. R. Bergmann, Dr. B. M. Kirstein and Dr. W. Näther helped by reading parts of the manuscript. They and many other people willingly informed me about new and unpublished papers and moreover made important suggestions to me.

I am very much indebted to my wife for her untiring cooperation in producing the manuscript.

Finally I thank the Akademie-Verlag Berlin, and especially Miss Helle, for their cooperation concerning my wishes on the organisation of the book: it was thus possible to include at the proof reading stage recently published papers and thereby to extend in part the manuscript finished in September 1975.

Dietrich Stoyan

Freiberg, December 1976

---

[1] Some problems from the German edition have been solved; such problems are omitted in this revised edition and their solutions incorporated into the text.

Foreword to the English Edition

This monograph is the revised version of an English translation of a book initially published in German (1977) and Russian (1979). It presents unified methods for obtaining approximations and bounds for stochastic models, in particular for queueing models, though it is to be hoped that readers will find the methods of use for other models as well. While several important approximation methods are presented, coverage is not comprehensive, there being for example no discussion of diffusion approximations and heavy traffic methods.

This English edition differs from the original German in many ways. Of course, misprints and errors have been corrected. I am happy that I had Dr. D. J. DALEY as an editor, for he has worked through many of the proofs and derivations, and in several cases given much improved versions of them. Also the organisation of the book and its style have been much enhanced by his work. Clearly, I am indebted to Dr. V. V. KALASHNIKOV, too, for his suggestions made while editing the Russian edition also led to improvements in the text.

Preparation of the English translation has been a good opportunity to include new results and to redraft parts of the monograph. Thus, this English edition contains about 200 more references than the original in German. Chapter One has been redrafted and includes, for example, results by KAMAE, KRENGEL and O'BRIEN on partial orderings of distributions in Polish spaces. Elsewhere there are new bounds for GI/GI/1 and GI/GI/s queues, renewal processes, while there are new sections on queueing networks, branching processes and metrics for distributions. Chapter Eight has been much altered as a result of fruitful discussion with Dr. V. V. KALASHNIKOV and Dr. G. SH. TSITSIASHVILI who wrote an excellent appendix on robustness properties of queueing systems for the Russian edition.

It should also be remarked that many papers referenced in the text or bibliographic notes are included merely for bibliographic purposes.

The process of revising the text has convinced me that the field of "qualitative properties of stochastic models and their applications" is still flourishing, and may be expected to continue to do so, in spite of my having been pessimistic about it.

I am indebted to many people for information on recent research so that this edition does contain references to papers in press at the time of the manuscript being sent to the publisher.

I thank Professor DAVID KENDALL very much for supporting the publication of this book in the Wiley monograph series. I also take this opportunity of thanking Professor BORIS V. GNEDENKO for his early interest for my work, for it helped me to find the interest of a greater circle of mathematicians. Then I am grateful to Dr. HÖPPNER of Akademie-Verlag Berlin for a good co-operation and for giving us the opportunity to include some new material in the text finished in June 1981.

Last but not least I thank very much my wife HELGA, for her help in producing the initial rough English translation of the book and those who typed the final manuscript.

February, 1983

DIETRICH STOYAN

Translation Editor's Note

As indicated by Dr. Stoyan, he and his wife jointly produced a rough English version from which I worked, making sometimes no change, sometimes a less or more extensive paraphrase, and occasionally recasting the argument. This version has then been the basis of discussion by post with Dr. Stoyan. I must pay tribute both to his willingness to read and comment on this version, and to the openness of his reaction: such a spirit has made this joint venture a pleasure.

I thank Professor P. A. P. Moran for his support in the project — and for a vital conversation in 1966 that did much to kindle my interest in what led to my work on stochastic monotonicity. Beryl Cranston, Jan Williams and Sue Watson all shared in typing the final manuscript, and both Dietrich Stoyan and I are most grateful to them for their efforts in coping with the various modifications to Helga Stoyan's typescript. Finally, I thank Nola, John, Geoffrey and Alan for their support and tolerance.

Daryl J. Daley

Table of contents

1. Partial orderings for distributions and random variables

1.0 Introduction

The study of models in applied probability theory provided much of the motivation for the subject matter of this monograph. Some of the simpler models can be formulated in terms of one or two distributions, usually in a nonlinear fashion, so that the study of changes in the properties of the model as the constituent distributions vary need not be a simple matter. One type of question which can be asked is the extent to which the model may vary under constraints on the constituent distributions; more succintly, what are the extremes of the model as the components range over a specified class of distributions?

Another use of the methods below concerns the approximation of a model either by a simpler model or by a model with simpler constituent components than the given model. In stochastic analysis, limit theorems, bounds on approximations, and rate of convergence results, are all used to provide information on different facets of this last question.

The emphasis of this monograph is on bounds and extremes rather than limit results, and to answer such questions it is necessary to discuss inequalities. Accordingly, since at the simplest level the stochastic components of the models involve random variables (or, their distributions), we must start by reviewing tools for comparing these entities. This chapter is therefore devoted to partial orderings for distribution functions (equivalently, of random variables on a suitably defined probability space).

We start by studying briefly some general properties of partial orderings for distribution functions (d.f.s) of univariate random variables (r.v.s), and proceed to consider the four most important orderings, namely, stochastic ordering, convex ordering and its concave counterpart, and an ordering based on the Laplace-Stieltjes transform. Some criteria for comparisons both within parametric families and within special general classes of d.f.s are presented. Extremal d.f.s within certain classes with respect to a given ordering can sometimes be identified. Finally it is shown how it is possible to extend definitions of partial orderings to multidimensional r.v.s or distributions on more general spaces.

For readers interested in applications and formulae for various stochastic models, it may be sufficient to concentrate on Sections 1.2, 1.3 and 1.5 unless specific aspects of the chapter are required.

1.1 General properties of partial orderings of distribution functions

We start by recalling that a *partial ordering* $\prec$ on any given set $\mathscr{D}$ of elements $a, b, c, \ldots$ is a binary relation satisfying the three axioms

PO(i) $a \prec a$ (reflexivity),

PO(ii) if $a \prec b$ and $b \prec c$ then $a \prec c$ (transitivity),

PO(iii) if $a \prec b$ and $b \prec a$ then $a = b$ (anti-symmetry).

It is sometimes convenient to write $a \prec b$ in the equivalent form $b \succ a$.

Our main concern in this opening section is with desirable properties of partial orderings defined on the set $\mathscr{D}$ (or suitable subsets thereof) of all distribution functions (d.f.s) of real-valued random variables (r.v.s). Since the concepts concerning random variables are more intuitive than the purely analytic language of distributions, we introduce the

Convention. *If the r.v.s X, Y have d.f.s F, G for which $F \prec G$, then we write $X \prec Y$.*

However it must be noted here that, even assuming that X and Y have been defined on the same probability space, we can have the anti-symmetry property of PO(iii) holding for F and G without it necessarily being the case that X and Y are the same r.v.

Nevertheless, some desirable properties of any partial ordering $\prec$ on $\mathscr{D}$ come directly from considering the order properties of real numbers when expressed in the context of possibly degenerate random variables. To this end, we shall use the Dirac d.f., denoted by $\Theta_c(.)$, and defined for all real numbers by

$$\Theta_c(x) = \begin{cases} 0 & (x < c), \\ 1 & (x \geqq c). \end{cases}$$

Definition 1.1.1. *Given a partial ordering $\prec$ on (a subset of) the space $\mathscr{D}$ of d.f.s,*

(a) *it has property* (R) *if, for all real a and b for which $a \leqq b$, $\Theta_a \prec \Theta_b$;*

(b) *it has property* (E) *if, whenever $F \prec G$ and $m_F = \int_{-\infty}^{\infty} x \, \mathrm{d}F(x)$ and m_G are well-defined and finite, $m_F \leqq m_G$;*

(c) *it has property* (M) *if, whenever $F \prec G$, $F^c \prec G^c$ for all real $c > 0$, where the d.f. F^c is defined by $F^c(x) = F(x/c)$ (all x);*

(d) *it has property* (C) *if, whenever $F_1 \prec F_2$ and F_1, F_2, and $G \in \mathscr{D}$, the convolutions $F_i * G$ ($i = 1, 2$), defined by $(F_i * G)(x) = \int_{-\infty}^{\infty} F_i(x - y) \, \mathrm{d}G(y)$, are elements of $\mathscr{D}$ and $F_1 * G \prec F_2 * G$;*

(e) *it has property* (W) *if, whenever $\{F_n\}$ and $\{G_n\}$ are subsets of $\mathscr{D}$ for which F_n and G_n converge weakly to limits F and G also in $\mathscr{D}$, $F_n \prec G_n$ (all n) implies $F \prec G$.*

In terms of r.v.s X, Y, the scalar multiplication property (M) ensures that $X < Y$ is equivalent to $cX < cY$ (all $0 < c < \infty$), while the convolution property (C) is equivalent to having $X_1 < X_2$ imply $X_1 + Y < X_2 + Y$ whenever Y is independent of X_1 and X_2. The expectation property (E) ensures that $X < Y$ implies $\mathsf{E}X \leqq \mathsf{E}Y$. That this expectation property may be inherited from other properties, comes from

Proposition 1.1.1. *A partial ordering $<$ on (a subset of) $\mathcal{D}$ having the properties* (R), (M), (C) *and* (W), *also has the property* (E).

Proof. Let F and G be two elements of $\mathcal{D}$ having the means m_F and m_G and for which $F < G$. Let $\{X_n\}$ and $\{Y_n\}$ be two sequences of independent identically distributed (i.i.d.) r.v.s with d.f.s F and G respectively. From the weak law of large numbers the d.f.s F_n and G_n of the r.v.s $S_n = (X_1 + \cdots + X_n)/n$ and $T_n = (Y_1 + \cdots + Y_n)/n$ converge weakly to Θ_{m_F} and Θ_{m_G} respectively. From (C) and (M) it follows that $F_n < G_n$ for all n; from (W) it then follows that $\Theta_{m_F} < \Theta_{m_G}$, and finally (R) implies that $m_F \leqq m_G$.

It transpires that many partial orderings on $\mathcal{D}$ that are of interest, can be defined in terms of classes of functions as follows.

Definition 1.1.2. *For a given class $\mathcal{F}_<$ of real functions, the partial ordering $<$ on (a subset of) $\mathcal{D}$ is said to be* generated by $\mathcal{F}_<$ *if $F < G$ is equivalent to the inequality*

$$\int_{-\infty}^{\infty} f(x)\,\mathrm{d}F(x) \leqq \int_{-\infty}^{\infty} f(x)\,\mathrm{d}G(x)$$

holding for all f in $\mathcal{F}_<$ for which the integrals exist.

Indeed, for each of the orderings considered later in this chapter there is a corresponding class of functions $\mathcal{F}_<$. As may be anticipated, certain procerties of $\mathcal{F}_<$ are then equivalent to properties of $<$ as we now illustrate.

Definition 1.1.3. *A class $\mathcal{G}$ of real functions on the real line* $\mathbb{R}$ *(respectively, the positive half-line* $\mathbb{R}_+$) *is* invariant with respect to translations *if for all $a \in \mathbb{R}$ (all $a \in \mathbb{R}_+$), whenever $f \in \mathcal{G}$, $f_a \in \mathcal{G}$ also, where the function f_a is defined by $f_a(x) = f(x + a)$ (all $x \in \mathbb{R}$ (all $x \in \mathbb{R}_+$)).*

Proposition 1.1.2. *Let the partial ordering $<$ on $\mathcal{D}$ be generated by the class $\mathcal{F}_<$ which is invariant with respect to translations. Then $<$ has the property* (C).

Proof. For arbitrary F_1, F_2 and G in $\mathcal{D}$ for which the integrals below exist and $F_1 < F_2$,

$$\int_{-\infty}^{\infty} f(x)\,\mathrm{d}(F_1 * G)(x) = \int_{-\infty}^{\infty}\int_{-\infty}^{\infty} f(x + y)\,\mathrm{d}F_1(x)\,\mathrm{d}G(y)$$

$$\leqq \int_{-\infty}^{\infty}\int_{-\infty}^{\infty} f(x + y)\,\mathrm{d}F_2(x)\,\mathrm{d}G(y) = \int_{-\infty}^{\infty} f(x)\,\mathrm{d}(F_2 * G)(x).$$

Observe that if in this proof we associate r.v.s X_1, X_2 and Y with the d.f.s F_1, F_2 and G, this chain of relations can be expressed as

$$\begin{aligned}\mathsf{E}f(X_1 + Y) &= \mathsf{E}\big(\mathsf{E}(f_Y(X_1) \mid Y)\big)\\ &\leqq \mathsf{E}\big(\mathsf{E}(f_Y(X_2) \mid Y)\big) = \mathsf{E}f(X_2 + Y).\end{aligned}$$

In general, the class $\mathscr{R}_{<}$ of all functions f for which

$$\int_{-\infty}^{\infty} f(x)\,\mathrm{d}F(x) \leqq \int_{-\infty}^{\infty} f(x)\,\mathrm{d}G(x)$$

whenever $F, G \in \mathscr{D}$ and $F \prec G$, may be larger than a class $\mathscr{F}_{<}$ of functions generating $\prec$ in the sense of Definition 1.1.2; this class $\mathscr{R}_{<}$ may be called the class of $\prec$-monotone functions. For example, if $\mathscr{F}_{<}$ is a class of functions $f_{(\alpha)}$ with index $\alpha \in A \subseteqq \mathbb{R}$, and if $\mathscr{H}_{<}$ is the class of all functions of the form

$$f(t) = \int_{-\infty}^{\infty} f_{(\alpha)}(t)\,\mu_f(\mathrm{d}\alpha) + c_f$$

where μ_f is a measure on A and c_f is real we then have

Proposition 1.1.3. $\mathscr{H}_{<} \subseteqq \mathscr{R}_{<}$.

Proof. For $f \in \mathscr{H}_{<}$, $F_k \in \mathscr{D}$ $(k = 1, 2)$, and assuming the integrals below exist,

$$\int_{-\infty}^{\infty} f(t)\,\mathrm{d}F_k(t) = \int_{-\infty}^{\infty} \left[\int_{-\infty}^{\infty} f_{(\alpha)}(t)\,\mu_f(\mathrm{d}\alpha) + c_f\right] \mathrm{d}F_k(t)$$

$$= \int_{-\infty}^{\infty} \int_{-\infty}^{\infty} f_{(\alpha)}(t)\,\mu_f(\mathrm{d}\alpha)\,\mathrm{d}F_k(t) + c_f.$$

Then if $F_1 \prec F_2$, the non-negativity of the measure μ_f yields

$$\int_{-\infty}^{\infty} f(t)\,\mathrm{d}F_1(t) \leqq \int_{-\infty}^{\infty} f(t)\,\mathrm{d}F_2(t).$$

1.2 Stochastic ordering for d.f.s ($\leqq_d$)

Definition 1.2.1. *The* r.v. X *is* stochastically smaller (*or* smaller in distribution) *than the* r.v. Y, *equivalently, their respective d.f.s* F *and* G *satisfy*[1] $F \leqq_d G$, *if*

$$F(x) \geqq G(x) \qquad (\textit{all real } x), \tag{1.2.1}$$

equivalently,

$$X \leqq_d Y. \tag{1.2.1'}$$

It is obvious that $\leqq_d$ satisfies the partial order axioms; this can also be established via the sample path property of the following

Proposition 1.2.1. *If* $F_1 \leqq_d F_2$, *then there exist r.v.s* X_1 *and* X_2 *defined on the same probability space* $(\Omega, \mathscr{A}, \mathsf{P})$ *for which*

$$X_1(\omega) \leqq X_2(\omega) \qquad (\textit{all } \omega \in \Omega)$$

and $\mathsf{P}(\{\omega\colon X_k(\omega) \leqq x\}) = F_k(x) \qquad (k = 1, 2)$.

---

[1] The notation $\leqq_d$ in this edition replaces the less descriptive $\overset{(1)}{\leqq}$ of the German original and other earlier papers of the author and co-workers.

Proof. Let Ω be the unit interval [0, 1], $\mathscr{A}$ the σ-algebra of Borel subsets on [0, 1], and P Lebesgue measure on [0, 1]. Then set

$$X_k(\omega) = F_k^{-1}(\omega) \equiv \inf\{x \geqq 0 : F_k(x) \geqq \omega\},$$

from which it follows that

$$F_1^{-1}(\omega) \leqq F_2^{-1}(\omega) \quad \text{for} \quad 0 \leqq \omega \leqq 1, \quad \text{and}$$

$$\mathsf{P}(\{\omega : X_k(\omega) \leqq x\}) = \mathsf{P}(\{\omega : F_k^{-1}(\omega) \leqq x\}) = F_k(x).$$

From this proposition it follows that $\leqq_d$ has properties (R) and (M). To see that $\leqq_d$ has the expectation property (E), we can again use the proposition in writing

$$m_{F_1} = \mathsf{E}X_1 = \int_\Omega X_1(\omega)\, \mathsf{P}(\mathrm{d}\omega) \leqq \int_\Omega X_2(\omega)\, \mathsf{P}(\mathrm{d}\omega) = m_{F_2}.$$

Alternatively, since

$$m_{F_k} = \int_{-\infty}^{\infty} x \,\mathrm{d}F_k(x) = -\int_{-\infty}^{0} F_k(x)\,\mathrm{d}x + \int_0^\infty \big(1 - F_k(x)\big)\,\mathrm{d}x,$$

$m_{F_1} \leqq m_{F_2}$ follows from the definition of $F_1 \leqq_d F_2$ at (1.2.1), with equality of m_{F_k} here only if $F_1 = F_2$.

Since an equivalent form for (1.2.1) is

$$\int_{-\infty}^{\infty} \Theta_x(t)\,\mathrm{d}F_1(t) \leqq \int_{-\infty}^{\infty} \Theta_x(t)\,\mathrm{d}F_2(t) \qquad \text{(all } x\text{)}, \tag{1.2.2}$$

the relation $\leqq_d$ is seen to be generated in the sense of Definition **1.1.2** by the family $\mathscr{F}_d \equiv \{\Theta_x : x \in \mathbb{R}\}$. This family is clearly invariant with respect to translations, so $\leqq_d$ also has the property (C).

What is the class $\mathscr{R}_{\leqq_d}$ of $\leqq_d$-monotone functions? This question is answered by the next result which shows that

$$\mathscr{R}_{\leqq_d} = \mathfrak{R}_d(\mathbb{R}) \tag{1.2.3}$$

where the latter denotes the class of all nondecreasing real functions on $\mathbb{R}$. The proof also shows that the class $\mathscr{H}_{\leqq_d}$ of (linear) mixtures of $\mathscr{F}_d$ (cf. Proposition 1.1.3) coincides with $\mathscr{R}_{\leqq_d}$.

Theorem 1.2.2. *The inequality*

$$\int_{-\infty}^{\infty} f(t)\,\mathrm{d}F_1(t) \leqq \int_{-\infty}^{\infty} f(t)\,\mathrm{d}F_2(t) \tag{1.2.4}$$

holds for all functions $f \in \mathfrak{R}_d(\mathbb{R})$ for which the integrals are defined, if and only if $F_1 \leqq_d F_2$.

For given f, (1.2.4) holds for all F_1, F_2 with $F_1 \leqq_d F_2$ only if f is nondecreasing.

2*

Proof. $\Theta_x \in \mathfrak{K}_d(\mathbb{R})$ so if (1.2.4) holds over this class, (1.2.2) follows. Conversely, suppose first that $f(t) \downarrow 0$ $(t \to -\infty)$ and $\uparrow K < \infty$ $(t \to \infty)$. Then the integrals at (1.2.4) are well-defined,

$$f(t) = \int_{-\infty}^{t} \mathrm{d}f(x) = \int_{-\infty}^{\infty} \Theta_x(t)\, \mathrm{d}f(x) \tag{1.2.5}$$

and so

$$\int_{-\infty}^{\infty} f(t)\, \mathrm{d}F_k(t) = \int_{-\infty}^{\infty} \mathrm{d}F_k(t) \int_{-\infty}^{\infty} \Theta_x(t)\, \mathrm{d}f(x) = \int_{-\infty}^{\infty} \big(1 - F_k(x)\big)\, \mathrm{d}f(x). \tag{1.2.6}$$

The monotonic nature of f ensures that the increments $\mathrm{d}f$ are non-negative, and the sufficiency of the result for such f follows. Use of a truncation argument extends the result to unbounded non-negative f, and simple extension yields the validity of (1.2.4) for all $f \in \mathfrak{K}_d(\mathbb{R})$ when $F_1 \leqq_d F_2$.

The last part of the theorem is equivalent to the statement that if f is not non-decreasing everywhere, there exist F_1 and F_2 with $F_1 \leqq_d F_2$ for which (1.2.4) is not satisfied. This is readily realized by taking $F_1 = \Theta_x$ and $F_2 = \Theta_y$ for some $x < y$ for which $f(x) > f(y)$.

Corollary 1.2.2a. *For non-negative r.v.s X, Y with $X \leqq_d Y$,*

$$\mathsf{E}X^r \leqq \mathsf{E}Y^r \qquad (r \geqq 0), \tag{1.2.7a}$$

$$\mathsf{E}X^r \geqq \mathsf{E}Y^r \qquad (r < 0) \tag{1.2.7b}$$

whenever the expectations exist, and, provided again that the expectations are well defined,

$$\mathsf{E}X^r \leqq \mathsf{E}Y^r \qquad (r = 1, 3, 5, \ldots) \tag{1.2.7c}$$

without the non-negativity constraint.

Proposition 1.2.3. $\leqq_d$ *has the weak convergence property* (W).

Proof. Suppose $\{F_n\}$ and $\{G_n\}$ converge weakly to F and G and that $F_n \leqq_d G_n$. Then if y is a continuity point of both F and G, it follows that $F(y) \geqq G(y)$. Thus, $F(y) < G(y)$ is possible only if y is a discontinuity point of either F or G. Such discontinuity points are at most countable, so there exist continuity points $x_n(y)$ of F and G for which $x_n(y) \downarrow y$ $(n \to \infty)$. Consequently, appealing to the right-continuity property of d.f.s,

$$F(y) = \lim_{n\to\infty} F\big(x_n(y)\big) \leqq \lim_{n\to\infty} G\big(x_n(y)\big) = G(y),$$

whence a contradiction.

We conclude this section with four examples of classes of d.f.s together with criteria for the stochastic ordering of their elements.

Example 1.2(a). For *exponential distributions* with the parameters λ_1, λ_2, set

$$F_k(t) = 1 - \exp\,(-\lambda_k t) \equiv \mathrm{Exp}\,(\lambda_k)\,(t) \qquad (0 \leqq t < \infty;\ k = 1, 2)$$

for some finite $\lambda_k > 0$. Then

$$F_1 \leqq_d F_2 \text{ if and only if } \lambda_2 \leqq \lambda_1, \text{ i.e., } m_1 = \lambda_1^{-1} \leqq \lambda_2^{-1} = m_2. \tag{1.2.8}$$

Example 1.2(a′). Let the distribution F on $(0, \infty)$ have density F' and *failure rate* (also called *hazard rate*)

$$\lambda(t) \equiv F'(t)/\big(1 - F(t)\big) \qquad \big(0 \leqq t < \inf\{t : F(t) = 1\}\big),$$

so that, since $\lambda(t) = -(\mathrm{d}/\mathrm{d}t) \ln \big(1 - F(t)\big)$, $F(t) = 1 - \exp\left(-\int_0^t \lambda(u)\, \mathrm{d}u\right)$. Then

$$F \leqq_d \operatorname{Exp}(\lambda_0) \text{ provided that } \int_0^t \lambda(u)\, \mathrm{d}u \geqq \lambda_0 t \quad (\text{all } t \geqq 0). \tag{1.2.9}$$

Example 1.2(b). For *Erlang distributions* with the parameters λ_k, α_k $(k = 1, 2)$ defined for positive integers α_k and $\lambda_k > 0$ by

$$F_k(t) = \big((\alpha_k - 1)!\big)^{-1} \int_0^t \lambda_k^{\alpha_k} x^{\alpha_k - 1} \exp(-\lambda_k x)\, \mathrm{d}x \tag{1.2.10a}$$

$$= 1 - \sum_{i=0}^{\alpha_k - 1} \exp(-\lambda_k t)\, (\lambda_k t)^i / i!, \tag{1.2.10b}$$

it then follows from (1.2.8) and property (C) that $\lambda_2 \leqq \lambda_1$ and $\alpha_1 \leqq \alpha_2$ imply $F_1 \leqq_d F_2$. We denote such d.f.s F by $\operatorname{Erl}(\lambda, \alpha)$.

Example 1.2(b′). More generally, replacing $(\alpha_k - 1)!$ at (1.2.10a) by $\Gamma(\alpha_k)$, the equation then defines for all $\alpha_k > 0$ the d.f. of a *gamma r.v.*, for which it is not difficult to check that $F_1 \leqq_d F_2$ if $\lambda_2 \leqq \lambda_1$ and $\alpha_1 \leqq \alpha_2$.

Example 1.2(c). For *Poisson distributions* with the parameters m_1 and m_2, so that

$$p_{k,n} = (m_k^n / n!) \exp(-m_k) \quad (n = 0, 1, \ldots;\ k = 1, 2),$$

$$m_1 \leqq m_2 \text{ implies } F_1 \leqq_d F_2 \tag{1.2.11}$$

as follows from (1.2.10b) with $t = 1$ and $\lambda_k = m_k$.

Example 1.2(d). For *normal distributions* with the parameters μ_k, σ_k $(k = 1, 2)$ so that the density function equals

$$F_k'(t) = (2\pi)^{-1/2} \sigma_k^{-1} \exp\big(-(t - \mu_k)^2 / 2\sigma_k^2\big) \quad (-\infty < t < \infty),$$

$$\mu_1 \leqq \mu_2 \text{ and } \sigma_1 = \sigma_2 \text{ implies } F_1 \leqq_d F_2. \tag{1.2.12}$$

Further criteria are collected in Appendix 1 which is based on Lisek (1978) and Taylor (1983).

1.3 Convex ordering for d.f.s ($\leqq_c$)

Using $x_+ = \max(0, x)$ we indroduce now the

Definition 1.3.1. *The r.v. X is* smaller in mean residual life *than the r.v. Y, written*

$$X \leqq_c Y,$$

or equivalently, their respective d.f.s F and G satisfy $F \leqq_c G$, if

$$\mathsf{E}(X - x)_+ = \int_x^\infty (t - x)\,\mathrm{d}F(t) = \int_x^\infty \big(1 - F(t)\big)\,\mathrm{d}t \leqq \int_x^\infty \big(1 - G(t)\big)\,\mathrm{d}t$$

$$= \mathsf{E}(Y - x)_+ \quad (\textit{all real } x) \tag{1.3.1}$$

provided these expectations (equivalently, integrals) are finite.[1]

It is an immediate consequence of the definition that

$$\text{if } F \leqq_d G \text{ and } \mathsf{E}Y_+ < \infty \text{ then } F \leqq_c G. \tag{1.3.2}$$

The integrals at (1.3.1) can be written more succinctly, using $\overline{F}(t) = 1 - F(t)$, in the form

$$\int_x^\infty \overline{F}(t)\,\mathrm{d}t \leqq \int_x^\infty \overline{G}(t)\,\mathrm{d}t, \tag{1.3.3}$$

while another equivalent form of this inequality is that

$$\mathsf{E}\max(x, X) = x + \mathsf{E}(X - x)_+ \leqq x + \mathsf{E}(Y - x)_+ = \mathsf{E}\max(x, Y) \quad (\text{all } x). \tag{1.3.4}$$

Provided $\mathsf{E}(x - X)_+$ and $\mathsf{E}(x - Y)_+$ are finite, $X \leqq_c Y$ implies $\mathsf{E}X \leqq \mathsf{E}Y$, because

$$\mathsf{E}Y - \mathsf{E}X = \mathsf{E}(Y - x)_+ - \mathsf{E}(X - x)_+ - \mathsf{E}(x - Y)_+ + \mathsf{E}(x - X)_+$$

$$\geqq \mathsf{E}(x - X)_+ - \mathsf{E}(x - Y)_+ \to 0 \quad (x \to -\infty).$$

From Theorem 1.3.1 below it emerges that it is appropriate to call $\leqq_c$ a *convex ordering* relation, and hence the notation; Bessler and Veinott (1966) used the term "stochastically smaller in mean". From (1.3.1) the transitivity and anti-symmetry properties of $\leqq_c$ are evident, so it is a partial ordering on the subset of $\mathscr{D}$ for which $\int_0^\infty t\,\mathrm{d}F(t) < \infty$; without this finiteness property the anti-symmetry property could fail.

For any r.v. X with finite mean m, it is clear that $\mathsf{E}\max(x, X) \geqq \max(x, \mathsf{E}X) = \max(x, m)$, so from (1.3.4)

$$m \leqq_c X. \tag{1.3.5}$$

It also follows from (1.3.4) that $\leqq_c$ has the property (R) as well as (M).

---

[1] The notation $\leqq_c$ in this edition replaces $\overset{(2)}{\leqq}$ used in the German original.

From (1.3.1) we see that $\leqq_c$ is generated by the family

$$\mathscr{F}_c \equiv \{e_x: -\infty < x < \infty\} \tag{1.3.6}$$

of functions e_x defined by

$$e_x(t) = (t - x)_+ = \int_{-\infty}^{t} \Theta_x(u)\, du. \tag{1.3.7}$$

Since $\mathscr{F}_c$ is invariant with respect to translations, $\leqq_c$ has property (C).

The next result, the main part of which is analogous to Theorem 1.2.1, identifies the class $\mathscr{R}_{\leqq_c}$ of $\leqq_c$-monotone functions as the class $\mathfrak{K}_c(\mathbb{R})$ of non-decreasing convex real functions, and the proof shows that the class $\mathscr{H}_{\leqq_c}$ of (linear) mixtures of $\mathscr{F}_c$ coincides with $\mathscr{R}_{\leqq_c}$ (cf. Proposition 1.1.3).

Theorem 1.3.1. *The inequality*

$$\int_{-\infty}^{\infty} f(t)\, dF_1(t) \leqq \int_{-\infty}^{\infty} f(t)\, dF_2(t) \tag{1.3.8}$$

holds for all functions $f \in \mathfrak{K}_c(\mathbb{R})$ *for which the integrals are defined if and only if* $F_1 \leqq_c F_2$. *For given* f, *the inequality* (1.3.8) *holds for all* F_1, F_2 *for which* $F_1 \leqq_c F_2$ *only if* f *is non-decreasing and convex.*

If $F_1 \leqq_c F_2$ *and their means exist and are equal,* (1.3.8) *holds for all convex* f.

Proof. The functions $e_x \in \mathfrak{K}_c(\mathbb{R})$ so if (1.3.8) holds, $F_1 \leqq_c F_2$ by definition. Conversely, suppose first that for given $f \in \mathfrak{K}_c(\mathbb{R})$ we also have $f(t) \downarrow 0$ $(t \to -\infty)$. Then f is expressible

$$f(t) = \int_{-\infty}^{\infty} e_x(t)\, dg(x) = \int_{-\infty}^{t} g(x)\, dx \tag{1.3.9}$$

where g is non-decreasing and $\lim_{t \to -\infty} g(t) = 0$. Consequently

$$\int_{-\infty}^{\infty} f(t)\, dF_k(t) = \int_{-\infty}^{\infty} dF_k(t) \int_{-\infty}^{\infty} e_x(t)\, dg(x) = \int_{-\infty}^{\infty} dg(x) \int_{-\infty}^{\infty} e_x(t)\, dF_k(t) \quad (k = 1, 2), \tag{1.3.10}$$

and from (1.3.1) the inequality (1.3.8) now follows for such non-negative f. A simple extension argument establishes (1.3.8) for all $f \in \mathfrak{K}_c(\mathbb{R})$.

For the second part of the theorem, the non-decreasing nature of f follows from Theorem 1.2.2 and the convexity of f follows by putting

$$F_1 = \Theta_x,\ F_2 = \gamma\Theta_{x_1} + (1 - \gamma)\, \Theta_{x_2}$$
$$\text{where } 0 < \gamma < 1 \text{ and } x = \gamma x_1 + (1 - \gamma)\, x_2.$$

Finally, if f is any given convex function, either f is everywhere non-increasing, or there exists some (not necessarily unique) finite t_0 such that f

is expressible in the form

$$f(t) = f(t_0) + f_1(t) + f_2(t)$$

where $f_1(t) = f(t) - f(t_0)$ for $t < t_0$, zero elsewhere, and $f_2(t) = f(t) - f(t_0)$ for $t > t_0$, zero elsewhere. Then f_k are non-negative monotone functions, and by the earlier part of the theorem, (1.3.8) holds for $f(t) = f(t_0) + f_2(t)$ when $F_1 \leqq_c F_2$. It remains to show that (1.3.8) holds for $f(t) = f_1(t)$ when also $\int_{-\infty}^{\infty} t\,\mathrm{d}F_k(t) = m$ for $k = 1, 2$. To this end let X_k denote r.v.s with d.f.s F_k for which $\mathsf{E}X_k = m$ $(k = 1, 2)$ and $F_1 \leqq_c F_2$, and let the r.v.s $Y_k \equiv -X_k$ have d.f.s G_k. Then since for all x

$$\begin{aligned} \mathsf{E}(X_k - x)_+ &= m - x + \mathsf{E}(x - X_k)_+ \\ &= m - x + \mathsf{E}\big(Y_k - (-x)\big)_+ \quad (k = 1, 2) \qquad (1.3.11) \end{aligned}$$

we must also have $G_1 \leqq_c G_2$. Consequently, setting $f_3(t) = f_1(-t)$, being a non-decreasing convex function, $\int_{-\infty}^{\infty} f_1(t)\,\mathrm{d}F_k(t) = \int_{-\infty}^{\infty} f_3(t)\,\mathrm{d}G_k(t)$, and so (1.3.8) holds as required.

Corollary 1.3.1a. *For non-negative r.v.s X, Y with $X \leqq_c Y$,*

$$\mathsf{E}X^r \leqq \mathsf{E}Y^r \qquad (r \geqq 1) \qquad (1.3.12\text{a})$$

whenever the expectations exist; for general r.v.s X, Y with $\mathsf{E}X = \mathsf{E}Y$ *and* $X \leqq_c Y$,

$$\mathsf{E}X^r \leqq \mathsf{E}Y^r \qquad (r = 2, 4, \ldots). \qquad (1.3.12\text{b})$$

Recall that Jensen's inequality states that for any convex function f and any r.v. X with finite mean $\mathsf{E}X$,

$$f(\mathsf{E}X) \leqq \mathsf{E}f(X).$$

Reference to (1.3.5) then shows that the last part of Theorem 1.3.1 is a generalization of this inequality.

It is worth remarking that for non-negative r.v.s X and Y for which $X \leqq_c Y$, the equality $\mathsf{E}X^r = \mathsf{E}Y^r$ for any $r > 1$ implies $X =_d Y$, because

$$\mathsf{E}X^r = \int_0^\infty r x^{r-1}\big(1 - F(x)\big)\,\mathrm{d}x = \int_0^\infty r(r-1)\,x^{r-2}\,\mathrm{d}x \int_x^\infty \big(1 - F(y)\big)\,\mathrm{d}y.$$

This property is the analogue of $X \leqq_d Y$ and $\mathsf{E}X = \mathsf{E}Y$ implying $X =_d Y$ noted above (1.2.2).

Proposition 1.3.2. *The limit r.v.s X and Y of weakly convergent sequences of r.v.s $\{X_n\}$ and $\{Y_n\}$ for which $X_n \leqq_c Y_n$ have the property $X \leqq_c Y$ if* $\mathsf{E}X_+$ *and* $\mathsf{E}Y_+$ *are finite and* $\mathsf{E}(X_n)_+ \to \mathsf{E}X_+$ *and* $\mathsf{E}(Y_n)_+ \to \mathsf{E}Y_+$ $(n \to \infty)$.

Remark. To see that without some constraint it can be the case that $X \geqq_c Y$, let $X_n = 1$ (all n), and $Y_n = 0$ with probability $1 - n^{-1}$, $= n$

with probability n^{-1}, so that $X_n \leqq_c Y_n$ by (1.3.5) and Y_n converges weakly to a r.v. $Y = 0$ with probability 1.

Proof. It suffices to show that the assumptions of the proposition ensure that $\mathsf{E}(X_n - x)_+ \to \mathsf{E}(X - x)_+$ $(n \to \infty)$ for all finite x, and similarly for $\mathsf{E}(Y_n - x)_+$. This follows from writing

$$\mathsf{E}(X_n - x)_+ = \int_x^\infty \bar{F}_n(t)\,\mathrm{d}t = \mathsf{E}(X_n)_+ - \int_0^x \bar{F}_n(t)\,\mathrm{d}t$$

and applying the bounded convergence theorem to the last term in conjunction with the assumed convergence to $\mathsf{E}X_+$ of $\mathsf{E}(X_n)_+$.

1.4 Concave ordering for d.f.s ($\leqq_{cv}$)

The following partial ordering is a counterpart of the convex ordering introduced in the previous section. The proofs of the various statements proceed in an analogous fashion to those for convex ordering, and accordingly they are omitted.

Definition 1.4.1.[1] *The r.v. X is* smaller in mean used life *than the r.v. Y, written*

$$X \leqq_{cv} Y,$$

or equivalently, their respective d.f.s F and G satisfy $F \leqq_{cv} G$, if

$$\mathsf{E}(x - X)_+ = \int_{-\infty}^x (x - t)\,\mathrm{d}F(t) = \int_{-\infty}^x F(t)\,\mathrm{d}t \geqq \int_{-\infty}^x G(t)\,\mathrm{d}t$$

$$= \mathsf{E}(x - Y)_+ \quad (\textit{all real } x) \tag{1.4.1}$$

provided these expectations (equivalently, integrals) are finite.

Immediately therefore,

$$\text{if } F \leqq_d G \text{ and } \mathsf{E}X_- \equiv \mathsf{E}\max(0, -X) < \infty, \text{ then } X \leqq_{cv} Y, \tag{1.4.2}$$

and as at (1.3.4), $X \leqq_{cv} Y$ is the same as

$$\mathsf{E}\min(x, X) \leqq \mathsf{E}\min(x, Y) \quad (\text{all } x). \tag{1.4.3}$$

We check that $\leqq_{cv}$ is a partial ordering on the subset of $\mathscr{D}$ for which $\int_{-\infty}^0 |t|\,\mathrm{d}F(t) < \infty$ much as for $\leqq_c$.

Observe from (1.4.1) that

$$X \leqq_{cv} Y \text{ if and only if } -Y \leqq_c -X. \tag{1.4.4}$$

---

(3)

[1] The notation $\leqq_{cv}$ in this edition replaces $\leqq$ used in the German original.

Recalling (1.3.11),

$$\text{if } \mathsf{E}X = \mathsf{E}Y, \text{ then } X \leqq_{cv} Y \text{ if and only if } Y \leqq_c X. \qquad (1.4.5)$$

Using $\mathfrak{K}_{cv}(\mathbb{R})$ to denote the class of non-decreasing concave functions f, we have

Theorem 1.4.1. *The inequality*

$$\int_{-\infty}^{\infty} f(t)\, \mathrm{d}F(t) \leqq \int_{-\infty}^{\infty} f(t)\, \mathrm{d}G(t) \qquad (1.4.6)$$

holds for all functions $f \in \mathfrak{K}_{cv}(\mathbb{R})$ *for which the integrals are defined if and only if* $F \leqq_{cv} G$. *For given* f, *the inequality* (1.4.6) *holds for all* F, G *for which* $F \leqq_{cv} G$ *only if* f *is non-decreasing and concave.*

If $F \leqq_{cv} G$ *and their means exist and are equal, then* (1.4.6) *holds for all concave* f.

Corollary 1.4.1a. *For non-negative r.v.s* X, Y *with* $X \leqq_{cv} Y$

$$\mathsf{E}X^r \leqq \mathsf{E}Y^r \; (0 \leqq r \leqq 1) \qquad (1.4.7)$$

provided the moments exist.

For a d.f. F or a r.v. X with F as its d.f., define

$$\text{ess sup } X = \inf \{x\colon \mathsf{P}(X > x) = 0\},$$

$$\text{ess inf } X = \sup \{x\colon \mathsf{P}(X < x) = 0\},$$

these bounds on X being not necessarily finite. It is clear from the definition of $\leqq_d$ that, if $X \leqq_d Y$, then ess sup $X \leqq$ ess sup Y. Since $X \leqq_c Y$ means that $\mathsf{E}(X - t)_+ \leqq \mathsf{E}(Y - t)_+$ and $E(X - t)_+ = \int_t^{\infty} \mathsf{P}(X > x)\, \mathrm{d}x$, we do in fact have the stronger result of

Proposition 1.4.2.

$$X \leqq_c Y \textit{ implies } \text{ess sup } X \leqq \text{ess sup } Y, \qquad (1.4.8\text{a})$$

$$X \leqq_{cv} Y \textit{ implies } \text{ess inf } X \leqq \text{ess inf } Y. \qquad (1.4.8\text{b})$$

1.5 Criteria for the convex ordering $\leqq_c$

1.5.1 The cut criterion of Karlin and Novikoff (1963)

Definition 1.5.1. *R.v.s X and Y with d.f.s F and G and finite first moments* satisfy the cut criterion *if* $\mathsf{E}X \leqq \mathsf{E}Y$ *and there exists some finite* ξ *for which*

$$F(x) \leqq G(x) \quad (\textit{all } x < \xi), \qquad (1.5.1\text{a})$$

$$F(x) \geqq G(x) \quad (\textit{all } x > \xi). \qquad (1.5.1\text{b})$$

These inequalities may hold for ξ in a non-degenerate interval, I say, in which case $F(x) = G(x)$ for x in I.

Proposition 1.5.1. *R.v.s X and Y satisfying the cut criterion of Definition 1.5.1 have $X \leqq_c Y$.*

Proof. It follows from (1.5.1b) that $\mathsf{E} \max(x, X) \leqq \mathsf{E} \max(x, Y)$ for $x \geqq \xi$, and from (1.5.1a) that $\mathsf{E} \min(x, X) \geqq \mathsf{E} \min(x, Y)$ for $x \leqq \xi$. Since for any r.v. X with finite first moment

$$\mathsf{E} \max(x, X) = \mathsf{E}X + x - \mathsf{E} \min(x, X) \quad \text{(all } x\text{)},$$

it now follows from the assumptions that $\mathsf{E} \max(x, X) \leqq \mathsf{E} \max(x, Y)$ (all x), so by (1.3.4) $X \leqq_c Y$.

This result has immediate application to some well known d.f.s.

Example 1.5.1(a). Let X and Y be *normally distributed.* Then it follows from Proposition 1.5.1 that

$$\text{if } \mathsf{E}X \leqq \mathsf{E}Y \text{ and } \operatorname{var} X \leqq \operatorname{var} Y, \quad X \leqq_c Y. \tag{1.5.2}$$

Example 1.5.1(b). For non-negative r.v.s X_k with $\mathsf{E}X_1 \leqq \mathsf{E}X_2$, let the d.f.s F_k have *hazard rates* $\lambda_k(.)$ (cf. Example 1.2(a')) satisfying

$$\lambda_1(t) \leqq \lambda_2(t) \text{ for } t < \tau, \lambda_1(t) \geqq \lambda_2(t) \text{ for } t > \tau \tag{1.5.3}$$

for some real $\tau \geqq 0$. Then it follows that $X_1 \leqq_c X_2$, because by the representation in Example 1.2(a') for the tails $\bar{F}_k$ of the d.f. in terms of the hazard rates, $F = F_1$ and $G = F_2$ satisfy (1.5.1). (Note that when $\tau = 0$, the stronger relation $F_1 \leqq_d F_2$ holds.)

Example 1.5.1(c). From the preceding example we deduce that for *Weibull r.v.s* X_k whose distributions have parameters α_k, λ_k, so that the d.f. is given by

$$F_k(t) = 1 - \exp(-\lambda_k t^{\alpha_k}) \qquad (t \geqq 0)$$

and the hazard rates equal $\alpha_k \lambda_k t^{\alpha_k - 1}$ $(k = 1, 2)$,

$$\text{if } \mathsf{E}X_1 \leqq \mathsf{E}X_2 \text{ and } \alpha_1 \geqq \alpha_2, \text{ then } F_1 \leqq_c F_2. \tag{1.5.4}$$

Example 1.5.1(d). A sufficient condition for two d.f.s F and G to cut one another at most once, i.e. satisfy Definition 1.5.1 for some (possibly infinite) ξ, is that the function ψ defined by $\psi(x) = G^{-1}(F(x))$ be *star-shaped*, i.e.

$$\psi(cx) \leqq c\psi(x) \qquad \text{(all } 0 < c < 1, \text{ all } x \geqq 0\text{)}. \tag{1.5.5}$$

To see this rewrite (1.5.5) in the form

$$\psi(cx)/cx \leqq \psi(x)/x \qquad \text{(all } 0 < c < 1, \text{ all } x \geqq 0\text{)},$$

so that $\psi(x)/x$ is a non-decreasing function, and hence $G^{-1}(F(x))$ crosses any line $X(x) = Kx$ at most once, and from below, and, having crossed it, never touches it again. Taking $K = 1$ and using the non-decreasing nature of G^{-1}, it follows that, if there is such a crossing point ξ say, $F(x) <$ or $> G(x)$ as $x <$ or $> \xi$.

It is clear that if ψ is convex then it is starshaped. BARLOW and PROSCHAN (1975) studied the partial orderings on the space $\mathcal{D}_+$ of d.f.s of non-negative r.v.s, denoted by $<_c$ and $<_*$, defined by

$$F <_c G \text{ if and only if } G^{-1}(F) \text{ is convex,}$$

$$F <_* G \text{ if and only if } G^{-1}(F) \text{ is starshaped.}$$

Of course $F <_c G$ implies $F <_* G$, and the preceding discussion shows that if the r.v.s X and Y have d.f.s F and G, than

$$\mathsf{E}X \leqq \mathsf{E}Y \quad \text{and} \quad F <_* G \quad \text{imply} \quad F \leqq_c G. \tag{1.5.6}$$

LANGBERG, LEON and PROSCHAN (1978) studied $<_c$ and $<_*$ and included some criteria for these partial orderings.

Example 1.5.1(e). VAN ZWET (1964) showed that for *gamma distributions* F_k with parameters α_k, λ_k, so that

$$F_k(t) = \big(\Gamma(\alpha_k)\big)^{-1} \int_0^t (\lambda_k x)^{\alpha_k - 1} \lambda_k \, \mathrm{e}^{-\lambda_k x} \, \mathrm{d}x \qquad (t \geqq 0, \ k = 1, 2),$$

the function $F_2^{-1}\big(F_1(x)\big)$ is convex if and only if $\alpha_1 \geqq \alpha_2$. Since F_k has first moment α_k/λ_k, it follows that

$$\text{if} \quad \alpha_1 \geqq \alpha_2 \quad \text{and} \quad \alpha_1/\lambda_1 \leqq \alpha_2/\lambda_2, \quad \text{then} \quad F_1 \leqq_c F_2. \tag{1.5.7}$$

Example 1.5.1(f). It is clear that d.f.s F_α with the common mean m and differing from one another only by the *scale parameter* α, so that

$$\begin{gathered} F_\alpha(t - m) = F_1\big((t - m)/\alpha\big) \qquad (\alpha > 0, \text{ all } t), \\ \alpha_1 \leqq \alpha_2 \quad \text{implies} \quad F_{\alpha_1} \leqq_c F_{\alpha_2} \end{gathered} \tag{1.5.8}$$

because F_{α_1} and F_{α_2} satisfy Definition 1.5.1 with $\xi = m$; see also NÄTHER (1975b).

Further examples are presented in Appendix 1.

1.5.2 Criteria for the comparison of d.f.s related to normal d.f.s

Example 1.5.2(a). Let two r.v.s X_k have the *log normal distributions* F_k with parameters μ_k and σ_k, so that $X_k = \exp(Y_k)$ where the r.v.s Y_k have normal distributions with means μ_k and variances σ_k^2 $(k = 1, 2)$. From (1.5.2) and the monotonic convex nature of the exponential function it follows (see Theorem 2.2.2a) that

$$\text{if} \quad \mu_1 \leqq \mu_2 \quad \text{and} \quad \sigma_1 \leqq \sigma_2, \quad \text{then} \quad F_1 \leqq_c F_2. \tag{1.5.9}$$

Example 1.5.2(b). Suppose the d.f.s F_k correspond to normal r.v.s with a common positive mean μ, variances σ_k^2, conditioned to be positive, so that the corresponding density functions f_k can be expressed in the form

$$f_k(x) = \begin{cases} 0 & (x < 0) \\ \phi(x;\, \mu,\, \sigma_k^2)/\big(1 - \Phi(0;\, \mu,\, \sigma_k^2)\big) & (x \geqq 0) \end{cases}$$

where $\phi(\cdot\,;\mu,\sigma^2)$ and $\Phi(\cdot\,;\mu,\sigma^2)$ are respectively the density function and d.f. of a normal r.v. with mean μ and variance σ^2. Then we assert that

$$\sigma_1 \leqq \sigma_2 \text{ implies } F_1 \leqq_c F_2. \tag{1.5.10}$$

To see this, recall from (1.5.2) that the normal d.f.s Φ_k defined by $\Phi(\cdot\,;\mu,\sigma_k^2)$ satisfy $\Phi_1 \leqq_c \Phi_2$, while by the cut property $\Phi(0;\mu,\sigma_1^2) \leqq \Phi(0;\mu,\sigma_2^2)$. Then (1.5.10) follows on using these inequalities in the relation

$$\int_x^\infty \bar{F}_k(t)\,\mathrm{d}t = \big(1 - \Phi(0;\mu,\sigma_k^2)\big)^{-1} \int_x^\infty \big(1 - \Phi(t;\mu,\sigma_k^2)\big)\,\mathrm{d}t.$$

Example 1.5.2(c). NÄTHER (1975b) considered d.f.s $F_{\alpha,\sigma}$ with densities of the form

$$f(t;\alpha,\sigma) = w(\alpha,\sigma)\exp\left(-{}^1\!/_2\,|t/\sigma|^\alpha\right) \quad (\text{all real } t,\ \alpha > 0,\ \sigma > 0), \tag{1.5.11}$$

showing that

$$\sigma_1 \leqq \sigma_2 \text{ implies } F_{\alpha,\sigma_1} \leqq_c F_{\alpha,\sigma_2}, \tag{1.5.12}$$

$$\alpha_1 \geqq \alpha_2 \text{ implies } F_{\alpha_1,\sigma} \leqq_c F_{\alpha_2,\sigma}. \tag{1.5.13}$$

1.5.3 Criteria for comparison with exponential distributions

Comparison of certain classes of d.f.s with the exponential d.f. are considered later in this chapter. Thus, Proposition 1.5.2 states that for every d.f. F which is NBUE (see the next section) and has mean m,

$$F \leqq_c \operatorname{Exp}(m^{-1}). \tag{1.5.14}$$

Other criteria, ensuring that (1.5.14) is satisfied, but requiring knowledge only of the mean (and, perhaps, variance) are given in STOYAN (1972e). For example, when $m = 1$, we have (1.5.14) if for given $0 < \alpha < 1$, $F(\alpha) = 0$ and $F\big(\beta(\alpha)\big) = 1$ where the parameter $\beta \equiv \beta(\alpha)$ is the root in $(1, \infty)$ of

$$\beta = 1 - \ln\left[(1 - \alpha)/(\beta - \alpha)\right]; \tag{1.5.15}$$

Table 1.1. Values of $\beta(\alpha)$ in (1.5.15)

α	.1	.2	.3	.4	.5	.6	.7	.8	.9
$\beta(\alpha)$	1.20	1.41	1.69	1.95	2.26	2.63	3.07	3.66	4.61

some values are indicated in the Table 1.1. (a more extensive table could be found from DAVID, KENDALL and BARTON (1966)).

If we know not only that $m = 1$ but σ^2 is given, then (1.5.14) is satisfied by any d.f. F for which $F(\alpha(\sigma)) = 0$ and $F(\beta(\sigma)) = 1$ where the values of α and β are as indicated in the Table 1.2.

Table 1.2. Values of $\alpha(\sigma)$ and $\beta(\sigma)$

σ^2	.1	.2	.3	.4	.5	.6
$\alpha(\sigma)$	.05	.11	.17	.24	.32	.44
$\beta(\sigma)$	6.66	5.74	4.96	4.40	3.83	3.18

If $\sigma^2 > 0.6477$, we need further constraints or information on F in order to establish (1.5.14).

1.6 Some classes of d.f.s (IFR/DFR, NBU/NWU, DMRL/IMRL, NBUE/NWUE, γ-MRLA/γ-MRLB)

The classes of d.f.s considered in this section are important for comparison of "new" and "residual" life times; they arose in the context of models in reliability theory and have been used in queueing theory as well (see e.g. Sections 5.4—5.7 and 7.2 below).

Let X be a positive r.v. with d.f. F and denote by X_τ a generic r.v. with d.f. F_τ given by

$$\bar{F}_\tau(x) = 1 - F_\tau(x) = \mathsf{P}(X_\tau > x) = \mathsf{P}(X > x + \tau \mid X > \tau)$$
$$= \bar{F}(x + \tau)/\bar{F}(\tau) \qquad (x \geqq 0, \tau > 0),$$

assuming $F(\tau) < 1$; if not set $F_\tau = \Theta_0$. If we interpret X as the lifetime of a component, then X_τ is the residual lifetime of a component which has already survived the time τ. Note the equivalent expression, for F_τ, namely

$$F_\tau(x) = \big(F(x + \tau) - F(\tau)\big)/\big(1 - F(\tau)\big).$$

We shall note in this section some results for the following four pairs of concepts which compare the d.f.s F and F_τ, though it is just as easy to formulate the definition in terms of X and X_τ.

Definition 1.6.1. *The d.f. F of the non-negative r.v. X*

(a) *has an* increasing failure rate (IFR) *if*

$$F_y \leqq_d F_x \quad (all\ 0 \leqq x < y < \infty); \tag{1.6.1}$$

(b) *has* decreasing mean residual life (DMRL) *if*

$$\mathsf{E}X_\tau = \int_\tau^\infty \big(1 - F(u)\big)\,du\big/\big(1 - F(\tau)\big) \downarrow (0 < \tau \uparrow \infty); \tag{1.6.2}$$

(c) *is* new better than used (NBU) *if*

$$F_\tau \leqq_d F \quad (all\ 0 < \tau < \infty); \tag{1.6.3}$$

(d) *is* new better than used in expectation (NBUE) *if*

$$\mathsf{E}X_\tau \leqq \mathsf{E}X \quad (all\ 0 < \tau < \infty); \tag{1.6.4}$$

If the inequalities are reversed in (1.6.1)–(1.6.4), *then we speak of*

(a') decreasing failure rate (DFR),
(b') increasing mean residual life (IMRL),
(c') new worse than used (NWU), *and*
(d') new worse than used in expectation (NWUE),
respectively.

If follows from this definition that IFR implies DMRL and NBU, and that either of these implies NBUE. An equivalent form of (1.6.1) is that

$$\big(F(\tau + x) - F(\tau)\big)/\big(1 - F(\tau)\big) \text{ is non-increasing in } \tau \text{ for each fixed } x > 0. \tag{1.6.5}$$

It is not difficult to verify that an equivalent form of (1.6.5) is that

$$\ln\big(1 - F(x)\big) \text{ is concave in } x > 0, \tag{1.6.5'}$$

while yet another criterion, which is applicable only when F has a density function f, is that

$$\text{the hazard rate } f(x)/\overline{F}(x) \text{ is non-decreasing in } x. \tag{1.6.5''}$$

Example 1.6(a). Gamma distributions with index $\alpha \geqq 1$, in particular the exponential and Erlang distributions, are IFR (cf. Examples 1.2(a') and (b), and 1.5.1(b)).

Example 1.6(b). The Weibull distribution with $\alpha \geqq 1$ is IFR (cf. Example 1.5.1(c)).

Example 1.6(c). Uniform distributions are IFR.

We now prove some properties of r.v.s having one (or more) of these properties.

Proposition 1.6.1. *For independent (not necessarily identically distributed) non-negative r.v.s X and Y, let X_Y be the generic r.v. with d.f. F_Y given by*

$$F_Y(x) = \mathsf{P}(X \leqq x + Y \mid X > Y) \quad (x \geqq 0)$$
$$= \int_0^\infty \big(F(x + y) - F(y)\big)\,dG(y) \Big/ \int_0^\infty \big(1 - F(y)\big)\,dG(y) \tag{1.6.6}$$

where X, Y have d.f.s F, G. Then if F is IFR (DFR) distribution,

$$F_Y \leqq_d (\geqq_d)\, F; \tag{1.6.7}$$

if F has DMRL (IMRL),

$$\mathsf{E}X_Y^2/\mathsf{E}X_Y \leqq (\geqq)\ \mathsf{E}X^2/\mathsf{E}X; \tag{1.6.8}$$

if F is NBUE (NWUE), then

$$\mathsf{E}X_Y^2/\mathsf{E}X_Y \leqq (\geqq)\ 2\mathsf{E}X. \tag{1.6.9}$$

Proof. It will suffice to consider the IFR, DMRL, and NBUE cases. Let $X_{Y+\tau}$ be the generic r.v. with d.f.

$$\mathsf{P}(X \leqq x + Y + \tau \mid X > Y + \tau) \qquad (\text{all } x \geqq 0,\ \tau > 0);$$

we start by showing the property, at first sight more general than (1.6.7) but in fact no more general because X_τ has IFR whenever X has IFR, that

$$X_{Y+\tau} \leqq_d X_\tau. \tag{1.6.7'}$$

To this end, consider $\bar{F}(\tau + x)/\bar{F}(\tau) - \bar{F}_Y(\tau + x)/\bar{F}_Y(\tau)$, which has the same sign as

$$\int_0^\infty \big(\bar{F}(\tau + x)\,\bar{F}(\tau + y) - \bar{F}(\tau + x + y)\,\bar{F}(\tau)\big)\,\mathrm{d}G(y),$$

in which the integrand is non-negative by the IFR property of F. Thus (1.6.7′) follows.

To prove (1.6.8) we shall show that

$$\mathsf{E}(X - y)_+^2/\mathsf{E}(X - y)_+ \ \textit{decreases with increasing } y,$$

for then we have the weaker property

$$\mathsf{E}(X - y)_+^2 \leqq (\mathsf{E}X^2/\mathsf{E}X)\ \mathsf{E}(X - y)_+$$

from which (1.6.8) follows on replacing y by Y, taking expectations, and division by $\mathsf{P}(X > Y)$. Now

$$(\mathrm{d}/\mathrm{d}y)\big(\mathsf{E}(X - y)_+^2/\mathsf{E}(X - y)_+\big) = -2 + \mathsf{E}(X - y)_+^2\ \mathsf{P}(X > y)/\mathsf{E}(X - y)_+^2$$

which has the same sign as $\mathsf{E}(X - y)_+^2\ \mathsf{P}(X > y) - 2\mathsf{E}(X - y)_+^2$.

Since X has DMRL, we have for all $x, y > 0$ that

$$\mathsf{E}(X - x - y)_+\ \mathsf{P}(X > y) \leqq \mathsf{E}(X - y)_+\ \mathsf{P}(X > x + y),$$

and integrating x on $(0, \infty)$ leads to

$${}^1\!/_2\mathsf{E}(X - y)_+^2\ \mathsf{P}(X > y) \leqq \big(\mathsf{E}(X - y)_+\big)^2,$$

so the assertion is proved.

Finally, when X is NBUE, $\mathsf{E}(X - x)_+ \leqq \mathsf{E}X\ \mathsf{P}(X > x)$, from which we have by integration on $y < x < \infty$ that

$$\mathsf{E}(X - y)_+^2 \leqq 2\mathsf{E}X\ \mathsf{E}(X - y)_+.$$

Replacing y by Y, taking expectations, and dividing by $\mathsf{P}(X > Y)$, leads to (1.6.9).

We turn now to the NBUE property for which the defining relation (1.6.4) can be expressed as

$$\int_0^\infty \bar{F}(\tau + x)\,\mathrm{d}x \leqq m\bar{F}(\tau)\ (0 < \tau < \infty;\ m \equiv \mathsf{E}X), \tag{1.6.10}$$

and for which we have an alternative interpretation as follows. Consider a stationary renewal process with lifetime (or, inter-event) d.f. F (see e.g. Cox (1962)). Let F_R be the stationary forward (or, backward) recurrence time d.f., being given by

$$F_R(x) = m^{-1} \int_0^x \bar{F}(t)\,\mathrm{d}t$$

where $m = \int_0^\infty \bar{F}(t)\,\mathrm{d}t$. Then (1.6.10) is equivalent to

$$F_R \leqq_d F. \tag{1.6.10'}$$

Proposition 1.6.2. *Let the r.v. X with d.f. F have finite mean m. If F is NBU (NWU), then*

$$F \leqq_d (\geqq_d)\ \mathrm{Exp}\,(\lambda) \tag{1.6.11}$$

for some $\lambda \leqq (\geqq)\ m^{-1}$, with equality possible only if $F = \mathrm{Exp}\,(m^{-1})$. If F is NBUE (NWUE), then

$$F \leqq_c (\geqq_c)\ \mathrm{Exp}\,(m^{-1}). \tag{1.6.12}$$

Proof. We start by remarking that the implication of the assertion about $\lambda = m^{-1}$ following (1.6.11), is that none of the properties given in Definition 1.6.1 can lead to the assertion, stronger than either (1.6.11) or (1.6.12), that

$$F \leqq_d \mathrm{Exp}\,(m^{-1}) \tag{1.6.13}$$

without then having $F = \mathrm{Exp}\,(m^{-1})$. To see this, (1.6.13) is equivalent to $1 - F(x) \leqq \mathrm{e}^{-x/m}$ (all $x > 0$), which in view of the relations

$$m = \int_0^\infty \big(1 - F(x)\big)\,\mathrm{d}x = \int_0^\infty \mathrm{e}^{-x/m}\,\mathrm{d}x$$

means that equality must hold for all $x > 0$.

Assuming F to be NBU, introduce the function $r(x) = -\ln\big(1 - F(x)\big)$ and write the defining relation (1.6.3) for NBU in the form

$$r(x + y) \geqq r(x) + r(y) \qquad (\text{all } x, y > 0).$$

Then elementary properties of super-additive functions and the finiteness of $r(y)$ for some positive y ensures that

$$r(x)/x \geqq \lim_{y \downarrow 0} r(y)/y = \inf_{y>0} r(y)/y \equiv \lambda < \infty.$$

Reverting to the d.f. F, we have $1 - F(x) \leqq e^{-\lambda x}$ (all $x > 0$), which is the same as (1.6.11), and integrating on $0 < x < \infty$ shows that $m \leqq \lambda^{-1}$ if $\lambda > 0$, while $\lambda < m^{-1}$ is trivially true for $\lambda = 0$ which is the case if $F(x) = 0$ for any $x > 0$.

The proof for F NWU is similar, except that the finiteness of λ can no longer be ensured.

Now suppose that F is NBUE. From the discussion following (1.6.10), F_R has $m^{-1}\bar{F}(\cdot)$ as its density function, and hence its hazard rate λ_R is given for all $t > 0$ by

$$\lambda_R(t) = m^{-1}\bar{F}(t)/m^{-1}\int_t^\infty \bar{F}(u)\,du = 1/\mathsf{E}(X - t \mid X > t) \geqq m^{-1}.$$

Thus $\bar{F}_R(x) = \exp\left(-\int_0^x \lambda_R(t)\,dt\right) \leqq e^{-x/m}$, i.e. $F_R \leqq_d \operatorname{Exp}(m^{-1})$. Since $\bar{F}_R(x) = \mathsf{E}(X - x)_+/m$ and $m\,e^{-x/m} = \int_x^\infty e^{-u/m}\,du$, (1.6.12) now follows.

Barlow and Proschan (1975), in their section 4.6, established a result that also indicates why the inequality (1.6.13) cannot hold, at least under the stronger assumption that F is IFR with mean m. They used the property that such an F is convex with respect to Exp (m^{-1}) (i.e., $F <_c \operatorname{Exp}(m^{-1})$ in the notation of Example 1.5.1(d)), and deduced that then

$$F(x) \leqq 1 - e^{-x/m} \quad \text{for} \quad 0 \leqq x < m. \qquad (1.6.13')$$

Corollary 1.6.2a. *If the r.v. X has mean m, is NBUE (NWUE), and has finite r^{th} moment $\mu_r = \mathsf{E}X^r$, then*

$$\mu_r \leqq (\geqq)\, r!\, m^r \qquad (r = 1, 2, \ldots).$$

In particular therefore, the variance σ^2 is smaller (greater) than m^2, and the coefficient of variation $v \equiv \sigma/m$ is smaller (greater) than one.

A further weakening of the concept of NBUE is that of the mean residual life being bounded above by some constant $\gamma > m$, denoted γ-MRLA: X is γ-MRLA if

$$\mathsf{E}(X_\tau) \equiv \mathsf{E}(X - \tau \mid X > \tau) \leqq \gamma \qquad (\text{all } \tau > 0).$$

If $\gamma = m$, then X is NBUE. Integrating the equivalent inequality

$$\mathsf{E}(X - y)_+ \leqq \gamma \mathsf{P}(X \geqq y)$$

over $0 < y < \infty$ yields $\frac{1}{2}\mathsf{E}X^2 \leqq \gamma m$, and hence, using also the analogous concept γ-MRLB (mean residual life bounded below by γ), we have the first part of

Proposition 1.6.3. *If X is γ-MRLA (γ-MRLB), then $\mathsf{E}X^2 \leqq (\geqq)\, 2\gamma\mathsf{E}X$, and if Y is a non-negative r.v. independent of X, then $\mathsf{E}X_Y{}^2 \leqq (\geqq)\, 2\gamma\mathsf{E}X_Y$.*

The second part of this proposition is proved by the same argument as used in establishing (1.6.9).

We remark that the Erlang distribution Erl (λ, α) (see Example 1.2(b))

is (α/λ)-MRLB, and so too is the gamma distribution (cf. Example 1.2(b')).

KARPELEVICH and KREININ (1976) established the essence of the final result of this section, though under the stronger hypothesis of IFR (DFR).

Proposition 1.6.4. *For independent r.v.s X and Y with $\mathsf{E}X^2 < \infty$ and $Y \geqq 0$ a.s., define the constant ξ as the root of*

$$\mathsf{E}(X - \xi)_+ = \mathsf{E}(X - Y)_+. \tag{1.6.14}$$

Then if X has DMRL (IMRL),

$$\operatorname{var}(X - Y)_+ \geqq (\leqq) \operatorname{var}(X - \xi)_+. \tag{1.6.15}$$

Proof. The key to the proof is the observation that the curve

$$\Gamma \equiv \{(\mathsf{E}(X - t)_+, \mathsf{E}(X - t)_+^2) : 0 \leqq t < \infty\}$$

is convex (concave) when X is DMRL (IMRL). To see this, compute the slope of Γ at the parametric point t (assuming $\mathsf{E}(X - t)_+ > 0$) by

$$(\mathrm{d}/\mathrm{d}t)\, \mathsf{E}(X - t)_+^2 / (\mathrm{d}/\mathrm{d}t)\, \mathsf{E}(X - t)_+ = -2\mathsf{E}(X - t)_+/(-\mathsf{P}(X > t)) = 2\mathsf{E}X_t$$

in the notation at (1.6.2). The function $z(t) = \mathsf{E}(X - t)_+$ is continuous monotonic decreasing, and so has inverse $t(z)$ say for $0 \leqq z \leqq \mathsf{E}X_+$ that is one-one on $0 < z \leqq \mathsf{E}X_+$, while $\mathsf{E}(X - t)_+^2 = 0$ when $z = 0$. Thus

$$\Gamma = \{(z, \mathsf{E}(X - t(z))_+^2) : 0 \leqq z \leqq \mathsf{E}X_t\}$$

is convex or concave when $\mathsf{E}X_t = \mathsf{E}X_{t(z)}$ increases or decreases monotonically with increasing z. Now the r.v. Y defines a r.v. Z by $Z = z(Y)$, and in turn this defines the r.v. T by $T = t(Z)$, with $(X - Y)_+ = (X - T)_+$ a.s. and (cf. (1.6.14)) $\xi = t(\mathsf{E}Z)$. By Jensen's inequality, writing $v(t) = \mathsf{E}(X - t)_+^2$, if X has DMRL (IMRL),

$$\mathsf{E}(X - Y)_+^2 = \mathsf{E}v(t(Z)) \geqq (\leqq)\, v(t(\mathsf{E}Z)) = \mathsf{E}(X - \xi)_+^2,$$

which in view of (1.6.14) is the same as (1.6.15).

1.7 Weakenings of the convex and concave order relations

Write

$$\mathsf{E}(X - x)_+^\alpha = \int_x^\infty (t - x)^\alpha \,\mathrm{d}F(t)$$

for a r.v. X with d.f. F and having a finite moment of order α. For $\alpha > 1$, integrating by parts yields

$$\begin{aligned} \mathsf{E}(X - x)_+^\alpha &= \int_x^\infty \alpha(t - x)^{\alpha-1} (1 - F(t))\, \mathrm{d}t \\ &= \int_x^\infty \alpha(\alpha - 1)\, (t - x)^{\alpha-2} \left(\int_t^\infty (1 - F(u))\, \mathrm{d}u \right) \mathrm{d}t, \end{aligned} \tag{1.7.1}$$

whence inequality (1.7.2) below follows easily.

Proposition 1.7.1. *If r.v.s X and Y have finite moments of order $\alpha > 1$, and if $X \leqq_c Y$, then*

$$\mathsf{E}(X - x)_+^\alpha \leqq \mathsf{E}(Y - x)_+^\alpha \qquad (\text{all } x), \tag{1.7.2}$$

while if $X \leqq_{cv} Y$, then

$$\mathsf{E}(x - X)_+^\alpha \geqq \mathsf{E}(X - Y)_+^\alpha \qquad (\text{all } x). \tag{1.7.2'}$$

Proof. $X \leqq_{cv} Y$ means that $\mathsf{E}(x - X)_+ \geqq \mathsf{E}(x - Y)_+$ (all x); (1.7.2') follows from the appropriate analogue of (1.7.2).

It follows similarly that for all $\beta > 0$, $X \leqq_c^\alpha Y$ implies $X \leqq_c^{\alpha+\beta} Y$ and $X \leqq_{cv}^\alpha Y$ implies $X \leqq_{cv}^{\alpha+\beta} Y$, where $\leqq_c^\alpha$ and $\leqq_{cv}^\alpha$, based on the inequalities at (1.7.2), are weakenings of the convex and concave order relations as follows.

Definition 1.7.1. *For r.v.s X and Y with $\mathsf{E}X_+^\alpha$ and $\mathsf{E}Y_+^\alpha < \infty$ for some $\alpha > 1$ we write*

$$X \leqq_c^\alpha Y$$

when (1.7.2) holds; if $\mathsf{E}X^\alpha$ and $\mathsf{E}Y^\alpha < \infty$ and (1.7.2') holds, we write

$$X \leqq_{cv}^\alpha Y.$$

It is also appropriate to comment that, from the first integral expression for $\mathsf{E}(X - x)_+^\alpha$ at (1.7.1), a similar weakening of $\leqq_d$ is possible for all $\alpha > 0$ for which $\mathsf{E}X_+^\alpha$ and $\mathsf{E}Y_+^\alpha < \infty$. However the cases of interest are those with $\alpha = 1, 2, \ldots$

It is not difficult to verify that both $\leqq_c^\alpha$ and $\leqq_{cv}^\alpha$ have the properties (R), (E), (M) and (C).

1.8 Orderings by transforms and generating functions ($\leqq_L$ and $\leqq_g$)

Definition 1.8.1. *For d.f.s F and G of non-negative r.v.s X and Y, F is* smaller than G with respect *to $\leqq_L$, denoted $F \leqq_L G$, if for all positive s*

$$\mathsf{E}\exp(-sX) = \int_0^\infty \exp(-sx)\,\mathrm{d}F(x) \geqq \int_0^\infty \exp(-sx)\,\mathrm{d}G(x) = \mathsf{E}\exp(-sY). \tag{1.8.1}$$

It is clear that $\leqq_L$ is reflexive and transitive; it is also antisymmetric because the Laplace-Stieltjes transform (L.-S.T.) of a d.f. on $\mathbb{R}_+$ determines the d.f. uniquely.

Recall that integration by parts in (1.8.1) yields

$$\mathsf{E}\exp(-sX) = 1 - s\int_0^\infty \mathrm{e}^{-sx}\big(1 - F(x)\big)\,\mathrm{d}x,$$

so

$$F \leqq_d G \text{ implies } F \leqq_L G, \tag{1.8.2}$$

while, if the means exist, letting $s \downarrow 0$ in $\big(1 - \mathsf{E}\exp(-sX)\big)/s$ shows that

$$F \leqq_L G \text{ and } \mathsf{E}Y < \infty \text{ implies } \mathsf{E}X \leqq \mathsf{E}Y. \tag{1.8.3}$$

A second integration by parts yields

$$\mathsf{E}\exp(-sX) = 1 - s^2 \int_0^\infty e^{-sx}\,dx \int_0^x \big(1 - F(y)\big)\,dy$$

$$= 1 - s^2 \int_0^\infty e^{-sx}\mathsf{E}\min(x, X)\,dx,$$

so that

$$F \leqq_{cv} G \text{ implies } F \leqq_L G; \tag{1.8.4a}$$

observe that (1.8.4a) may hold irrespective of the finiteness or otherwise of $\mathsf{E}X$ and $\mathsf{E}Y$.

More generally, for non-negative r.v.s X and Y,

$$X \leqq_{cv}^{\alpha} Y \text{ (some } \alpha \geqq 1) \text{ implies } X \leqq_L Y. \tag{1.8.4b}$$

To see this, we interchange the order of integration in

$$\int_0^\infty e^{-sx}\,dx \int_0^x (x-u)^\alpha\,dF(u) = \int_0^\infty e^{-su}\,dF(u) \int_u^\infty e^{-s(x-u)}(x-u)^\alpha\,dx,$$

and cancellation of the common factor yields (1.8.4b).

The converse implication at (1.8.4) need not hold, that is, there exist non-negative r.v.s X and Y for which $X \leqq_L Y$ but it is not true that $X \leqq_{cv} Y$. For example, suppose $X = 0$ or β with probabilities $1 - \beta^{-1}$ and β^{-1} respectively, so that $\mathsf{E}X = 1$, $\mathsf{E}X^2 = \beta$, and that the r.v. Y also has $\mathsf{E}Y = 1$ and $\mathsf{E}Y^2 = \beta$ but $Y \neq_d X$. Then (cf. e.g. Rolski (1972)) $\mathsf{E}\,e^{-sX} \geqq \mathsf{E}\,e^{-sY}$ (all $s \geqq 0$) so that $X \leqq_L Y$, while from the remark above Proposition 1.3.2, $X \leqq_{cv} Y$ can only hold if also $X \leqq_d Y$, contrary to assumption.

The class $\mathcal{F}_L$ of functions f on $\mathbb{R}_+$ of the form $f(x) = a\,e^{-bx}$ $(a, b \geqq 0)$ is translation invariant here, so $\leqq_L$ has property (C). It also has properties (R) and (M) by inspection, and it inherits property (W) from the closure under weak convergence of the space of L-S.T.s in the sense that, if a sequence $\{F_n\}$ of d.f.s converges weakly to the d.f. F, then the L-S.T.s of F_n converge to the L-S.T. of F.

A characterization of the class $\mathcal{R}_L$ of all $\leqq_L$-monotonic functions can be given by means of the following class of functions.

Definition 1.8.2. *A real function f on $\mathbb{R}_+$ is completely monotonic if all its derivatives $f^{(n)}$ exist and satisfy*

$$(-1)^n f^{(n)}(t) \geqq 0 \quad (\textit{all } t > 0,\ n = 1, 2, \ldots). \tag{1.8.5a}$$

A well known equivalent of (1.8.5a) is the existence of a measure μ on $\mathbb{R}_+$ with

$$f(t) = \int_0^\infty e^{-xt}\mu(dx) \tag{1.8.5b}$$

(see e.g. XIII.4 of FELLER (1966)). REUTER and RIEDRICH (1981) proved

Theorem 1.8.1. *Let the function f have a completely monotonic derivative. Then $F \leqq_L G$ implies*

$$\int_0^\infty f(t)\,dF(t) \leqq \int_0^\infty f(t)\,dG(t). \tag{1.8.6}$$

Conversely, any real function f for which (1.8.6) *holds for all F and G for which $F \leqq_L G$, has a completely monotonic derivative.*

Thus, $\mathscr{R}_{\leqq_L}$ is the set of all real functions f on $\mathbb{R}_+$ for which

$$(-1)^{n+1} f^{(n)}(t) \geqq 0 \qquad (n = 1, 2, \ldots, t \geqq 0).$$

If the non-negative r.v.s of Definition 1.8.1 are moreover *integer-valued*, it is more usual to use in place of their L-S.T.s the *probability generating functions* (p.g.f.s), defined for all $|z| \leqq 1$ by

$$g(z) = \mathsf{E}z^X = \sum_{n=0}^\infty \mathsf{P}(X = n)\, z^n.$$

Since for $0 < z < 1$ we can use $s = -\ln z$ to write $\mathsf{E}z^X = \mathsf{E}\,e^{-sX}$, it follows that the partial ordering $\leqq_g$ defined for non-negative integer-valued r.v.s X, Y by

$$X \leqq_g Y \text{ if } \mathsf{E}z^X \geqq \mathsf{E}z^Y \qquad (\text{all } 0 < z < 1) \tag{1.8.1'}$$

inherits the properties of $\leqq_L$, notably

$$X \leqq_d Y \text{ implies } X \leqq_g Y; \tag{1.8.2'}$$

$$X \leqq_{cv} Y \text{ implies } X \leqq_g Y; \tag{1.8.4'}$$

$$X \leqq_g Y \text{ and } \mathsf{E}Y < \infty \text{ imply } \mathsf{E}X < \infty.$$

Since when $\mathsf{E}X = \mathsf{E}Y$, $X \leqq_c Y$ and $Y \leqq_{cv} X$ are equivalent, it also follows that

$$X \leqq_c Y \text{ and } \mathsf{E}X = \mathsf{E}Y \text{ imply } Y \leqq_L (\leqq_g)\, X. \tag{1.8.7}$$

Observe that by letting $s \to \infty$ in (1.8.1) or $z \to 0$ in (1.8.1'), it also follows that

$$X \leqq_L Y \text{ or } X \leqq_g Y \text{ implies } \mathsf{P}(X = 0) \geqq \mathsf{P}(Y = 0). \tag{1.8.8}$$

1.9 Extremal elements of sets of d.f.s

Let $\mathscr{M}$ be a set of d.f.s and $\prec$ a partial ordering on a class of d.f.s $\mathscr{D}$ with $\mathscr{M} \subseteq \mathscr{D}$. The maximum (respectively, minimum) of $\mathscr{M}$ with respect to $\prec$, if it exists, is the d.f. $F_{\max}(F_{\min})$ in $\mathscr{M}$ for which

$$F \prec F_{\max} \ (F_{\min} \prec F) \qquad (\text{all } F \in \mathscr{M}).$$

Similarly, the supremum $F_{\sup}$ (infimum $F_{\inf}$) of $\mathscr{M}$ with respect to $\prec$, if it exists, is the $\prec$-smallest (greatest) d.f. $F_{\sup}(F_{\inf})$, not necessarily in $\mathscr{M}$, for which

$$F \prec F_{\sup} \ (F_{\inf} \prec F) \qquad (\text{all } F \in \mathscr{M}).$$

The purpose of this section is to note below the extremal elements (when they exist) for various classes of d.f.s. For brevity we speak of such elements as $\prec$-extremal elements.

Example 1.9(a). For the class $\mathscr{M}_m$ of d.f.s with mean m,

(i) Θ_m is its $\leqq_c$-minimum and $\leqq_c^\alpha$-minimum ($\alpha > 1$);
(ii) there is no $\leqq_c$-supremum;
(iii) Θ_0 is its $\leqq_L$-infimum;
(iv) Θ_m is its $\leqq_L$-maximum and $\leqq_{cv}^\alpha$-maximum ($\alpha \geqq 1$).

Example 1.9(b). For the class $\mathscr{M}_m^{[a,b]}$ of d.f.s with mean m and $F(a-0) = 0$, $F(b+0) = 1$,

(i) Θ_m is its $\leqq_c$-minimum;
(ii) $\Theta_{a,b}^m \equiv (b-a)^{-1}\big((b-m)\,\Theta_a + (m-a)\,\Theta_b\big)$ is its $\leqq_c$-maximum;
(iii) $\Theta_{a,b}^m$ is its $\leqq_L$-minimum and $\leqq_{cv}$-minimum;
(iv) Θ_m is its $\leqq_L$-maximium and $\leqq_{cv}$-maximum.

Example 1.9(c). For the class $\mathscr{M}_m^\sigma$ of d.f.s with mean m and finite variance σ^2, and restricted to $\mathscr{M}_m^\sigma \cap \mathscr{M}_m^{[0,\infty)}$ for (iii) and (iv),

(i) Θ_m is its $\leqq_c$-infimum;
(ii) The d.f. $\tfrac{1}{2}\big[1 + (x-m)\big((x-m)^2 + \sigma^2\big)^{-1/2}\big]$ is its $\leqq_c$-supremum;
(iii) $\omega\Theta_0 + (1-\omega)\,\Theta_{m+\sigma^2/m}$, where $\omega = \sigma^2/(m^2+\sigma^2)$, is its $\leqq_L$-minimum, $\leqq_{cv}^\alpha$-minimum, and $\leqq_c^\alpha$-maximum ($\alpha \geqq 2$);
(iv) Θ_m is its $\leqq_L$-supremum.

Example 1.9(d). For the class $\mathscr{M}_m^I$ of all d.f.s on $[0, \infty)$ with mean m that are NBUE,

(i) Θ_m is its $\leqq_c$-minimum;
(ii) Exp (m^{-1}) is its $\leqq_c$-maximum;
(iii) Exp (m^{-1}) is its $\leqq_L$-minimum;
(iv) Θ_m is its $\leqq_L$-maximum.

Example 1.9(e). For the class of all probability density functions of the form (1.5.11) with $0 \leqq \alpha_1 \leqq \alpha \leqq \alpha_2 < \infty$, $0 < \sigma_1 \leqq \sigma \leqq \sigma_2 < \infty$,

(i) the $\leqq_c$-minimum has parameters α_2, σ_1;
(ii) the $\leqq_c$-maximum has parameters α_1, σ_2.

1.10 Partial orderings for general distributions and multidimensional random variables

In this section where proofs are mostly suppressed, we consider distributions P on a σ-algebra $\mathcal{M}$ of subsets of a set E which is partially ordered by a relation $\leqq$. In particular if $E = \mathbb{R}^n$ with generic element $x = (x_1, \ldots, x_n)$ we write $x \leqq y$ to mean $x_i \leqq y_i$ $(i = 1, \ldots, n)$.

We shall mainly work with the definition of the partial ordering $F < G$ by means of Definition 1.1.2 for suitable classes of functions f in the inequality

$$\int_E f(t)\, F(\mathrm{d}t) \leqq \int_E f(t)\, G(\mathrm{d}t).$$

In particular, the analogues of $\leqq_c$ (respectively, $\leqq_{cv}$) are defined by functions f that are convex (concave) on E in the sense that

$$f\big(\lambda x + (1 - \lambda)\, y\big) \leqq (\geqq)\, \lambda f(x) + (1 - \lambda)\, f(y) \quad (\text{all } 0 < \lambda < 1)$$

for all x, y in E. Such a definition requires that E be a linear space, to which we shall add the requirement that the linear operations and $\leqq$ on E be compatible in the sense that for x, y, z in E,

$$0 \leqq x \ \text{ implies } \ 0 \leqq \lambda x \ \text{ if } \ \lambda > 0,$$

$$x \leqq y \ \text{ implies } \ x + z \leqq y + z.$$

Denote by $\mathfrak{K}_c(E)$ $\big(\mathfrak{K}_{cv}(E)\big)$ the set of all monotonic and convex (concave) functions on E, and by $P_{\mathcal{M}}$ the set of all probability measures on $(E, \mathcal{M})$.

Definition 1.10.1. *For P, $Q \in P_{\mathcal{M}}$, $P \leqq_d Q$, $P \leqq_c Q$, $P \leqq_{cv} Q$ according as the inequality*

$$\int_E f(t)\, P(\mathrm{d}t) \leqq \int_E f(t)\, Q(\mathrm{d}t) \tag{1.10.1}$$

holds for all $f \in \mathfrak{K}_d(E)$, $\mathfrak{K}_c(E)$, $\mathfrak{K}_{cv}(E)$ respectively for which the integrals are well-defined.

Here, $\mathfrak{K}_d(E)$ consists of all $\leqq$-monotonic functions f: $f \in \mathfrak{K}_d(E)$ whenever the condition

$$x \leqq y \ \text{ implies } \ f(x) \leqq f(y) \qquad (x, y \in E)$$

is satisfied. For two E-valued r.v.s X and Y with distributions P and Q, $X \leqq_d, \leqq_c, \leqq_{cv} Y$ when $P \leqq_d, \leqq_c, \leqq_{cv} Q$. When $X \leqq_d Y$ we can speak of X being stochastically smaller than Y. The antisymmetry property of $\leqq_d$ can be verified easily, at least for $E = \mathbb{R}^n$, by the use of particular functions f, and hence $\leqq_d$ is then a partial ordering; for general E see Kamae and Krengel (1978) and Wieczorek (1978). In the sequel we need only the transitivity property of the various relations, so on occasions our use of the term "partial ordering" may not strictly be justified in that the antisymmetry axiom may not hold.

In the case $E = \mathbb{R}^n$, so that $P_{\mathcal{M}}$ is the space of n-dimensional d.f.s, generically $F, G, \ldots$, a seemingly natural relation $\leqq_D$ is defined by

$$F \leqq_D G \quad \text{whenever} \quad F(x) \geqq G(x) \quad \text{for all } x. \tag{1.10.2}$$

VEINOTT (1965), TONG (1967, 1980a), and BERGMANN (1978) used this definition, but $\leqq_D$ is weaker than $\leqq_d$, i.e.

$$F \leqq_d G \text{ implies } F \leqq_D G \text{ but not necessarily conversely,} \tag{1.10.3}$$

as we now show. Let $f(t) = -1$ when $t \leqq x$, $f(t) = 0$ otherwise so that

$$\int_E f(t)\, F(\mathrm{d}t) = -F(x),$$

and $F \leqq_d G$ then implies that $G(x) \leqq F(x)$, i.e. $F \leqq_D G$. As a counter example to the converse, let F and G be the d.f.s of the r.v.s X, Y as specified in the Table 1.3.

Table 1.3. Distributions of the r.v.s X and Y

i	(0,0)	(0,1)	(1,0)	(1,1)
$\mathsf{P}(X = i)$	0.4	0.2	0.3	0.1
$F(i)$	0.4	0.6	0.7	1.0
$\mathsf{P}(Y = i)$	0.3	0.3	0.4	0
$G(i)$	0.3	0.6	0.7	1.0

By inspection, $F(i) \geqq G(i)$ at all points of increase of F or G, so $F \leqq_D G$, while for the function $f((x_1, x_2)) = \max(0, x_1 + x_2 - 1)$, $\mathsf{E}f(X) = 0.1 > 0 = \mathsf{E}f(Y)$, and since $f \in \mathfrak{R}_d(\mathbb{R}^2)$, we cannot have $X \leqq_d Y$.

We note also that some authors (see e.g. BERGMANN (1978), CAMBANIS, SIMONS and STOUT (1976)) have used the analogous relation for $E = \mathbb{R}^n$,

$$X \leqq_K Y \text{ if } \mathsf{P}(x \leqq X) \leqq \mathsf{P}(x \leqq Y) \quad (\text{all } x \text{ in } \mathbb{R}^n).$$

Proposition 1.10.1. *For the random n-vectors* $X = (X_1, \ldots, X_n)$ *and* $Y = (Y_1, \ldots, Y_n)$, $X \leqq_D Y$ *(respectively,* $X \leqq_K Y$*) if and only if*

$$\mathsf{E}\big(f_1(X_1) \ldots f_n(X_n)\big) \geqq (\leqq)\ \mathsf{E}\big(f_1(Y_1) \ldots f_n(Y_n)\big) \tag{1.10.4}$$

for all non-increasing (non-decreasing) positive functions $f_1, \ldots, f_n$.
We omit details of the proof which can be constructed via the expression of f_i as limits of non-decreasing jump functions.

In the case $E = \mathbb{R}^2$, another class of $\prec$-generating functions is given by the *quasi-monotone* functions f defined as those for which

$$f(x, y) + f(x', y') - f(x, y') - f(x', y) \geqq 0$$

for all reals $x \leqq x'$ and $y \leqq y'$.

For the random n-vector $X = (X_1, \ldots, X_n)$, let $X_{(j)}$ denote the $(n-1)$-vector X with X_j omitted.

Proposition 1.10.2. *Let the random n-vectors X and Y have $X_{(j)} =_d Y_{(j)}$ for each $j = 1, \ldots, n$. Then $X \leqq_K Y$ (or, equivalently $X \leqq_D Y$ if n even, $Y \leqq_D X$ if n odd) implies the inequality* (1.10.4) *with $\leqq$ for all non-decreasing functions $f_1, \ldots, f_n$.*

Further, for $n = 2$,

$$\mathsf{E}f(X) \geqq \mathsf{E}f(Y)$$

for all quasi-monotone right-continuous functions f if either f is symmetric with $\mathsf{E}f(X_1, X_1)$ and $\mathsf{E}f(X_2, X_2)$ finite, or $\mathsf{E}f(X_1, y_2)$ and $\mathsf{E}f(y_1, X_2)$ are finite for some $y = (y_1, y_2)$.

For partial orderings $<$ on $P_{\mathscr{M}}$ for general σ-algebras $\mathscr{M}$ similar considerations to those given in Section 1.1 for d.f.s are possible. In particular, let $\mathscr{R}_<$ be the set of all functionals on E for which (1.10.1.) holds for all P, Q in $P_{\mathscr{M}}$ for which $P < Q$.

The following result is the composite analogue of Theorems 1.2.2, 1.3.1 and 1.4.1 which treated the case $E = \mathbb{R}$.

Theorem 1.10.3. $\mathfrak{R}_d(E) = \mathscr{R}_{\leqq_d}$, $\mathfrak{R}_c(E) = \mathscr{R}_{\leqq_c}$, $\mathfrak{R}_{cv} = \mathscr{R}_{\leqq_{cv}}$.

Proof. By definition we have $\mathfrak{R}_d(E) \subseteqq \mathscr{R}_{\leqq_d}$ etc. Let $f \in \mathscr{R}_{\leqq_d}$; to show that f is $\leqq$-monotone, i.e. $f \in \mathfrak{R}_d(E)$, take $x, y \in E$ with $x \leqq y$, and continue to denote by Θ_x the Dirac distribution which has unit mass at x. Then $\Theta_x \leqq_d \Theta_y$ by definition of $\leqq_d$, and therefore $f(x) \leqq f(y)$, i.e. $f \in \mathfrak{R}_d(E)$. Similarly, by using two-point distributions the convexity (concavity) of f with $\leqq_c$ ($\leqq_{cv}$) can be proved.

By using results of Strassen (1965), Kamae, Krengel and O'Brien (1977) proved the following generalization of Proposition 1.2.1 (we omit the proof).

Proposition 1.10.4. *When E is a Polish space (i.e. a complete separable metric space) endowed with a closed partial ordering and $\mathscr{M}$ is the corresponding Borel σ-algebra, given two distributions $P_k \in P_{\mathscr{M}}$ with $P_1 \leqq_d P_2$, there exists a probability space $(\Omega, \mathscr{A}, \mathsf{P})$ and E-valued random variables X_k such that $\mathsf{P}(X_k \in B) = P_k(B)$ (all $B \in \mathscr{M}$) and $X_1(\omega) \leqq X_2(\omega)$ (all $\omega \in \Omega$).*

Just as $F \leqq_d G$ is established via (1.10.1) in the case $E = \mathbb{R}$ by considering merely the functions $\Theta_x(.)$ (all x), so Levhari, Paronsh and Peleg (1975) and Kamae, Krengel and O'Brien (1977) showed that so establish whether or not $P \leqq_d Q$ holds for more general E, it is sufficient to consider indicator functions $f_S(x) = 1$ or 0 as $x \in$ or $\notin S$ where the sets S are closed so-called "increasing" subsets of E, i.e. sets for which $x \in S$, $y \in E$, $x \leqq y$, implies $y \in S$. Whitt (1980a) pointed out that another class of functions with the same property is that of all continuous bounded nondecreasing functions. This ensures that $\leqq_d$ is closed under weak convergence.

Nevertheless, it is not necessarily a simple matter to establish, for given P, Q whether $P \leqq_d Q$ or $P \leqq_c Q$ or $P \leqq_{cv} Q$. These particular order relations are useful for applications because many distributions of interest can be generated from simpler distributions by mappings or constructions which preserve the comparability property concerned. Instances of this may be found in Chapter 2.

Criteria for the comparability of random vectors

Let $X = (Y_1, \ldots, X_n)$ and $Y = (Y_1, \ldots, Y_n)$ be n-vectors whose components are E-valued r.v.s. When each of X and Y consists of sets of mutually independent r.v.s,
for each of the partial orderings $\leqq_d$, $\leqq_c$, *and* $\leqq_{cv}$,

$$X < Y \quad \textit{if and only if} \quad X_j < Y_j \quad (j = 1, \ldots, n). \tag{1.10.5}$$

That the same assertion is true for infinite sequences of r.v.s is given at Theorem 4.1.2.

VEINOTT (1965) introduced a different criterion for $X \leqq_d Y$ involving possibly dependent sequences of r.v.s in terms of the conditional distribution

$$P_j(B \mid x_1, \ldots, x_{j-1}) \equiv \mathsf{P}(X_j \in B \mid X_1 = x_1, \ldots, X_{j-1} = x_{j-1}),$$

$$Q_j(B \mid y_1, \ldots, y_{j-1}) \equiv \mathsf{P}(Y_j \in B \mid Y_1 = y_1, \ldots, Y_{j-1} = y_{j-1}).$$

Then

$$X \leqq_d Y \tag{1.10.6}$$

if

$$X_1 \leqq_d Y_1 \tag{1.10.7}$$

and

$$P_j(\cdot \mid x_1, \ldots, x_{j-1}) \leqq_d Q_j(\cdot \mid y_1, \ldots, y_{j-1}) \tag{1.10.8}$$

for all $x_i \leqq y_i$ $(i = 1, \ldots, j-1)$, $j = 2, \ldots, n$.

Proof. We show inductively that

$$(X_1, \ldots, X_k) \leqq_d (Y_1, \ldots, Y_k) \tag{1.10.7'}$$

under the conditions (1.10.7) and (1.10.8). It is true for $k = 1$ by (1.10.7). Supposing it is true for $k = j - 1$, consider for any $f \in \mathfrak{R}_d(E^k)$

$$\begin{aligned}&\mathsf{E}f(X_1, \ldots, X_k)\\ &= \int_{E^{k-1}} \int_E f(x_1, \ldots, x_{k-1}, x_k)\, P_k(\mathrm{d}x_k \mid x_1, \ldots, x_{k-1})\, \mathsf{P}(X_1 \in \mathrm{d}x_1, \ldots, X_{k-1} \in \mathrm{d}x_{k-1}).\end{aligned}$$

It follows from (1.10.8) that for all $x_i \leqq y_i$ $(i = 1, \ldots, k-1)$,

$$\begin{aligned} g_1(x_1, \ldots, x_{k-1}) &\equiv \int_E f(x_1, \ldots, x_{k-1}, x_k)\, P_k(\mathrm{d}x_k \mid x_1, \ldots, x_{k-1})\\ &\leqq \int_E f(y_1, \ldots, y_{k-1}, y_k)\, Q_k(\mathrm{d}y_k \mid y_1, \ldots, y_{k-1})\\ &\equiv g_2(y_1, \ldots, y_{k-1}).\end{aligned}$$

Since the function

$$g(x_1, \ldots, x_{k-1}) \equiv \inf_{t_i \geqq x_i} g_2(t_1, \ldots, t_{k-1})$$

satisfies

$$g_1 \leqq g \leqq g_2 \quad \text{and} \quad g \in \mathfrak{R}_d(E^{k-1})$$

the definition of $\leqq_d$ and the induction assumption now ensure that (1.10.7′) holds for $k = j$, so (1.10.7) and (1.10.8) imply (1.10.6) as asserted.

Franken and Kirstein (1977) used a similar procedure to prove criteria for $\leqq_c$ and $\leqq_{cv}$.

Just as for the one-dimensional case at (1.3.5) it follows by Jensen's inequality that for $X = (X_1, \ldots, X_n) \in E = \mathbb{R}^n$,

$$(X_1, \ldots, X_n) \leqq_{cv} (m_1, \ldots, m_n) \leqq_c (X_1, \ldots, X_n) \tag{1.10.9}$$

where $m_j = \mathsf{E}X_j$ $(j = 1, \ldots, n)$.

Comparability of point processes

We turn briefly to the question of comparability of Palm distributions of stationary point processes on $\mathbb{R}$ (see Matthes, Kerstan, and Mecke (1978)). Intuitively the Palm distribution of a stationary point process describes its behaviour conditional on there being a point at the origin. We consider the sequence of n-dimensional d.f.s P_n $(n = 1, 2, \ldots)$ of distances between the origin and the $1^{\text{st}}, \ldots, n^{\text{th}}$ points to the right of the origin.

Definition 1.10.2. *Given two stationary point processes $\mathscr{P}$ and $\mathscr{Q}$, $\mathscr{P}$ is* $<$-thicker than $\mathscr{Q}$ *if for each n the corresponding n-dimensional Palm distributions P_n and Q_n satisfy*

$$P_n < Q_n \quad (< \text{ is one of } \leqq_d, \leqq_c, \leqq_{cv}). \tag{1.10.10}$$

In certain special cases criteria can be given for $\mathscr{P}$ to be $<$-thicker than $\mathscr{Q}$.

Example 1.10(a). For a *renewal process* $\mathscr{P}$ the interpoint distances are i.i.d. so, when $\mathscr{Q}$ is also a renewal process, (1.10.10) is true for all n iff $P_1 < Q_1$ where P_1, Q_1 are the d.f.s of the interpoint distances.

Example 1.10(b). *A semi-Markov point process* (SMPP) results from a semi-Markov process (SMP) on interpreting the jump points of the SMP as the epochs of a point process. The n-dimensional Palm distribution of a SMPP is given by

$$P_n(x_1, \ldots, x_n) = \sum_{k, k_1, \ldots, k_n} p_k p_{kk_1} \ldots p_{k_{n-1}k_n} F_{kk_1}(x_1) \ldots F_{k_{n-1}k_n}(x_n) \tag{1.10.11}$$

where p_{jk} are the transition probabilities, $F_{jk}(\cdot)$ the sojourn time d.f.s, and p_k the stationary distribution of the embedded jump Markov chain, assumed to be ergodic, with one-step transition probabilities p_{jk}. If $\{q_k\}$, $\{q_{jk}\}$, $\{G_{jk}(\cdot)\}$ denote the corresponding quantities for another semi-Markov point process, $\mathcal{Q}$ say, it is clear that if $p_{jk} = q_{jk}$ (all j, k), then $p_k = q_k$ (all k), and if also $F_{jk} \prec G_{jk}$ (all j, k), then $\mathcal{P}$ is $\prec$-thicker than $\mathcal{Q}$.

As an application of this set-up, suppose there are only finitely many states in the SMPP. Then we can find renewal processes $\mathcal{R}_1$ and $\mathcal{R}_2$ such that

$$\mathcal{R}_1 \leqq_d \mathcal{P} \leqq_d \mathcal{R}_2,$$

for it suffices to have $\mathcal{R}_1$, $\mathcal{R}_2$ determined by the d.f.s

$$G_1(x) = \max_{(j,k)} F_{jk}(x), \quad G_2(x) = \min_{(j,k)} F_{jk}(x) \quad (\text{all } x \geqq 0),$$

and then $G_1 \leqq_d F_{jk} \leqq_d G_2$ (all (j, k)). Of course we then have also $G_1 \leqq_c F_{jk} \leqq_c G_2$ (all (j, k)), but because the maximum of a set of convex functions is convex and minimum of a set of concave functions is concave, if not all the F_{jk} are the same, we can find extremal d.f.s $G_{cv} \neq G_1$ and $G_c \neq G_2$ satisfying

$$G_1 \leqq_{cv} G_{cv} \leqq_{cv} F_{jk} \leqq_c G_c \leqq_c G_2 \quad (\text{all } (j, k)),$$

and hence extremal renewal processes $\mathcal{R}_{cv}$ and $\mathcal{R}_c$ for which $\mathcal{R}_{cv} \leqq_{cv} \mathcal{P} \leqq_c \mathcal{R}_c$. With

$$\Delta_{jk} = \int_0^\infty t \, \mathrm{d}F_{jk}(t) = \int_0^\infty \bigl(1 - F_{jk}(u)\bigr) \, \mathrm{d}u,$$

these processes have

$$\int_0^\infty \bigl(1 - G_{cv}(u)\bigr) \, \mathrm{d}u = \min_{(j,k)} \Delta_{jk}, \quad \int_0^\infty \bigl(1 - G_c(u)\bigr) \, \mathrm{d}u = \max_{(j,k)} \Delta_{jk}.$$

If we dispense with the requirement that $p_{jk} = q_{jk}$ (all (j, k)), some other requirement like an order structure for the state space is needed. For example, taking the state space to be the non-negative integers, and assuming the Markov chain to satisfy

$$\sum_{k=1}^n p_{jk} \geqq \sum_{k=1}^n q_{jk} \quad (\text{all } i, j, n \text{ with } i \leqq j),$$

the chain has comparable n-dimensional distributions (this follows from Proposition 4.2.7), so that if also

$$F_{jk} \prec F_{mn} \text{ or } G_{jk} \prec G_{mn} \quad (\text{all } j \leqq m, \text{ all } k \leqq n) \text{ and } F_{jk} \prec G_{jk} \ (\text{all } (j, k)),$$

then the SMPPs satisfy $\mathcal{P} \prec \mathcal{Q}$ where $\prec$ is one of $\leqq_d$, $\leqq_c$ and $\leqq_{cv}$. The proof is similar to that of (1.10.6) given Veinott's criteria at (1.10.7) and (1.10.8).

Finally, we mention the following comparability property of interpoint distances of stationary point processes. Let P_1 be the one-dimensional d.f. of the distance of two subsequent points, and let P_0 be the d.f. of the distance of the two points which enclose the origin. Then

$$P_0(t) = 1 - \int_t^\infty x \,\mathrm{d}P_1(x)/m, \quad m = \int_0^\infty x \,\mathrm{d}P_1(x)$$

and it is easy to show that

$$P_1 \leqq_d P_0.$$

Notes and bibliographic comments to Chapter 1

1.1

Methods of functional analysis can be used to construct the class $\mathscr{R}$ of $\prec$-monotone functions given the class of functions $\mathscr{F}_{\prec}$ generating $\prec$. There are some connections with the bipolar theorem concerning topological vector spaces (see for example section 8.1 of Edwards (1965)). The essence of the problem appears to be to find suitable topological vector spaces E and E' which contain f and F as elements and are related by duality.

Rolski (1976a, b) considered the relations that may exist between the partial orderings $\leqq_d$, $\leqq_c$ and $\leqq_{cv}$ and properties of partial orderings for d.f.s. He showed for example that the convex ordering $\leqq_c$ is characterized by the properties (R) and (M) and the two additional properties that

$$F_1 \prec F_2 \text{ implies } F_1(\cdot + c) \prec F_2(\cdot + c) \quad \text{(all real } c\text{)},$$

and there exists an extremal element in $\mathscr{M}_m$ (see Example 1.9(a)).

1.2

The stochastic order relation $\leqq_d$ has long been used, at least since Mann and Whitney (1947) and Lehmann's work on some statistical problems (see e.g. Lehmann (1959) for references). Karlin (1960) introduced its use into inventory problems (see also Section 7.4 below), using it to prove some external monotonicity properties.

Kalmykov (1962) studied comparison problems for Markov processes. The first explicit use of $\leqq_d$ to establish certain monotonicity properties of queueing systems appears in Gaede (1965).

There are now many papers using $\leqq_d$ for the study of stochastic models (cf. also Daley (1968) for some other early references).

1.3

The relation $\leqq_c$ has not been used as long as $\leqq_d$, essentially because it is more complicated, has a smaller field of applications, and unlike $\leqq_d$ does not have an intuitive sample path property as at Proposition 1.2.1. At first $\leqq_c$ was studied in the case of equal means: see KARLIN and NOVIKOFF (1963). BESSLER and VEINOTT (1966) used $\leqq_c$ for approximations in inventory theory. The first use of $\leqq_c$ for the investigation of waiting times in the GI/GI/1 queueing system is in STOYAN and STOYAN (1969), with similar results being obtained by BOROVKOV (1970) (see Section 5.2 below). A further early use of $\leqq_c$, for d.f.s on $\mathbb{R}_+$, is in MARSHALL and PROSCHAN (1970) who used it to obtain bounds for reliability systems (see Section 7.2.4 below).

Theorem 1.3.1 can also be proved by integration by parts; for this and related problems, see also CHONG (1974).

WHITT (1980b) showed an analogue of Proposition 1.2.1: $X_1 \leqq_c X_2$ holds if and only if there exists a r.v. Z jointly distributed with X_1 such that $\mathsf{P}(X_1 + Z \leqq x) = \mathsf{P}(X_2 \leqq x)$ for all real x and $\mathsf{E}(Z \mid X_1) \geqq 0$ almost surely. The "if"-part also holds for random vectors.

1.4

The first use of $\leqq_{cv}$ was probably in MARSHALL and PROSCHAN (1970) where it appeared as a particular relation useful for obtaining bounds for the mean life time of series systems in reliability theory see p. 136.

1.5

The criteria in Examples 1.5.1(b), (c), (d) and (e) are given by MARSHALL and PROSCHAN (1970) and those in 1.5.2 by STOYAN (1972c, e) and BANDEMER and STOYAN (1971). LISEK (1978) and TAYLOR (1983) have made a thorough study of comparison properties for d.f. classes; Appendix 1 is drawn from these two papers.

ROLSKI (1977) studied the class $\{F : F \leqq_c \text{Exp}\,(m)\}$ of d.f.s that are $\leqq_c$-smaller than a given exponential d.f.

1.6

The terms NBU, ..., NWUE are defined by MARSHALL and PROSCHAN (1972) who used them for the comparison of replacement policies in reliability theory models. They considered the preservation of these properties of life time d.f.s under such operations as convolution, mixture, etc., and also studied the question as to sufficient conditions for the d.f. of system life time to be of a similar type to that of the life time d.f.s of the components; see also BARLOW and PROSCHAN (1975) and ESARY, MARSHALL and PROSCHAN (1970).

ROSS (1979), EL-NEWEIHI, PROSCHAN and SETHURAMAN (1978), EL-NEWEIHI (1981), EL-NEWEIHI and PROSCHAN (1978), ESARY and MARSHALL (1979), MARSHALL and SHAKED (1979), and BLOCK and SAVITS (1980) generalized the notions NBU and IFRA so as to include multivariate distributions and random processes. Further criteria for IFR, NBU, ... based on order statistics are given in LANGBERG, LEON and PROSCHAN (1980).

MARSHALL (1968a) proved bounds for GI/GI/1 with interarrival d.f.s of type IFR, γ-MRLA, DFR and γ-MRLB. Proposition 1.6.1 is contained in his paper with the more stringent IFR assumption for (1.6.8); DALEY and TRENGOVE (1977) showed that the weaker DMRL property suffices.

A typical example of the application of the NBU idea in queueing can be found in Proposition 5.5.1 below where residual service of interarrival times are replaced by "new" ones.

BERGMANN (1974, 1979b) defined and studied NBU, NBUE, ... properties on the basis of $\leqq_c$ instead of $\leqq_d$.

WHITT (1980a) suggested an interesting partial ordering $\leqq_{d|\mathscr{C}}$ of distributions, namely

$$P_1 \leqq_{d|\mathscr{C}} P_2 \quad \text{iff} \quad P_1(.\mid A) \leqq_d P_2(.\mid A)$$

for all $A \in \mathscr{C}$ with $P_1(A) > 0$, $P_2(A) > 0$, where $\mathscr{C}$ is a class of elements of the σ-algebra of subsets of $\mathbb{R}$ on which P_1 and P_2 are defined. In particular, if $\mathbb{R} \in \mathscr{C}$, then $P_1 \leqq_{d|\mathscr{C}} P_2$ implies $P_1 \leqq_d P_2$, while for a non-negative r.v. X with the DFR property and

$$P_2(.) = P_1(. + t \mid X > t),$$

$$P_1 \leqq_{d|\mathscr{C}} P_2 \quad \text{for}$$

$$\mathscr{C} = \{(x, \infty); 0 \leqq x < \infty\}.$$

This ordering is related to the monotone likelihood ratio ordering which is also of use, see also FERGUSON (1967), KARLIN and RINOTT (1980), KEILSON and SUMITA (1983), SMITH and WHITT (1981) and WHITT (1979, 1982b).

1.7

ROLSKI and STOYAN (1974) (see also ROLSKI (1976a) and, in similar vein, BERGMANN (1974, 1979b), defined and studied $\leqq_c^{\alpha}$ and $\leqq_{cv}^{\alpha}$ (using the notation $\leqq_{1,\alpha}$ and $\leqq_{2,\alpha}$ respectively); for the case of equal means, see also KARLIN and NOVIKOFF (1963). For these relations a generalized cut criterion holds.

STOYAN and STOYAN (1980) considered relations which facilitate the comparison of d.f.s with given fixed n^{th} moment with the aim of comparing random sets.

1.8

Of potential interest are the relations $\leqq_E$ and $\leqq_M$ defined by

$$X \leqq_E Y \quad \text{if} \quad \mathsf{E}(e^{xX}) \leqq \mathsf{E}(e^{xY}) \qquad (\text{all } x \geqq 0),$$

$$X \leqq_M Y \quad \text{if} \quad \mathsf{E}X^n \leqq \mathsf{E}Y^n \qquad (\text{all } n = 1, 2, \ldots).$$

Expansion of the exponential function shows that $X \leqq_M Y$ implies $X \leqq_E Y$, but the converse need not be true as shown by the r.v.s X and Y

for which $X = 0$ or 2 with probability 0.5 each, $Y = 1$ or 3 with probabilities 0.9 and 0.1, so that $\mathsf{E}X = 1 < \mathsf{E}Y = 1.2$, $\mathsf{E}X^2 = 2 > \mathsf{E}Y^2 = 1.8$, and $10(\mathsf{E}z^Y - \mathsf{E}z^X) = z^3 - 5z^2 + 9z - 5 = (z - 1)(z^2 - 4z + 5) > 0$ (all $z > 1$).

1.9

The nature of the $\leqq_L$ and $\leqq_c$ extremal elements of the set $\mathcal{M}_m^\sigma$ is established in Kozlov and Vasil'ev (1969) and Stoyan (1973b). The proofs rely on the fact that under certain conditions, the extreme values of the integrals of the form $\int f(t)\, \mathrm{d}F(t)$ occur for two- or three-point distributions (see, for example, Harris (1962), Hartley and David (1954) and Hoeffding (1955)). Eckberg (1977) studied bounds for Laplace-Stieltjes transforms in the case of known moments and applied them to queueing theory.

For the study of variational problems in the set of d.f.s Karlin and Studden (1966) is an important reference.

The extremal elements (in the sense of extreme values of some expectations) of sets of discrete probability distributions $\{p_n\}$ with constraints on the p_n, play an important role in the theory of decision making under uncertainty (see e.g. Kofler and Menges (1976)).

1.10

Lehmann (1955) and Veinott (1965) seem to be the first to have studied the relation $\leqq_d$ in the case of $E = \mathbb{R}^n$; the general definition of $\leqq_d$, $\leqq_c$ and $\leqq_{cv}$ has been given by Stoyan (1972b). Franken and Kirstein (1977) (see also Arjas and Lehtonen (1978)) proved Veinott's criterion in the form presented here; Veinott's original condition is a special case.

A relation similar to $\leqq_c$ was studied by Blackwell (1953), Brumelle and Vickson (1975), Hardy, Littlewood and Polya (1929), Rothchild and Stiglitz (1970, 1971), and Strassen (1965); it is defined analogously to (1.10.1), but with convex functions f. Bergmann (1978) discussed a further variant of generalizations of the relations $\leqq_d$ and $\leqq_c$ for $E = \mathbb{R}^n$ and more general spaces; his relations are useful for the proof of monotonicity properties of mixed moments and covariances. More recently, there have appeared the monographs by Marshall and Olkin (1979) and Tong (1980a) with further discussion and references on partial orderings for random vectors. They show that partial orderings of the type (1.10.1) with f Schur-convex have important applications in statistics and reliability. Further general properties of partial orderings of random vectors and random variables in general spaces are studied in Kemperman (1977) and Cambanis and Simons (1982). The relation $\leqq_D$ has also been studied by Whitt (1976) and Tchen (1980).

Hoffman-Jørgensen (1977) proved certain limit theorems by using the partial orderings for random vectors which result from $\leqq_d$ and $\leqq_c$ if the r.v.s. X are replaced by $|X| \equiv (|X_1|, \ldots, |X_n|)$. Cambanis and Simons (1980) considered $\leqq_D$ with the same modification.

For comparing stationary point processes Schmidt (1976) considered both Definition 1.10.2 and a definition based on the comparison of numbers of points in intervals of given lengths. WHITT (1981a) and HAEZENDONCK and VYLDER (1980) also studied partial orderings of point processes. For certain renewal processes MILLER (1979) showed how to construct inhomogeneous Poisson processes which are respectively "thicker" and "thinner" than the given renewal process. STOYAN and STOYAN (1980) presented some variants of stochastic orderings of random closed sets.

Open problems:

1. The relation $\leqq_c$ is closed under weak convergence in the case $E = \mathbb{R}$ provided that the means converge (see Propositions 1.2.3 and 1.3.2). What is the analogue for $E = \mathbb{R}^n$ and more general spaces? (see around Proposition 1.10.4 for $\leqq_d$; also GOROFF and WHITT (1980) is related to this matter).

2. Can simple criteria for the $\leqq_c$-comparison of stationary point processes with Poisson processes be given? Such criteria would be of considerable importance for applications in queueing theory, reliability, etc. (cf. Section 5.7 and Chapter 7).

Supplementary material added in proof

NOEBELS (1981) showed that, under certain conditions on E, any non-empty subset $\{P \in P_{\mathcal{M}} : Q_1 \leqq_d P \leqq_d Q_2\}$ defined for distributions $Q_1, Q_2 \in P_{\mathcal{M}}$ with $Q_1 \leqq_d Q_2$, is weakly compact. GOROFF and WHITT (1980) is also related to problems of convergence and comparison of distributions. Some relations even sharper than $\leqq_d$ have been considered in the literature, as for example comparisons of failure rates and of likelihood ratios (see e.g. PFANZAGL (1964) and the supplementary material referring to 1.6 on p. 35).

Concerning properties of system lifetimes, see also BIRNBAUM, ESARY and MARSHALL (1966). Concerning generalizations of NBU etc. see also ARJAS (1981a, b) and EL-NEWEIHI (1981). It is observed in CHANDRA and SINGPURWALLA (1981) that the Lorenz curve, occurring in economics contexts, can be used to characterize DFR d. f. s.

KARR (1982) studies extremal elements (in the sense of extremal elements of convex sets) of sets of distribution functions having prescribed values for a finite number of moments.

For partial orderings based on Schur-convex functions, see also RÜSCHENDORF (1981a). EATON (1982) gives an excellent survey of multivariate probability inequalities and partial orderings of multivariate distributions.

ARJAS, E. (1981a) A stochastic process approach to multivariate reliability systems: notions based on conditional stochastic order. *Math. Operat. Res.* **6**, 263—276.

— (1981b) The failure and hazard processes in multivariate reliability systems. *Math. Operat. Res.* **6**, 551—562.

BIRNBAUM, Z. W., ESARY, J. D., & MARSHALL, A. W. (1966) A stochastic characterization of wear out for components and systems. *Ann. Math. Statist.* **37**, 816—825.

CHANDRA, M., & SINGPURWALLA, N. D. (1981) Relationships between some notions which are common to reliability theory and economics. *Math. Operat. Res.* **6**, 113–121.

EATON, M. L. (1982) A review of selected topics in multivariate probability inequalities. *Ann. Statist.* **10**, 11–43.

KARR, A. F. (1982) Extreme points of certain sets of probability measures, with applications. *Math. Operat. Res.* **7**.

NOEBELS, A. (1981) A note on stochastic order of probability measures and application to Markov processes. *Zeit. Operat. Res.* **A 25**, 35–43.

RÜSCHENDORF, L. (1981 a) Ordering of distributions and rearrangement of functions. *Ann. Prob.* **9**, 276–283.

2. Stochastic models and monotonicity

2.0 General considerations concerning stochastic models

We are not aware of any general mathematical definition of the term *stochastic model* as we wish to use it, namely, that it should embrace the mathematical models of queueing, reliability, inventory, and other areas of applied probability theory in which random influences act. We have in mind the following two basic types of stochastic models:

(1°) "Finite" models. Suppose given N r.v.s $X_1, \ldots, X_N$ which are sufficient to describe the behaviour of the system. Depending on the structure of the model these X_i may denote quantities that run down simultaneously, or one after the other, or in some other fashion, and when all are so run down then the system has finished its life. The general problem is to consider the properties of a r.v. Z describing the behaviour of the system and having the form

$$Z = \Phi(X_1, \ldots, X_n) \tag{2.0.1}$$

for some suitable mapping Φ.

Example 2.0(a). In a reliability model without renewal, X_i = life time of ith element, Z = life time of the system (see Section 7.2 below).

Example 2.0(b). In a planning network, X_i = duration of ith activity, Z = duration of the full project (see Section 7.3).

(2°) "Recursive" models. Suppose given an infinite sequence of r.v.s (for example, random vectors) $\{X_n\}$, describing random influences on the state of the system at the (deterministic or random) instants t_n (with $t_n < t_{n+1}$) for $n = 1, 2, \ldots$ The system state at time t_{n+1} is described by a r.v. Z_{n+1} which is given by the recursion formula

$$Z_{n+1} = \phi_n(Z_n, X_n) \tag{2.0.2}$$

where the ϕ_n are suitable mappings. Such processes have been called "chains with complete connections" (see Iosifescu and Theodurescu (1969)).

Example 2.0(c). In a queueing model take X_n = $\{n^{\text{th}}$ inter-arrival time, n^{th} service time$\}$, Z_n = waiting time(s) associated with n^{th} customer arrival (see Sections 5.0 and 6.0). Loss systems can also be described by such recursion formulae as for example in Section 27 of Borovkov (1976a).

Example 2.0(d). In an inventory model X_n = demand(s) during n^{th} period, Z_n = inventory at the beginning of nth period (see Section 7.4).

Many other examples could be given from branching processes, learning theory, two-armed bandit problems, etc.

In what follows the distributions F_i of X_i (and of analogous r.v.s appearing in such stochastic models that do not lie within the framework of (2.0.1) and (2.0.2)) will be called *constituent distributions* (WHITT (1974a) used the terminology "initial data" for quantities having the nature of the X_i). Thus, in the case of Example 2.0(c), if the X_i are independent and identically distributed, then the constituent distribution would be just that of X_1, but if the X_i form a dependent sequence, then the constituent distribution may consist of the distribution of the whole sequence.

The mappings Φ or ϕ_n are determined for the system under consideration by its structure, i.e. by the rules for transitions between the states of the system. Just as the term stochastic model has not been described precisely, neither is the term structure: what is important in the sequel is the convention of referring to two systems as being identical in structure if the relevant mappings Φ (or, $\{\phi_n\}$) are the same (i.e. differences in constituent distributions and different stochastic dependencies are allowed). For example, two queueing systems having equal numbers of servers, the same queue discipline, the same waiting room capacity etc., have the same structure notwithstanding any differences in constituent distributions or any different stochastic dependencies in those distributions. The behaviour of a stochastic model of given structure is unambiguously determined once the constituent distributions (including their dependency properties, if any) and, in the case of recursive models, the initial distribution, are known.

The principal aim of studying stochastic models mathematically is to obtain statements about quantities which, for a given model Σ consisting of specified structure and constituent distribution(s), etc., describes its behaviour as a whole. These may be called *system quantities*, a term we use to cover real-valued quantities like means and probabilities (e.g. mean waiting time, loss probability) as well as d.f.s or distributions (e.g. of waiting time or the output process).

Let c_Σ be a system quantity in Σ and let C_Σ denote the set of possible values of c_Σ. For given structure and initial distribution $\mathfrak{A}$, c_Σ depends only on the F_i, symbolically,

$$c_\Sigma = c_\Sigma(F_1, F_2, \ldots) \in C_\Sigma. \tag{2.0.3}$$

In simpler models we may be able to deduce an explicit expression for c_Σ, but in many situations such mathematical work may not be feasible or may lead to complicated formulae that may not be useful from a practical viewpoint.

Such circumstances lead us to seek qualitative properties of c_Σ with respect to the F_i, by which we mean, roughly speaking, the way in which c_Σ is affected by variations in the F_i. Important qualitative properties of stochastic models include insensitivity (meaning, that for all F_i satisfying certain constraints, c_Σ is constant; see for example Chapter 6 of FRANKEN,

König, Arndt and Schmidt (1981)), monotonicity (if the F_i "increase" in some sense, so does c_Σ), and robustness (small perturbations in the F_i lead to small perturbations in the c_Σ).

It is by means of qualitative properties that bounding formulae can be obtained mathematically and approximations given a rigorous basis. Further, since in practice constituent distributions may not be known exactly but only through statistical data or (for projected models) through intelligent guesses, there is good reason to study qualitative properties in their own right, for on the basis of such incomplete information it may be possible at the very least to determine bounds on the system quantities of interest (see also Section 2.3.2).

2.1 Monotonicity properties

2.1.1 *Internal monotonicity properties*

Let Σ be a stochastic model with constituent distributions $(\mathfrak{A}, F_1, F_2, \ldots) = (\mathfrak{A}, \{F_i\})$, and let c_Σ be a system quantity which is time dependent, being defined either for all t in $\mathbb{R}$ or $\mathbb{R}_+$ (for example, with Σ a queueing system, c_Σ could be the mean virtual waiting time at t, or the d.f. of the number of customers in Σ at t) or for a sequence of (deterministic or random) instants t_n $(n = 1, 2, \ldots)$ (for example, with the same Σ, the waiting time d.f. or the loss probability of the nth arrival). We shall write c_Σ more precisely as

$$c_\Sigma(t) = c_\Sigma(t, \mathfrak{A}, \{F_i\}) \quad \text{or} \quad c_\Sigma(n) = c_\Sigma(n, \mathfrak{A}, \{F_i\}).$$

Let $\prec$ denote a partial ordering on the set C_Σ of possible values of c_Σ.

Definition 2.1.1. *The system quantity* $c_\Sigma(.)$ *is non-decreasing* (*non-increasing*) *with respect to the initial distribution* $\mathfrak{A}$ *if either for all* $t < u$,

$$t \leqq u \;\; \textit{implies} \;\; c_\Sigma(t) \prec (\succ)\, c_\Sigma(u), \tag{2.1.1}$$

or for integers $m < n$,

$$m < n \;\; \textit{implies} \;\; c_\Sigma(m) \prec (\succ)\, c_\Sigma(n), \tag{2.1.2}$$

which property is called internal monotonicity.

Another possible term for (2.1.1) or (2.1.2) is *temporal* or *intrinsic* monotonicity; the essential feature is that it does not depend in any way on the constituents $\{F_i\}$ other than the initial distribution. If $c_\Sigma(t)$ or $c_\Sigma(n)$ converges as t or $n \to \infty$ to some limit $c_\Sigma \equiv c_\Sigma(\{F_i\})$ independent of $\mathfrak{A}$, then it follows, in the case of (2.1.1), that for all t

$$c_\Sigma(t, \mathfrak{A}, \{F_i\}) \prec (\succ)\, c_\Sigma(\{F_i\}) \tag{2.1.3}$$

provided $\prec$ is closed under the mode of convergence used. The discrete analogue of (2.1.3) is obvious. Thus, bounds on c_Σ may follow from the choice of suitable initial distributions for internally monotonic system quantities.

We shall see in Proposition **2.2.8** and Chapter 4 how internal monotonicity properties are useful when the $c_\Sigma(.)$ are (a function of) the one-dimensional distributions of certain homogeneous Markov chains.

2.1.2 External monotonicity properties

With Σ, $\{F_i\}$ and c_Σ as in Section **2.1.1**, write $\mathscr{D}_k$ for sets of distributions F_k partially ordered by $<_k$, and let $<_c$ be a partial ordering on the set C_Σ of all possible c_Σ.

Definition 2.1.2. *The system quantity c_Σ is non-decreasing over $\mathscr{D}_k$ with respect to $<_k$ if for F, G in $\mathscr{D}_k$ and all other constituent distributions constant,*

$$F <_k G \text{ implies } c_\Sigma(F_1,\ldots,F_{k-1},F,F_{k+1},\ldots) <_c c_\Sigma(F_1,\ldots,F_{k-1},G,F_{k+1},\ldots), \tag{2.1.4}$$

which property is called external monotonicity.

When a system has the property of external monotonicity, system quantities of stochastic models having the same structure but different (and comparable) constituent distributions can be compared. Consequently the essence of external monotonicity can be expressed as follows:

Let Σ_1 and Σ_2 be stochastic models with the same structure and initial distributions. Then the models are externally monotonic if the relation $F < G$ for F and G constituent distributions in Σ_1 and Σ_2 implies $c_\Sigma(F) <_c c_\Sigma(G)$ for the system quantity c_Σ.

External monotonicity is of great value for constructing bounds on system quantities. If for example a given constituent distribution F_k can be bounded by distributions $G_1, G_2 \in \mathscr{D}_k$ for which

$$G_1 <_k F_k <_k G_2, \tag{2.1.5}$$

then for the corresponding system quantities c_Σ

$$c_\Sigma(G_1) <_c c_\Sigma(F_k) <_c c_\Sigma(G_2) \tag{2.1.6}$$

when external monotonicity holds. Many examples of external monotonicity are given in Chapters 5, 6, and 7; in the next section and Chapter 4 general methods for proving them are described.

2.2 Methods for establishing monotonicity properties

2.2.1 The functional method

Monotonicity properties of stochastic models can often be proved on the basis of monotonicity properties of functionals v_f defined over sets of d.f.s or distributions by

$$v_f(F) = \int_{-\infty}^{\infty} f(t)\,\mathrm{d}F(t) \quad \text{or} \quad v_f(P) = \int_E f(x)\,P(\mathrm{d}x)$$

for suitable classes of functions f. Such functionals occur in many known formulae for system quantities of stochastic models, and if these have the form

$$c_\Sigma(F_1, F_2, \ldots) = g\big(F_1, \ldots, F_{k-1}, v_f(F_k), F_{k+1}, \ldots\big) \tag{2.2.1}$$

where the function g is real-valued and non-decreasing in its kth argument for all other fixed F_j ($j \neq k$), then we have

Proposition 2.2.1. *Let $<_k$ be a partial ordering on the set $\mathscr{D}_k$ of all possible constituent distributions F_k, and let $f \in \mathscr{R}_{<_k}$. Then for $G_1, G_2 \in \mathscr{D}_k$,*

$$G_1 <_k G_2 \text{ implies } c_\Sigma(F_1, \ldots, F_{k-1}, G_1, \ldots) \leqq c_\Sigma(F_1, \ldots, F_{k-1}, G_2, \ldots). \tag{2.2.2}$$

To apply this functional method it is clear that g must be known, which fact restricts its application to those models in which the formulae are known, and then, in principle, an exact computation of the system quantity concerned is possible. However, these formulae may be rather complicated, and both then and in cases of incomplete information (see Section 2.3.2) the method may be of considerable value. Applications of the method may be found in Chapters 3, 5, 6 and 7.

2.2.2 *The mapping method*

The idea of the mapping method consists in using relations of the form (2.0.1) or (2.0.2); because (2.0.2) can be reduced to (2.0.1) with an appropriate function Φ, the discussion below covers both cases.

In the case of recursive models, we obtain external monotonicity properties in this way, first for nonstationary system quantities (for example, in a queueing system, the d.f. of the waiting time of the nth customer), and then, if the corresponding stationary system quantities exist and provided the nonstationary quantities converge to their stationary counterparts in a manner that is suitably "consistent" with the partial ordering used, we may be able to use Propositions 1.2.3 and 1.3.2 and its $\leqq_{cv}$ analogue to deduce external monotonicity properties for the stationary quantities.

We start by considering generally the comparison of random variables Y of the form

$$Y = f(X) \quad \text{or} \quad Y = \Phi(X)$$

as regards their dependence on X and through the functional f or mapping Φ. In the following, X_k are r.v.s and f_k measurable functions for $k = 1, 2$, and $<$ denotes one of $\leqq_d$, $\leqq_c$, $\leqq_{cv}$. When comparing the r.v.s $f_k(X_k)$ by $\leqq_c$ or $\leqq_{cv}$, we assume the finiteness of the moments $\mathsf{E} \max\big(0, f_k(X_k)\big)$, $\mathsf{E} \min\big(0, f_k(X_k)\big)$ respectively.

Theorem 2.2.2a. *If there exists $f \in \mathfrak{R}_<(\mathbb{R})$ with*

$$f_1(t) \leqq f(t) \leqq f_2(t) \qquad (\textit{all } t \in \mathbb{R}), \tag{2.2.3}$$

then

$$X_1 < X_2 \tag{2.2.4}$$

implies

$$f_1(X_1) \prec f_2(X_2). \tag{2.2.5}$$

Conversely, if (2.2.5) *holds for all* X_1 *and* X_2 *satisfying* (2.2.4), *then there exists* $f \in \mathfrak{K}_{\prec}(\mathbb{R})$ *satisfying* (2.2.3).

Proof. Suppose the existence of f satisfying (2.2.3), and depending on $\prec$ being $\leqq_d$, $\leqq_c$, $\leqq_{cv}$, let

$$h_{\prec,x}(t) = \Theta_x\big(f(t)\big),\ \max\big(x, f(t)\big),\ \min\big(x, f(t)\big)$$

respectively, so that for all x, the function $h_{\prec,x}(t) \in \mathfrak{K}_{\prec}(\mathbb{R})$. Using (2.2.4) and appealing to Theorem 1.2.2, 1.3.1 or 1.4.1 as appropriate, it follows that $f(X_1) \prec f(X_2)$, while from (2.2.3) it is trivially true that $f_1(X_1) \prec f(X_1)$ and $f(X_2) \prec f_2(X_2)$. Then (2.2.5) is proved.

We now prove the converse when $\prec$ denotes $\leqq_d$ or $\leqq_c$; the proof for $\leqq_{cv}$ is similar. We do this by constructing an $f \in \mathfrak{K}_{\prec}(\mathbb{R})$ satisfying (2.2.3). Suppose then that (2.2.4) and (2.2.5) hold as required when $\prec$ denotes $\leqq_d$, and consider

$$f(t) = \sup\nolimits_{x \leqq t} f_1(x).$$

Then f is non-decreasing and $f_1(t) \leqq f(t)$ (all t). Suppose that there exists v for which $f_2(v) < f(v)$. Then there is also some $u \leqq v$ for which $f_1(u) > f_2(v)$. Let $X_1 = u$ a.s., $X_2 = v$ a.s., so that $X_1 \leqq_d X_2$ but (2.2.5) fails. Consequently $f_2(t) \geqq f(t)$ (all t), and f as constructed satisfies (2.2.3).

Now suppose that (2.2.4) and (2.2.5) hold as required when $\prec$ denotes $\leqq_c$. We form the convex hull K of the set $\{(x, y): y \geqq f_2(x)\}$ and for each real t define

$$f^*(t) = \inf \{y: (t, y) \in K\} \tag{2.2.6}$$

so that, K being convex, $f^*(t)$ is convex in t, and by construction of K we have the alternative expression

$$f^*(t) = \inf_{\{t_i\}} \Big\{\sum_i \lambda_i f_2(t_i): \sum_i \lambda_i t_i = t, \sum_i \lambda_i = 1, \lambda_i \geqq 0\Big\}.$$

Suppose that for some real u, $f_1(u) > f^*(u)$, so that there will then exist a probability distribution $\{\lambda_i\}$ on the finite set of points $\{u_i\}$ for which $\sum_i \lambda_i u_i = u$ and $f_1(u) > \sum_i \lambda_i f_2(u_i) \geqq f^*(u)$. Let the r.v.s $X_1 = u$ a.s. and $X_2 = u_i$ with probability λ_i, so that $X_1 \leqq_c X_2$ and hence, by assumption, we should have $\mathsf{E} f_1(X_1) = f_1(u) < \sum_i \lambda_i f_2(u_i) = \mathsf{E} f_2(X_2)$, contradicting $f_1(u) > f^*(u)$. Thus, (2.2.3) holds with $f = f^*$.

If f^* is non-decreasing then $f^* \in \mathfrak{K}_{\leqq_c}(\mathbb{R})$. Otherwise, since $X_1 \leqq_d X_2$ implies $X_1 \leqq_c X_2$, we can use the argument given earlier for $\prec$ denoting $\leqq_d$ to conclude that $f_1(u) \leqq \inf_{t \geqq u} f_2(t)$. Now $\inf_{t \in \mathbb{R}} f^*(t) = \inf_{t \in \mathbb{R}} f_2(t)$ and so, if f^* is non-increasing, $f(u) = \inf_{t \geqq u} f_2(t)$ satisfies (2.2.3) and $f \in \mathfrak{K}_{\leqq_c}(\mathbb{R})$, or if f^* is convex with a minimum at t_0 say, then $f(t) = f^*(t)$ $(t \geqq t_0)$, $= f^*(t_0)$ $(t \leqq t_0)$ satisfies (2.2.3) and $\in \mathfrak{K}_{\leqq_c}(\mathbb{R})$. The converse is now proved.

Note that (2.2.3) is true for $\leqq_d$ if and only if

$$f_1(t_1) \leqq f_2(t_2) \qquad (t_1 \leqq t_2),$$

and it is true for $\leqq_c$ if and only if additionally

$$f_1\big(\lambda t_1 + (1-\lambda)\,t_2\big) \leqq \lambda f_2(t_1) + (1-\lambda)\,f_2(t_2).$$

For r.v.s X_1 and X_2 in the space $(E, \mathscr{M})$ as in Section 1.10 we have

Theorem 2.2.2b. *Given measurable functions f_1, f_2, if there exists $f \in \overline{\mathfrak{R}}_<(E)$ with*

$$f_1(t) \leqq f(t) \leqq f_2(t) \qquad (\textit{all } t \in E) \tag{2.2.3'}$$

then (2.2.4) *implies* (2.2.5). *Conversely, if* (2.2.5) *holds for all X_1, X_2 satisfying* (2.2.4) *in the case that $<$ denotes $\leqq_d$, then $f(t) \equiv \sup_{x<t} f_1(x)$ is a mapping satisfying* (2.2.3′), *provided it is measurable.*

The proof, being similar to that of Theorem 2.2.2a, is omitted, except to remark that in the construction for the converse, the partial ordering $\leqq$ on E is used. Also the measurability assumption is necessary; KEMPERMAN (1977) gives an example in $\mathbb{R}^2$ where measurability fails.

Using the same method it may also be proved that for two r.v.s X_1 and X_2 on $(E, \mathscr{M})$ with distributions P_1 and P_2, the relation $X_1 \leqq_d X_2$ is equivalent to

$$\int_E f_1(x)\,P_1(\mathrm{d}x) \leqq \int_E f_2(x)\,P_2(\mathrm{d}x) \tag{2.2.5'}$$

holding for all measurable functions $f_1, f_2 \colon E \to \mathbb{R}$ such that $f_1(x) \leqq f_2(y)$ when $x \leqq y$ on E. If (2.2.5′) is true some authors speak of P_2 being a *strong dilatation* of P_1 (see e.g. KEMPERMAN (1977)).

We now consider mappings Φ such as occur for example at (2.0.1); these mappings may be more general than functionals on E. For some positive integer N and for each $i = 1, \ldots, N$, let X_i be $(E_i, \mathscr{M}_i)$-valued r.v.s on the probability space $(\Omega, \mathscr{A}, \mathsf{P})$ and let Φ be a mapping from $E_1 \times \cdots \times E_N$ into some other space $(E, \mathscr{M})$ with appropriate measurability properties such that

$$Z \equiv \Phi(X_1, \ldots, X_N) \tag{2.0.1'}$$

is an $(E, \mathscr{M})$-valued r.v. on $(\Omega, \mathscr{A}, \mathsf{P})$. We assume that E and the E_i are vector spaces with partial orderings $\leqq_E$ and $\leqq_{E_i}$ respectively.

What emerges is that monotonicity, convexity or concavity properties of Φ facilitate the deduction of analogous properties of Z with respect to the X_i, according as $\leqq_d$, $\leqq_c$ or $\leqq_{cv}$ is used.

It should be noted that the statements below remain true if Z is a function of a possibly infinite sequence of r.v.s, with N possibly random also, as for example if $Z = \sum_{i=1}^{\nu} X_i$ for some random positive integer ν.

Definition 2.2.1. $\Phi \equiv \Phi(x_1, \ldots, x_N)$ *is* nondecreasing (nonincreasing) *in* $x_i (1 \leqq i \leqq N)$ *if for all fixed* x_j *for* $j \neq i$, *and all* $x_i', x_i'' \in E_i$, $x_i' \leqq_{E_i} x_i''$ *implies* $\Phi(x_1, \ldots, x_i', \ldots, x_N) \leqq_E (\geqq_E) \Phi(x_1, \ldots, x_i'', \ldots, x_N)$. Φ *is* convex *in* x_i $(1 \leqq i \leqq N)$ *if for all fixed* x_j *for* $j \neq i$, *and all* $x_i', x_i'' \in E_i$ *and* $0 < \lambda < 1$,

$$\Phi(x_1, \ldots, \lambda x_i' + (1 - \lambda) x_i'', \ldots, x_N) \leqq_E \lambda \Phi(x_1, \ldots, x_i', \ldots, x_N) + (1 - \lambda) \Phi(x_1, \ldots, x_i'', \ldots, x_N).$$

Concavity is defined similarly with $\leqq_E$ replaced by $\geqq_E$. Also, since the spaces E_i are vector spaces, so also are $E_{i_1} \times \cdots \times E_{i_m}$ and convexity and concavity can be defined in terms of the vector argument $x_{i_1}, \ldots, x_{i_m}$ of Φ. If Φ is defined only on a subset of $E_1 \times \cdots \times E_N$ then the conditions of Definition 2.2.1 are assumed to hold only on the domain of definition of Φ. According to the nature of Φ we have

Theorem 2.2.3. *Let the r.v.s* X_i', X_i'' *be independent of* $\{X_j: j \neq i\}$ *and set* $Z' = \Phi(X_1, \ldots, X_i', \ldots, X_N)$ *and similarly* Z''. *Then when* Φ *is nondecreasing in* x_i,

$$X_i' \leqq_d X_i'' \text{ implies } Z' \leqq_d Z'', \tag{2.2.7a}$$

while when Φ *is convex in* x_i

$$X_i' \leqq_c X_i'' \text{ implies } Z' \leqq_c Z''. \tag{2.2.7b}$$

The proof proceeds by the use of functions f in $\mathfrak{K}_d(E)$ or $\mathfrak{K}_c(E)$ as appropriate.

The theorem can be formulated more generally by writing (cf. Section 1.10) $\{X_1, \ldots, X_N\} \leqq_d \{Y_1, \ldots, Y_N\}$ when $\mathsf{E}f(X_1, \ldots, X_N) \leqq \mathsf{E}f(Y_1, \ldots, Y_N)$ for all $f \in \mathfrak{K}_d(E_1 \times \cdots \times E_N)$, the partial ordering $x \leqq y$ on this product space meaning $x_i \leqq_{E_i} y_i$ for each $i = 1, \ldots, N$. Then

Theorem 2.2.4. *For* $m \leqq N$ *let the r.v.s* $\{X_1', \ldots, X_m'\}$ *and* $\{X_1'', \ldots, X_m''\}$ *be independent of* $\{X_{m+1}, \ldots, X_N\}$. *Then when* Φ *is nondecreasing in* $\{x_1, \ldots, x_m\}$

$$\{X_1', \ldots, X_m'\} \leqq_d \{X_1'', \ldots, X_m''\} \text{ implies } Z' \leqq_d Z'', \tag{2.2.8a}$$

while for Φ *additionally convex in* $\{x_1, \ldots, x_m\}$,

$$\{X_1', \ldots, X_m'\} \leqq_c \{X_1'', \ldots, X_m''\} \text{ implies } Z' \leqq_c Z''. \tag{2.2.8b}$$

The following is an important example; as in Theorem 2.2.2a, $<$ denotes one of $\leqq_d$, $\leqq_c$, $\leqq_{cv}$.

Proposition 2.2.5. *Let* $\{X_n\}$, $\{Y_n\}$ *be sequences of non-negative stochastically independent r.v.s for which*

$$X_n < X_{n+1} \text{ and } Y_n < Y_{n+1} \quad (n = 1, 2, \ldots), \tag{2.2.9}$$

and let M, N *be non-negative integer-valued r.v.s independent of* $\{X_n\}$ *and* $\{Y_n\}$. *Then*

$$M < N \text{ and } X_n < Y_n \quad (n = 1, 2, \ldots)$$

imply

$$\sum_{n=1}^{M} X_n < \sum_{n=1}^{N} Y_n \tag{2.2.10}$$

where the empty sum equals zero.

Proof. We give the proof for $<$ denoting $\leqq_c$; the other cases are similar. Since $\leqq_c$ has the convolution property (C), it follows from the assumptions that for any fixed integer n, $\sum_{j=1}^{n} X_j < \sum_{j=1}^{n} Y_j$, i.e., that for any real t,

$$f_n(t) \equiv \mathsf{E}(X_1 + \cdots + X_n - t)_+ \leqq \mathsf{E}(Y_1 + \cdots + Y_n - t)_+ \equiv g_n(t).$$

Now (2.2.10) is equivalent to having

$$\sum_{n=0}^{\infty} \mathsf{P}(M = n)\, f_n(t) \leqq \sum_{n=0}^{\infty} \mathsf{P}(N = n)\, g_n(t) \qquad (\text{all } t) \tag{2.2.11}$$

in which we observe that $\{f_n(t)\}$ is a convex sequence because, with $S_n = X_1 + \cdots + X_n$ and using first the non-negativity of X_n and then that $X_{n+1} \geqq_c X_n$,

$$\begin{aligned} f_{n+1}(t) - f_n(t) &= \mathsf{E}(X_{n+1} + S_n - t)_+ - \mathsf{E}(S_n - t)_+ \\ &\geqq \mathsf{E}(X_{n+1} + S_{n-1} - t)_+ - \mathsf{E}(S_{n-1} - t)_+ \\ &\geqq \mathsf{E}(X_n + S_{n-1} - t)_+ - \mathsf{E}(S_{n-1} - t)_+ = f_n(t) - f_{n-1}(t). \end{aligned}$$

We can now rewrite the left-hand side of (2.2.11), as

$$\begin{aligned} &\sum_{n=0}^{\infty} \mathsf{E}(M - n)_+ \big(f_{n+1}(t) - 2f_n(t) + f_{n-1}(t)\big) \\ &\quad \leqq \sum_{n=0}^{\infty} \mathsf{E}(N - n)_+ \big(f_{n+1}(t) - 2f_n(t) + f_{n-1}(t)\big) \\ &\quad = \sum_{n=0}^{\infty} \mathsf{P}(N = n)\, f_n(t) \leqq \sum_{n=0}^{\infty} \mathsf{P}(N = n)\, g_n(t). \end{aligned}$$

In the case of recursive models it is of course desirable to express the monotonicity and/or convexity properties of Φ in terms of the properties of the ϕ_n. To illustrate the idea consider the following example.

Let the $\{Z_n\}$ constitute a stochastic process with values in $(E, \mathcal{M})$ for some vector space E on which $\leqq_E$ is a partial ordering and $\mathcal{M}$ is a σ-algebra of subsets as in Section 1.10. Let the $\{X_n\}$ be independent r.v.s on the probability space $(\Omega, \mathcal{A}, \mathsf{P})$ taking values in the vector space E_0 on which $\leqq_{E_0}$ is a partial ordering and $\mathcal{M}_0$ is a σ-algebra of subsets. Finally, let Z_1 be a r.v. on $(\Omega, \mathcal{A}, \mathsf{P})$ independent of all $\{X_n\}$ and let ϕ_n $\big(= \phi_n(x, y)\big)$ be measurable mappings of $E \times E_0$ into E. Clearly the sequence $\{Z_n\} = \{\phi_n(Z_{n-1}, X_{n-1})\}$ is then a Markov chain, and if the X_n are identically distributed and $\phi_n = \phi_1$ (all $n = 2, 3, \ldots$), it also has stationary transition probabilities.

For simplicity write $\boldsymbol{y}_n = \{y_1, \ldots, y_n\}$ and

$$\Phi_1(x_1, \boldsymbol{y}_1) = \phi_1(x_1, y_1),$$

$$\Phi_n(x_1, \boldsymbol{y}_n) = \phi_n\big(\Phi_{n-1}(x_1, \boldsymbol{y}_{n-1}), y_n\big) \quad (n = 2, 3, \ldots).$$

On $E_0{}^n$ we use the componentwise partial ordering $\leqq_{E_0{}^n}$. For the Φ_n we then have

Theorem 2.2.6. (a) *If all ϕ_n are nondecreasing in x and y, then for every n, Φ_n is nondecreasing in $\boldsymbol{y}_n$.*

(b) *If all ϕ_n are nondecreasing in x and convex (concave) in (x, y), then for every n, Φ_n is convex (concave) in $\boldsymbol{y}_n$.*

(c) *If all ϕ_n are nondecreasing in x and convex in x and y, then for every n, Φ_n is convex in y_j $(j = 1, \ldots, n)$.*

Proof. Φ_1 is nondecreasing in y_1 by the monotonicity of ϕ_1 in y. Since

$$\Phi_{k+1}(x, y_1, \ldots, y_{k+1}) = \phi_{k+1}\big(\Phi_k(x, y_1, \ldots, y_k), y_{k+1}\big),$$

an easy introduction step now establishes the monotonicity of Φ_k for all $k = 1, 2, \ldots$

Again, Φ_1 is convex in y_1 by the convexity of ϕ_1 in y. For $0 < \lambda < 1$, assuming Φ_k is convex in $\boldsymbol{y}_k$,

$$\Phi_{k+1}\big(\lambda\boldsymbol{y}_{k+1}' + (1-\lambda)\boldsymbol{y}_{k+1}''\big) = \phi_{k+1}\big(\Phi_k\big(\lambda\boldsymbol{y}_k' + (1-\lambda)\boldsymbol{y}_k''\big), \lambda y_{k+1}' + (1-\lambda)y_{k+1}''\big)$$

$$\leqq_E \phi_{k+1}\big(\lambda\Phi_k(\boldsymbol{y}_k') + (1-\lambda)\Phi_k(\boldsymbol{y}_k''), \lambda y_{k+1}' + (1-\lambda)y_{k+1}''\big)$$

$$\leqq_E \lambda\phi_{k+1}\big(\Phi_k(\boldsymbol{y}_k'), y_{k+1}'\big) + (1-\lambda)\phi_{k+1}\big(\Phi_k(\boldsymbol{y}_k''), y_{k+1}''\big)$$

$$= \lambda\Phi_{k+1}(\boldsymbol{y}_{k+1}') + (1-\lambda)\Phi_{k+1}(\boldsymbol{y}_{k+1}'')$$

by the properties assumed for Φ_k and ϕ_{k+1} at (b). The concave case is treated similarly, and induction completes the proof of (b). Part (c) is proved similarly.

The most important particular case of these results is part (a) of the following.

Theorem 2.2.7. (a) *For Markov processes $\{Z_n'\}$ and $\{Z_n''\}$ of the form (2.0.2) and which are defined by the same ϕ_n and sequences $\{X_n'\}$ and $\{X_n''\}$ of independent r.v.s for which $X_n' \prec X_n''$ (all n), where $\prec$ denotes one of $\leqq_d$, $\leqq_c$, $\leqq_{cv}$, if $Z_1' \prec Z_1''$, then when $\prec$ denotes $\leqq_d$ and for all n ϕ_n is nondecreasing in x and y, or when $\prec$ denotes $\leqq_c$ $(\leqq_{cv})$ and for all n ϕ_n is nondecreasing and convex (concave) in (x, y),*

$$Z_n' \prec Z_n''. \tag{2.2.12}$$

(b) *For stochastic processes $\{Z_n'\}$ and $\{Z_n''\}$ of the form (2.0.2) and which are generated by the same ϕ_n and sequences $\{X_n'\}$ and $\{X_n''\}$ of which the* *members are stochastically dependent but independent of Z_1' and Z_1'', and such that*

$$\{X_1', \ldots, X_n''\} < \{X_1'', \ldots, X_n''\} \qquad (n = 1, 2, \ldots) \tag{2.2.13}$$

where $<$ *is as in part* (a), *then under the same conditions as in part* (a)

$$\{Z_1', \ldots, Z_n'\} < \{Z_1'', \ldots, Z_n''\}. \tag{2.2.14}$$

By using a similar argument we also have the following statement.

Theorem 2.2.8. *Let $\{Z_n\}$ be a Markov chain with stationary transition probabilities of the form* (2.0.2) *with independent identically distributed $\{X_n\}$ and $\phi_n = \phi$ (all n). With $<$ denoting one of $\leqq_d$, $\leqq_c$, $\leqq_{cv}$, then $Z_1 < Z_2$ implies*

$$Z_n < Z_{n+1} \qquad (\text{all } n = 1, 2, \ldots) \tag{2.2.15}$$

if ϕ is nondecreasing in x and, additionally in the case $\leqq_c$ ($\leqq_{cv}$), convex (concave) as well.

For a general stationary sequence $\{X_n\}$ the final theorem of this section gives an internal monotonicity property with respect to $\leqq_d$ (cf. also (5.1.6)); it is a generalization of Theorem 2.2.8 for this partial ordering.

Theorem 2.2.9. *Let E and E_0 be partially ordered spaces with σ-algebras $\mathscr{M}$ and $\mathscr{M}_0$, and let $\phi(x, y)$: $E \times E_0 \to E$ be a measurable mapping which is nondecreasing in E. Let $\{X_n\}$ ($n = 0, \pm 1, \pm 2, \ldots$) be a stationary sequence of r.v.s in $(E_0, \mathscr{M}_0)$, and Z_0 an E-valued r.v. independent of $\{X_n\}$. If*

$$Z_0 \leqq \phi(Z_0, X_n) \text{ a.s.} \qquad (n = 0, \pm 1, \ldots), \tag{2.2.16}$$

then the sequence $\{Z_n\}$ defined by $Z_{n+1} = \phi(Z_n, X_n)$ satisfies

$$Z_n \leqq_d Z_{n+1} \qquad (n = 0, 1, \ldots). \tag{2.2.17}$$

Proof. For each $k = 0, 1, \ldots$ we construct the sequence $\{Z_n^k\}$ by

$$Z_n^k = \begin{cases} Z_0 & (n \leqq -k), \\ \phi(Z_{n-1}^k, X_{n-1}) & (n > -k), \end{cases}$$

so that $\{Z_n\} = \{Z_n^0\}$, and by stationarity of the sequence $\{X_n\}$ and its independence of Z_0, Z_n^0 has the same distribution as Z_0^n. We assert that for fixed n,

$$Z_n^k \leqq Z_n^{k+1} \text{ a.s.} \qquad (k = 0, \pm 1, \pm 2, \ldots). \tag{2.2.18}$$

To prove this, we have only to consider $k > -n$ since for $k < -n$ $Z_n^k = Z_n^{k+1} = Z_0$ by definition and for $k = -n$ it is true by assumption at (2.2.16). By monotonicity of ϕ, we have a.s.

$$Z_n^{-n+1} = \phi(Z_{n-1}^{-(n-1)}, X_{n-1}) \leqq \phi(Z_{n-1}^{-(n-2)}, X_{n-1}) = Z_n^{-n+2},$$

so (2.2.18) is now proved for $k \leqq -n + 1$. An induction proof now establishes (2.2.18) for all k. In particular then, $Z_0^n \leqq Z_0^{n+1}$ a.s., and since Z_0^n has the same distribution as $Z_n^0 = Z_n$, (2.2.17) now follows.

2.2.3 *The sample path method*

For the case of the partial ordering $\leqq_d$ another approach, nearly equivalent to the mapping method as in Section 2.2.2, is a method based on the comparison of sample paths and utilizing Theorems 1.2.1 and 1.10.4. In contrast to the mapping method which is based on the comparison of expectations, the idea now is to compare directly r.v.s that are suitably constructed. However, in both methods the same monotonicity properties must be proved, so it would appear that the fields of application of both methods are identical.

By way of illustration, we give an alternative proof of (2.2.8a) in Theorem 2.2.4; we shall show that

$$\Phi(X_1', \ldots, X_m', X_{m+1}, \ldots, X_N) \leqq_d \Phi(X_1'', \ldots, X_m'', X_{m+1}, \ldots, X_N) \qquad (2.2.19)$$

if (1) Φ is nondecreasing in $x_1, \ldots, x_m$, (2) the r.v.s $\{X_1', \ldots, X_m'\}$ and $\{X_1'', \ldots, X_m''\}$ are independent of $\{X_{m+1}, \ldots, X_N\}$, and (3) $\{X_1', \ldots, X_m'\}$ $\leqq_d \{X_1'', \ldots, X_m''\}$. We do so by appealing to Proposition 1.10.4 which ensures the existence of a probability space $(\Omega, \mathscr{A}, \mathsf{P})$ and r.v.s $\{Y_1', \ldots, Y_N'\}$ and $\{Y_1'', \ldots, Y_N''\}$ on it such that a.s. for $\omega \in \Omega$, $Y_j'(\omega) \leqq Y_j''(\omega)$ $(j = 1, \ldots, m)$, $Y_j'(\omega) = Y_j''(\omega)$ $(j = m + 1, \ldots, N)$, and $\{Y_1', \ldots, Y_N'\}$ distributed like $\{X_1', \ldots, X_m', X_{m+1}, \ldots, X_N\}$ and similarly for the Y_j''. Because of the monotonicity properties of Φ it then follows that

$$\Phi\big(Y_1'(\omega), \ldots, Y_N'(\omega)\big) \leqq \Phi\big(Y_1''(\omega), \ldots, Y_N''(\omega)\big) \quad \text{a.s. for } \omega \in \Omega,$$

from which (2.2.19) follows.

It is pertinent to observe that the sample path method is closely related to the way we may in theory construct simulations using uniform r.v.s and the inverses of d.f.s. Suppose that U is a generic r.v. uniformly distributed on $(0, 1)$ used to generate r.v.s X and Y with d.f.s F and G respectively. Then it is easily checked that of all bivariate r.v.s (X, Y) with specified marginals F and G, the r.v. $\big(F^{-1}(U), G^{-1}(U)\big)$ has a distribution for which $\mathsf{E}(X - Y)^2$ is least (assuming this moment to be finite). Such a construction, using inverse d.f.s, is most appropriate when it is sought to compare sample paths for two models having the same structure and differing only in the d.f.s of their constituent elements.

2.3 Extremal problems

2.3.1 *Solution of extremal problems by means of partial orderings*

By the term *extremal problem* for a stochastic model we mean a variational problem of the following type: for a given system quantity $c_\Sigma(F_1, F_2, \ldots)$ of a stochastic model Σ, determine its supremum or infimum when some (or all) of the F_k vary in specified classes $\mathscr{M}_k$ of constituent distributions while the others (if any) remain fixed, i.e., symbolically, determine for example

$$\sup_{F_k \in \mathcal{M}_k, k=k_1,\ldots,k_m} c_\Sigma(F_1, F_2, \ldots).$$

If for example c_Σ is a mean and the classes $\mathcal{M}_k$ are characterized by fixed moments, the results of HARRIS (1962), HARTLEY and DAVID (1954) and HOEFFDING (1955) can be used. Note that supremum and infimum are to be understood in the sense of the ordering of real numbers if c_Σ is real, and otherwise in the sense of a suitable partial ordering on c_Σ.

An extremal problem can certainly be solved by means of partial orderings when c_Σ is nondecreasing in the F_k with respect to $<_k$ and the $\mathcal{M}_k$ possess $<_k$-extremal elements $\sup_{\mathcal{M}_k} F_k$ or $\inf_{\mathcal{M}_k} F_k$. Then for example

$$\sup_{F_1 \in \mathcal{M}_1} c_\Sigma(F_1, F_2, \ldots) = c_\Sigma(\sup F_1, F_2, \ldots).$$

In Chapters 5, 6 and 7 we solve some extremal problems and establish some extremality properties in this way, making use of the extremal elements given in Section 1.9 and of external monotonicity.

To conclude this section it should be noted that extremality statements can sometimes be embedded in statements about external monotonicity. For example, it may be known that a system quantity is maximized for a constituent distribution in a given class by a certain distribution; if it can be shown that the system quantity has an external monotonicity property with respect to the partial ordering for which the extremal distribution has been chosen, then it is proper to regard the extremality statement as having been so embedded. It is possible that we may in this way obtain a deeper understanding of the original extremality statement. Examples of such embeddings are given in Sections 5.2 and 7.5.

2.3.2 Problems with incomplete information on distributions

An important area of application of extremality statements is the solution of problems with incomplete information on d.f.s or distributions. By such problems we mean the problem of determining a system quantity c_Σ when some constituent distribution F_k is not known precisely but only that it belongs to some class $\mathcal{M}_k$.

By way of example, consider the problem of determining the mean waiting time m_W in a GI/GI/1 queueing system when the service time d.f. B is given but only the mean inter-arrival time m_A is known. Letting Σ denote GI/GI/1, $c_\Sigma = m_W$, $F_1 = B$, F_2 = inter-arrival time d.f., $\mathcal{M}_2$ = {d.f.s on $\mathbb{R}_+$ with mean m_A}, we then have a problem with incomplete information in the sense just described, and, interpreting it as an extremality problem, it is solved by giving the bounds on c_Σ,

$$\inf_{F_k \in \mathcal{M}_k (k=1,2,\ldots)} (\sup) \; c_\Sigma(F_1, F_2, \ldots).$$

Notes and bibliographic comments to Chapter 2

2.2

A statement analogous to Theorem 2.2.2a and 2.2.2b can also be proved for other partial orderings $\prec$ for d.f.s. The "intermediate function" f must then have the property, in addition to (2.2.3) and $f \in \mathscr{R}_{<}$, that for all $g \in \mathscr{R}_{<}$, the function $g_f \in \mathbb{R}_{<}$ also where g_f is defined by $g_f(t) = g(f(t))$. General considerations concerning inequalities of the type $\mathsf{E}f_1(X_1) \leqq \mathsf{E}f_2(X_2)$ can be found in Kemperman (1977) and Cambanis and Simons (1982).

Theorems 2.2.3 and 2.2.4 can also be proved in a more general form, namely for mappings Φ_1 and Φ_2 and an "intermediate mapping" Φ analogous to the generalization just indicated for Theorem 2.2.2a. However, for the proof of external monotonicity the case of a fixed mapping is the more interesting, because this is equivalent to having models of the same structure.

The proof of Theorem 2.2.5 is from Borovkov (1976a). By theorems from measure theory homogeneous Markov chains can always be written in the form (2.0.2) with i.i.d. X_n if the state space E is a complete, separable, metric space and $\mathscr{M}$ is a corresponding Borel σ-algebra. It is possible moreover to assume that the X_n are uniformly distributed on $[0, 1]$; this is based on the isomorphism between non-atomic separable σ-algebras and the σ-algebra of all Borel sets in $[0, 1]$, and on the countability of the number of atoms of a probability measure; see Halmos (1958), §§ 39, 40.

The proof of Theorem 2.2.9 is a simple generalisation of the proof of internal monotonicity of waiting times in G/G/s; see Borovkov (1976a). The following example shows that it is not possible to weaken the assumption in Theorem 2.2.9 from a.s. inequality at (2.2.16) to $\leqq_d$. Let $\phi(x, y) = (x + y)_+$, $X_n = X_{n+5}$ a.s. (all n), $(X_0, X_1, X_2, X_3, X_4) = (1, 1, -1, -1, -1)$ and its four cyclic permutations with probability .2 each. When $\mathsf{P}(Z_0 = 0) = .8 = 1 - \mathsf{P}(Z_0 = 1)$, with Z_0 independent of $\{X_n\}$, it is easily checked that $\mathsf{P}(Z_2 = 3) = .04$ while for all $n \geqq 5$, $\mathsf{P}(Z_n \leqq 2) = 1$, and therefore (2.2.17) cannot hold.

Note that useful new partial orderings may arise by considering special problems: it was in this way that the author was led to the partial ordering $\leqq_c$, from investigations of GI/GI/1.

The sample path method has been used for the proof of monotonicity properties of stochastic models by Jacobs and Schach (1972), O'Brien (1975b), and others.

It would seem that, except for the functional method, it is a deficiency of the other methods known for proving external monotonicity that they do not make use of the identity of the X_k when this is assumed. It may be the case therefore that a method which does make use of this property of models, may lead to proofs of the hypotheses on the monotonicity properties of multiserver waiting and loss systems (see Sections 6.3.3 and 7.5) and deeper properties of other systems.

3. Some applications of the partial orderings $\leqq_d$, $\leqq_c$ and $\leqq_{cv}$ in experimental design

3.0 Introduction

In this chapter we will consider quite briefly two possible applications to statistical problems of the methods described in Chapter 2:

a) the solution of a problem in experimental design given incomplete information on distributions; and

b) the choice of optimal estimators.

We start by giving some convexity properties of commonly occurring sample functions; they can be a basis for qualitative investigations of the distributions of the sample functions with the help of the mapping method.

3.1 Convexity properties of sample functions

Some important sample functions are convex functions $Y = \Phi(X_1, \ldots, X_n)$ of the observations or measurement errors $X_1, \ldots, X_n$ which we assume to be i.i.d. With the help of the mapping method (Theorem 2.2.4) we can then prove comparison theorems of the following type for sample functions. Let F_1 and F_2 be the d.f.s of observations or measurement errors X_i' and X_i'' ($i = 1, 2, \ldots, n$) for which $\mathsf{E}X_i' = \mathsf{E}X_i''$. Then for the corresponding d.f.s G_1 and G_2 of the sample functions Y_1 and Y_2, where $Y_1 = \Phi(X_1', \ldots, X_n')$ and $Y_2 = \Phi(X_1'', \ldots, X_n'')$, it follows from Theorem 2.2.4 that

$$F_1 \leqq_c F_2 \text{ implies } G_1 \leqq_c G_2. \tag{3.1.1}$$

If Y_1 and Y_2 have the same mean, it then follows from (3.1.1) that

$$\mathsf{E}f(Y_1) \leqq \mathsf{E}f(Y_2) \tag{3.1.2}$$

for all convex f for which the means exist.

If Φ is concave in its arguments analogous relations hold for $\leqq_{cv}$.

Example 3.1(a). The sample mean $\bar{X} = \left(\sum_{i=1}^{n} X_i\right) \Big/ n$ is linear (and, consequently, convex) in the X_i.

Example 3.1(b). The sample variance $S^2 = \left(\sum_{i=1}^{n} (X_i - \bar{X})^2\right) \Big/ (n - 1)$ is convex in the X_i.

Example 3.1(c). The range $D = \max_{1 \leqq i \leqq n} X_i - \min_{1 \leqq i \leqq n} X_i$ is convex as it is the sum of two convex functions.

In estimation problems of statistics there often occur loss functions R which are functions of the deviations between a parameter θ and estimators $\hat{\theta}$ for θ. If the sample function $\theta = \Phi(X_1, \ldots, X_n)$ and R have suitable properties it may be possible to show that the risk $\mathsf{E}R(\theta, \hat{\theta})$ is monotonic or convex in its dependence on the X_i. For example, in estimating the population mean m by the sample mean $\overline{X}$, the loss $R(\overline{X} - m)$ is convex in the X_i if R is convex. So it follows, with $\overline{X}_k = \left(\sum_{i=1}^{n} X_{k,i}\right) / n$, $(k = 1, 2)$, that

$$X_{1,i} \leqq_c X_{2,i} \quad (i = 1, \ldots, n) \text{ implies } R(\overline{X}_1 - m) \leqq_c R(\overline{X}_2 - m) \tag{3.1.3}$$

if

$$\mathsf{E}X_{1,i} = \mathsf{E}X_{2,i}, \qquad i = 1, \ldots, n.$$

From (3.1.3) we then obtain for the risks

$$\mathsf{E}R(\overline{X}_1 - m) \leqq \mathsf{E}R(\overline{X}_2 - m). \tag{3.1.4}$$

3.2 Solving experimental design problems given incomplete information

The main goal of certain problems of experimental design is the estimation of the parameter θ of a d.f. $F(\cdot; \theta, \theta')$ which depends on both θ and a further parameter θ'. In this design problem the costs of obtaining the realizations $x_1, \ldots, x_n$ of a r.v. X are to be taken into account: the sample size n is to be so chosen that the sum of the costs of measurements $K(n)$ and of the risk is minimal. Here, "risk" describes the mean cost which arises from the use of the estimator $\hat{\theta}(n, \theta')$ instead of the (unknown) true parameter θ. The deviation of $\hat{\theta}(n, \theta')$ from θ is assessed by the loss function $R(\hat{\theta}(n, \theta'), \theta)$ so that the risk equals $\mathsf{E}R(\hat{\theta}(n, \theta'), \theta)$. We have then to solve the following optimization problem:

Determine n so as to minimize

$$K(n) + \mathsf{E}R(\hat{\theta}(n, \theta'), \theta). \tag{3.2.1}$$

We assume the following:

a) θ, θ' are real numbers, and for all n, $\hat{\theta}(n, \theta')$ are real-valued r.v.s;

b) R depends only on the difference $\hat{\theta}(n, \theta') - \theta$, i.e.,

$$R(\hat{\theta}(n, \theta'), \theta) = R(\hat{\theta}(n, \theta') - \theta). \tag{3.2.2}$$

c) The d.f. $G(\cdot; n, \theta')$ of $\hat{\theta}(n, \theta')$ is known.

If θ' is known we can compute the risk

$$\mathsf{E}R(\hat{\theta}(n, \theta'), \theta) = \int_{-\infty}^{\infty} R(x - \theta)\, \mathrm{d}G(x; n, \theta') \tag{3.2.3}$$

as a function of θ and solve the problem (3.2.1). There remains the problem of what to do when θ' is unknown.

A common way is to choose a Bayes prior d.f. $F_{\theta'}$, which expresses (subjective) suppositions on the probability of occurrence of different values for θ'. Because the choice of $F_{\theta'}$ is rather problematical (it may frequently require more information than is commonly available), the following approach may be useful. Instead of the prior d.f. $F_{\theta'}$, we start with a set $\mathcal{M}_{\theta'}$ of d.f.s in which the "most unfavourable" d.f. $F^0_{\theta'}$ is chosen and used as the prior d.f. For example, $\mathcal{M}_{\theta'}$ may be specified by giving bounds for θ' or bounds for the means of the d.f. $F_{\theta'} \in \mathcal{M}_{\theta'}$. The optimization problem (3.2.1) is then replaced by the minimax problem:
Find n to minimize

$$K(n) + \sup_{F_{\theta'} \in \mathcal{M}_{\theta'}} \int_{-\infty}^{\infty} \int_{-\infty}^{\infty} R(x - \theta) \, \mathrm{d}G(x; n, y) \, \mathrm{d}F_{\theta'}(y). \tag{3.2.4}$$

If we are able to show that

(1⁰) the function f defined by

$$f(y) = \int_{-\infty}^{\infty} R(x - \theta) \, \mathrm{d}G(x; n, y),$$

has suitable monotonicity and/or convexity properties; and

(2⁰) in $\mathcal{M}_{\theta'}$ there exists a partial ordering $\prec$ for the d.f.s with respect to which $F^0_{\theta'}$ is a maximum and which ensures that $f \in \mathcal{R}_{\prec}$, i.e. that

$$F_1 \prec F_2 \text{ implies } \int_{-\infty}^{\infty} f(y) \, \mathrm{d}F_1(y) \leq \int_{-\infty}^{\infty} f(y) \, \mathrm{d}F_2(y);$$

then (3.2.4) simplifies to:

Find n to minimize

$$K(n) + \int_{-\infty}^{\infty} \int_{-\infty}^{\infty} R(x - \theta) \, \mathrm{d}G(x; n, y) \, \mathrm{d}F^0_{\theta'}(y). \tag{3.2.5}$$

This is an optimization problem in the case of the prior d.f. $F^0_{\theta'}$.

Example 3.2(a). Consider estimating the mean μ of a normal distribution in the case of unknown variance σ^2 so that $\theta = \mu$, $\theta' = \sigma$. As estimator we use the sample mean $\overline{X}$. Then for fixed σ we have by (3.2.3) that the risk equals

$$\mathsf{E}R(\overline{X}(n, \sigma^2), \mu) = (2\pi)^{-1/2} \int_{-\infty}^{\infty} R(\sigma x n^{-1/2}) \exp(-x^2/2) \, \mathrm{d}x \tag{3.2.6}$$

where we have used the fact that $\overline{X} - \mu$ is normal with mean 0 and variance σ^2/n. If the functions f_n defined by

$$f_n(y) = (2\pi)^{-1/2} \int_{-\infty}^{\infty} R(y x n^{-1/2}) \exp(-x^2/2) \, \mathrm{d}x \quad (n = 1, 2, \ldots),$$

are convex and nondecreasing in y we can use the partial ordering $\leqq_c$ to conclude that

$$F_1 \leqq_c F_2 \text{ implies } \int_{-\infty}^{\infty} f_n(y)\,\mathrm{d}F_1(y) \leqq \int_{-\infty}^{\infty} f_n(y)\,\mathrm{d}F_2(y).$$

A sufficient condition for the convexity of f_n is the convexity of R. In this case we can show as follows that the f_n are also nondecreasing in $y > 0$. Let Z_y be a normal r.v. with mean 0 and variance y^2. Then we can write

$$f_n(y) = \mathsf{E}R(yZ_1n^{-1/2}) = \mathsf{E}R(Z_{yn^{-1/2}}).$$

Because Z_y is normal and $\mathsf{E}Z_y = 0$ (all y), it follows from (1.5.2) that

$$0 \leqq y_1 \leqq y_2 \text{ implies } Z_{y_1} \leqq_c Z_{y_2},$$

so using the convexity of R we also have $\mathsf{E}R(Z_{y_1}) \leqq \mathsf{E}R(Z_{y_2})$, i.e. f_n is nondecreasing in $y > 0$.

Suppose for illustrative purposes that $K(n) = n$, $R(x) = \alpha x^2$ for some $\alpha > 0$, and that, whatever the prior distribution of σ, $a \leqq \sigma \leqq b$ and $c \leqq \mathsf{E}\sigma \leqq d$ (so that $a \leqq c \leqq d \leqq b$). Then by Section 1.9 the maximal d.f. F_σ^0 for σ is a two-point distribution with atoms at a and b of masses $(b-d)/(b-a)$ and $(d-a)/(b-a)$ respectively. Thus the maximal risk

$$\int_{-\infty}^{\infty}\int_{-\infty}^{\infty} R(x-\mu)\,\mathrm{d}F_{\bar{X}}(x; n, \mu, y)\,\mathrm{d}F_\sigma^0(y)$$
$$= (2\pi)^{-1/2}(\alpha/n)\,[(b-d)\,a^2 + (d-a)\,b^2]/(b-a)$$

where $f_n(y) = \alpha y^2/n$ has been used. For example, if $\alpha = 0.31$, $a = 15$, $b = 25$, $c = 16$, $d = 23$, the optimal sample size $n = 13$.

3.3 Choice of optimal estimators

In some estimation problems optimal estimators can be determined or optimality properties of known estimators can be proved with the help of partial orderings. Let there be given a set $\mathscr{S}_\theta$ of sample functions

$$\hat{\theta} = \Phi_{\hat{\theta}}(X_1, \ldots, X_n),$$

any of which may be used as an estimator for the unknown parameter θ of the common d.f. of the X_i. The cost of deviations of the $\hat{\theta}$ from the θ are specified by a loss function $R(\theta, \hat{\theta})$. The object is to determine that estimator $\hat{\theta}^* \in \mathscr{S}_{\hat{\theta}}$ for which the risk is minimal, i.e.

$$\mathsf{E}R(\theta, \hat{\theta}^*) = \inf_{\theta \in \mathscr{S}_\theta} \mathsf{E}R(\theta, \hat{\theta}). \tag{3.3.1}$$

If the loss function R is convex in $\hat{\theta}$, if all $\hat{\theta}$ have the same mean (for example, if they are all unbiased) and if $\hat{\theta}^*$ has the property

$$\hat{\theta}^* \leqq_c \hat{\theta} \quad (\text{all } \hat{\theta} \in \mathscr{S}_\theta), \tag{3.3.2}$$

then (3.3.1) holds.

The statement at (3.3.2) can be also proved via the mapping method, by studying the properties of the mappings $\Phi_{\hat{\theta}}$ and the d.f.s generated by them.

Example 3.3(a) (Näther (1975a), Bandemer et al. (1976)). A proof of the extremality of the method of least squares (LS) in the linear model in the case of normally distributed errors and convex loss functions can be given by this method. Let the real-valued quantity y depend on the k-dimensional variable $\boldsymbol{x}$ according to the relation

$$y(\boldsymbol{x}) = \boldsymbol{f}(\boldsymbol{x})^T \boldsymbol{\theta}, \tag{3.3.3}$$

where $\boldsymbol{f}(\boldsymbol{x}) \in \mathbb{R}^r$, $\boldsymbol{\theta} \in \mathbb{R}^r$. Here $\boldsymbol{f}$ is a deterministic known mapping of $\mathbb{R}^k$ into $\mathbb{R}^r$, and $\boldsymbol{\theta}$ is an r-dimensional vector of unknown parameters. $\boldsymbol{\theta}$, and thereby $y(\boldsymbol{x})$, is to be estimated by measurements at the points $\boldsymbol{x}_1, \ldots, \boldsymbol{x}_n$. Measurement errors and similar random effects X_i occur at $\boldsymbol{x}_i$, so that the quantities y_i of the form

$$y_i = \boldsymbol{f}(\boldsymbol{x}_i)^T \boldsymbol{\theta} + X_i \qquad (i = 1, \ldots, n) \tag{3.3.4}$$

are observed. We assume that the X_i are i.i.d. with normal d.f. with mean 0 and variance σ^2. Writing

$$\boldsymbol{y} = (y_1, \ldots, y_n)^T, \quad \boldsymbol{X} = (X_1, \ldots, X_n)^T \text{ and } \boldsymbol{F} = \big(\boldsymbol{f}(\boldsymbol{x}_1), \ldots, \boldsymbol{f}(\boldsymbol{x}_n)\big)^T$$

(3.3.4) becomes

$$\boldsymbol{y} = \boldsymbol{F}\boldsymbol{\theta} + \boldsymbol{X}. \tag{3.3.5}$$

Starting from the y_i, estimators $\hat{\boldsymbol{\theta}}$ for $\boldsymbol{\theta}$ and $\hat{y}$ for $y(\boldsymbol{x})$ can be determined; we consider linear estimators only, i.e. estimators of the form

$$\hat{\boldsymbol{\theta}} = \boldsymbol{C}\boldsymbol{y} \tag{3.3.6}$$

where $\boldsymbol{C}$ is an $(r \times n)$ matrix. Then the estimator for $y(\boldsymbol{x})$ has the form $\hat{y}(\boldsymbol{x}) = \boldsymbol{f}(\boldsymbol{x})^T \hat{\boldsymbol{\theta}}$. Furthermore, we assume that $\hat{\boldsymbol{\theta}}$ is unbiased i.e. the relation $\mathsf{E}\hat{\boldsymbol{\theta}} = \boldsymbol{\theta}$ must be satisfied for all $\boldsymbol{\theta} \in \mathbb{R}^r$. A particular linear unbiased estimator that is often used is the LS estimator in which we determine $\hat{\boldsymbol{\theta}}$ by solving the optimization problem:

$$\text{Minimize } (\boldsymbol{y} - \boldsymbol{F}\boldsymbol{\theta})^T (\boldsymbol{y} - \boldsymbol{F}\boldsymbol{\theta}) \text{ over } \boldsymbol{\theta} \in \mathbb{R}^r. \tag{3.3.7}$$

When the rank of the matrix $\boldsymbol{F}$ is r, $\hat{\boldsymbol{\theta}}$ has the form

$$\hat{\boldsymbol{\theta}} = (\boldsymbol{F}^T\boldsymbol{F})^{-1} \boldsymbol{F}^T\boldsymbol{y}, \tag{3.3.8}$$

and we obtain

$$\hat{y} = \boldsymbol{f}(\boldsymbol{x})^T (\boldsymbol{F}^T\boldsymbol{F})^{-1} \boldsymbol{F}^T\boldsymbol{y}.$$

The accuracy of the estimator can be described by the loss function $R\big(y(\boldsymbol{x}), \hat{y}(\boldsymbol{x})\big)$. We than have

Theorem 3.3.1. *When the loss function $R(y, \hat{y})$ is convex in $\hat{y}$ for fixed y, the risk of the LS estimator is smaller than for all other linear unbiased estimators.*

Proof. The quantity $\hat{y}(x)$ is a linear function of the X_i, and therefore like the X_i it is normally distributed. Being an unbiased estimator, its mean is $y(\boldsymbol{x}) = \boldsymbol{f}(\boldsymbol{x})^T \boldsymbol{\theta}$. From the Gauss-Markov theorem (see e.g. RAO (1973) section 4a.2) it then follows that the variance of $\hat{y}(\boldsymbol{x})$ if the LS method is used, is smaller than for all other linear, unbiased estimators. (1.5.2) yields the theorem.

Notes and bibliographic comments to Chapter 3

Chapter 3 in no way purports to give a balanced account of applications in statistics of partial orderings in general nor of $\leqq_d$, $\leqq_c$, and $\leqq_{cv}$ in particular. For example, LEHMANN (see LEHMANN (1959) and WITTING (1965)) used $\leqq_d$ for the comparison for statistical tests. Problems which are similar to those considered above may be found in the monographs BANDEMER et al. (1976) and HUMAK (1977). Partial orderings play an important role in the books of BARLOW, BARTHOLOMEW, BREMNER, and BRUNK (1972), MARSHALL and OLKIN (1979), and TONG (1980a). HUBER (1976) suggested the use of capacities to determine bounds for probabilities in certain statistical problems; this concept could well be of some importance in the future.

3.2

The results in Section 3.2 can be found in BANDEMER and STOYAN (1971) for example. Similar results for univariate and multivariate linear models are also given by NÄTHER (1972, 1975a, 1975b, 1979) and RUDOLPH (1976). NÄTHER suggested the use of further partial ordering $\leqq_{t_0}$ which is suitable for the comparison of functionals of the form $\int_{-\infty}^{\infty} f(t)\,\mathrm{d}F(t)$ where the function $f(t)$ is nonincreasing for $t \leqq t_0$ and nondecreasing for $t \geqq t_0$. Another possible choice for "least favourable" prior d.f.s is the use of d.f.s with maximal entropy; see, for example, CHAN (1971). In general these d.f.s do not give the minimax solution of (3.2.4).

Monotonicity properties of the posterior distribution in Bayesian inference has been considered by WHITT (1979). He used the partial ordering $\leqq_{d|\mathscr{C}}$ (see notes to Section 1.6) which under some conditions is equivalent to the monotone likelihood ratio property which arises in the study of uniformly most powerful statistical tests. Also BARLOW and PROSCHAN (1980) used monotonicity properties of the posterior distribution, and used total positivity properties as in KARLIN (1968). An extension to the multivariate case give FAHMY, PEREIRA, PROSCHAN and SHAKED (1982). SIMONS

(1980) considered the conditional stochastic ordering of likelihood ratio quotients and the comparison of sequences of such quotients with respect to $\leqq_d$ (see also the references in SMITH and WHITT (1981)).

Added in proof. For monotone likelihood ratios and related partial orderings see the notes to § 1.6 on p. 35. RÜSCHENDORF (1981b), generalizing an earlier result of LEHMANN (1959), proved a general theorem on monotonicity of the OC-function of sequential probability ratio tests for stochastic processes.

RÜSCHENDORF, L. (1981b) Stochastically ordered distributions and monotonicity of the OC function of sequential probability ratio test. *Math. Operat.-forschung Statist., Ser. Statist* **14**, 327—338.

3.3

Section 3.3 is based on work of NÄTHER. It is clear that the linearity of the estimators plays an important role. It is therefore pertinent to ask whether it is possible to prove similarly the optimality of other more complicated estimators. For example, under what conditions can comparisons be made between $\overline{X}$ and the sample median $\tilde{X}$?

4. Monotonicity and comparability of stochastic processes

4.1 Introduction

In the following we consider stochastic processes $\{X_n\}$ or $\{X_t\}$ with discrete ($n = 1, 2, \ldots$) or continuous ($0 \leqq t < \infty$) time and with the state space $(E, \mathscr{M})$. Let $\prec$ be a partial ordering on $P_{\mathscr{M}}$ and, just as for the case of $\leqq_d$ for real-valued r.v.s and their d.f.s, we also let $\prec$ denote the corresponding relation for r.v.s with values in $(E, \mathscr{M})$.

Definition 4.1.1. *A stochastic process* $\{X_n\}$ *or* $\{X_t\}$ *is said to be* nondecreasing (nonincreasing) *with respect to* $\prec$ *if*

$$X_n \prec (\succ)\, X_{n+1} \quad (\textit{all } n) \tag{4.4.1'}$$

or

$$X_t \prec (\succ)\, X_{t+\tau} \quad (\textit{all } t, \textit{ all } \tau > 0). \tag{4.1.1''}$$

Letting $\prec$ *denote one of* $\leqq_d$, $\leqq_c$, $\leqq_{cv}$, *the process* $\{X_n\}$ *or* $\{X_t\}$ *is said to be* strictly nondecreasing (nonincreasing) *with respect to* $\prec$ *if for every* $m = 1, 2, \ldots$ *and all* $n_1', \ldots, n_m'$ *and* $n_1'', \ldots, n_m''$ *or* $t_1', \ldots, t_m'$ *and* $t_1'', \ldots, t_m''$ *satisfying* $n_1' \leqq n_1'', \ldots, n_m' \leqq n_m''$ *or* $t_1' \leqq t_1'', \ldots, t_m' \leqq t_m''$,

$$\{X_{n_1'}, \ldots, X_{n_m'}\} \prec \{X_{n_1''}, \ldots, X_{n_m''}\} \tag{4.1.2'}$$

or

$$\{X_{t_1'}, \ldots, X_{t_m'}\} \prec \{X_{t_1''}, \ldots, X_{t_m''}\}. \tag{4.1.2''}$$

In this book we study mainly homogeneous Markov processes (see Chapter 4.2, 5 and 6); "homogeneous" here and below is used to mean "having stationary transition probabilities". Of course, monotonicity properties can also be proved for other classes of processes and are interesting in some situations. For example, every submartingale is nondecreasing with respect to $\leqq_c$ and every supermartingale is nonincreasing with respect to $\leqq_{cv}$. This is a consequence of the fact (see e.g. VI. 12 of Feller (1966)) that if $\{X_t\}$ is a submartingale then so also is $\{f(X_t)\}$ for $f \in \mathfrak{K}_c(\mathbb{R})$ and hence $\mathsf{E} \max(c, X_t) \leqq \mathsf{E} \max(c, X_{t+\tau})$, i.e. $X_t \leqq_c X_{t+\tau}$. There is an analogous proof for supermartingales. The comparison of semi Markovian processes has been studied in Sonderman (1980) and Kirstein (1979), but the conditions obtained are not as succinct: for example, for stochastic comparability (i.e., $\leqq_d$) an assumption of the $\leqq_d$-monotonicity of the embedded Markov chain was made (compare also p. 31).

Definition 4.1.2. *Let* $\{X_t\}$ *and* $\{Y_t\}$ *for* t *in either* $\{0, 1, \ldots\}$ *or* $[0, \infty)$ *be stochastic processes with the state space* $(E, \mathscr{M})$. *Then the process* $\{X_t\}$ *is* smaller than $\{Y_t\}$ with respect to $\prec$, *symbolically*

$$\{X_t\} \prec \{Y_t\},$$

if for all t

$$X_t \prec Y_t. \tag{4.1.3}$$

The process $\{X_t\}$ *is* strictly smaller than $\{Y_t\}$ with respect to $\prec$ *if for every* $m = 1, 2, \ldots$ *and all* $\{t_1, \ldots, t_m\}$,

$$\{X_{t_1}, \ldots, X_{t_m}\} \prec \{Y_{t_1}, \ldots, Y_{t_m}\}, \tag{4.1.4}$$

symbolically,

$$\{X_t\} \text{ str-}\prec \{Y_t\}. \tag{4.1.5}$$

In Section 4.2 we shall consider methods for proving the comparability and strict comparability of homogeneous Markov processes. Here we discuss the relationship between strict comparability and the following stronger concept: we shall say that a process $\{X_t\}$ is *completely smaller* than the process $\{Y_t\}$ if the distributions on $E^{\mathbb{N}}$ or $E^{\mathbb{R}_+}$ generated by the processes are comparable with respect to a partial ordering $\prec$ for distributions on $E^{\mathbb{N}}$ or $E^{\mathbb{R}_+}$.

With $\prec$ denoting one of $\leqq_d$, $\leqq_c$ and $\leqq_{cv}$, the following theorem shows that (4.1.5) implies the validity of

$$f(\{X_t\}) \prec f(\{Y_t\}) \tag{4.1.6}$$

for certain functionals $f \in \mathfrak{R}(E^{\mathbb{R}_+})$, and for some others besides. Let $\mathscr{T} = \{\mathscr{T}_n\}$ denote sequences of non-negative real numbers $\mathscr{T}_n = \{t_{n0}, t_{n1}, \ldots, t_{nk_n}\}$ for which

$$0 = t_{n0} < t_{n1} < \cdots < t_{nk_n}, \quad \lim_{n\to\infty} t_{nk_n} = \infty, \quad \lim_{n\to\infty} \max_j (t_{n,j+1} - t_{nj}) = 0.$$

Further, given $\{X_t\}$ $(t \in \mathbb{R}_+)$ and $\mathscr{T}$, let $\{X_t^{(n)}\}$ denote the stochastic process

$$X_t^{(n)} = \begin{cases} X_{t_{nj}} & \text{if} \quad t_{nj} \leqq t < t_{n,j+1}, \quad j < k_n, \\ X_{t_{nk_n}} & \text{if} \quad t_{nk_n} \leqq t. \end{cases}$$

The d.f.s of the r.v.s $f(\{X_t\})$ and $f(\{X_t^{(n)}\})$ are denoted by F and $F^{(n)}$ respectively, and, similarly, G and $G^{(n)}$ for $f(\{Y_t\})$ and $f(\{Y_t^{(n)}\})$.

Theorem 4.1.1. *Let* $\prec$ *denote one of* $\leqq_d$, $\leqq_c$, $\leqq_{cv}$, *and let* $f \in \mathfrak{R}_\prec(E^{\mathbb{R}_+})$. *Then* (4.1.5) *implies* (4.1.6) *if the sequences* $\{F^{(n)}\}$ *and* $\{G^{(n)}\}$ *converge weakly for arbitrary* $\mathscr{T}$ *to F and G and if, additionally, in the case of* $\leqq_c$ *or* $\leqq_{cv}$ *that either* $\lim_{n\to\infty} \int_0^\infty \left(1 - F^{(n)}(t)\right) dt = \int_0^\infty \left(1 - F(t)\right) dt < \infty$ *or* $\lim_{n\to\infty} \int_0^\infty F^{(n)}(t)\, dt = \int_0^\infty F(t)\, dt < \infty$, *and similarly for* $\{G^{(n)}\}$, *respectively.*

Proof. From the strict comparability of the processes as at (4.1.5),

$$\{X_{t_{n1}}, \ldots, X_{t_{nk_n}}\} < \{Y_{t_{n1}}, \ldots, Y_{t_{nk_n}}\}$$

for all n and every $\mathscr{T}_n$. For f as stated, Theorem 2.2.2b then yields $f(\{X_t^{(n)}\}) < f(\{Y_t^{(n)}\})$ which is the same as $F^{(n)} < G^{(n)}$. The theorems of Chapter 1 giving conditions for the closure of $<$ under weak convergence now complete the proof of (4.1.6).

In the case of real-valued processes with paths in $D[0, \infty)$, the assumptions of the theorem are satisfied by all functionals on $D[0, \infty)$ which are continuous with respect to the Skorokhod metric. There are also other functionals for which the theorem is true, as for example,

$$f\{x(t)\} = \inf \{t: x(t) < a\} \qquad (-\infty < a < \infty).$$

If $\mathsf{P}(X_0 > a) = 1$ then $f(\{X_t\})$ can be interpreted as the first passage time of a from above.

In the case of $\leqq_d$ the complete comparability of $\{X_n\}$ and $\{Y_n\}$ follows from the strict comparability of $\{X_n\}$ and $\{Y_n\}$.

Theorem 4.1.2. *Let $E_1, E_2, \ldots$ be Polish spaces with Borel σ-algebras $\mathscr{M}_1, \mathscr{M}_2, \ldots$, and let $E^\infty = \prod_{i=1}^{\infty} E_i$ have the Borel σ-algebra $\mathscr{N}$. Let $P_1, P_2 \in P_{\mathscr{N}}$ and for $i = 1, 2$ let $P_i^{(n)}$ be the marginal distributions of the first n coordinates of P_i so that $P_i^{(n)} \in P_{\mathscr{M}_1 \times \cdots \times \mathscr{M}_n}$. If*

$$P_1^{(n)} \leqq_d P_2^{(n)} \qquad (n = 1, 2, \ldots), \tag{4.1.7}$$

then

$$P_1 \leqq_d P_2. \tag{4.1.8}$$

Proof. Let $(z_1, z_2, \ldots)$ be a fixed element of E^∞. By Proposition 1.10.4, there are for each n, random n-vectors $(X_{n1}, \ldots, X_{nn})$ and $(Y_{n1}, \ldots, Y_{nn})$ in $E_1 \times \cdots \times E_n$ with distributions $P_1^{(n)}$ and $P_2^{(n)}$ and such that $X_{nj} \leqq Y_{nj}$ for $j = 1, \ldots, n$. From these vectors we obtain random elements X_n and $Y_n \in E^\infty$ by defining $X_{nj} = Y_{nj} = z_j$ for $j = n + 1, n + 2, \ldots$. Let $P_{1,n}$ and $P_{2,n}$ be the distributions of X_n and Y_n respectively. Since $X_n \leqq Y_n$ we have $P_{1,n} \leqq_d P_{2,n}$. For fixed j the j-th one-dimensional marginal of $P_{1,n}$ is independent of n for $n \geqq j$; hence the sequence of such marginals is tight by theorem 1.4 of Billingsley (1968), and by Tykhonov's theorem, it follows that the sequences $\{P_{1,n}\}$ and $\{P_{2,n}\}$ are tight. Let U_n be the joint distribution of (X_n, Y_n) in $E^\infty \times E^\infty$. The sequence $\{U_n\}$ is also tight and hence has a subsequence $\{U_{n_i}\}$ which converges weakly to a probability measure $U \in P_{\mathscr{N} \times \mathscr{N}}$. The marginals (chosen appropriately) of U are $P_1^{(n)}$ and $P_2^{(n)}$ and P_1 and P_2. Also, each U_n has support in the closed set $H = \{(x, y) \in E^\infty \times E^\infty: x \leqq y\}$. By theorem 2.1 of Billingsley (1968), (see also Chapter 8 below),

$$U(H) = \limsup_{i \to \infty} U_{n_i}(H) = 1,$$

so that U has support in H. Thus $P_1 \leqq_d P_2$.

A similar statement holds for continuous time processes with paths in $D_E(\mathbb{R}_+)$, the space (with the Skorokhod metric) of all functions from $\mathbb{R}_+$ to E which are right continuous and have left limits at all $t \geqq 0$ (see KAMAE, KRENGEL and O'BRIEN (1977)).

Observe that the importance of Theorem 4.1.2 is that for $<$ denoting stochastic ordering, strictly smaller and completely smaller are equivalent. We do not know whether an analogous equivalence holds for $\leqq_c$ or $\leqq_{cv}$.

4.2 Comparability and monotonicity of homogeneous Markov processes

4.2.1 Monotone and comparable operators

Let $\boldsymbol{T}$, $\boldsymbol{T}_1$, $\boldsymbol{T}_2$ be operators on $P_{\mathcal{M}}$, i.e. mappings from $P_{\mathcal{M}}$ into itself, and let $<$ be a partial ordering on $P_{\mathcal{M}}$.

Definition 4.2.1. *An operator* $\boldsymbol{T}$ *is* $<$-monotone *if for all* $P_1, P_2 \in P_{\mathcal{M}}$ *with* $P_1 < P_2$, $\boldsymbol{T}P_1 < \boldsymbol{T}P_2$ *holds. The operator* $\boldsymbol{T}_1$ *is* smaller *than* $\boldsymbol{T}_2$ *if* $\boldsymbol{T}_1P < \boldsymbol{T}_2P$ *for all* $P \in P_{\mathcal{M}}$; *in symbols,* $\boldsymbol{T}_1 < \boldsymbol{T}_2$.

For applications to homogeneous Markov processes, we are interested in the comparability of the distributions $P_1^{(n)}$ and $P_2^{(n)}$ defined by

$$P_k^{(n)} = \boldsymbol{T}_k^n P_k \qquad (k = 1, 2;\, n = 1, 2, \ldots)$$

for two initial distributions P_k and operators $\boldsymbol{T}_k$.

Theorem 4.2.1. *Let* $\boldsymbol{T}_1$ *and* $\boldsymbol{T}_2$ *be operators on* $P_{\mathcal{M}}$, *and* $P_1, P_2 \in P_{\mathcal{M}}$. Then

$$P_1 < P_2 \tag{4.2.1}$$

implies

$$P_1^{(n)} < P_2^{(n)} \qquad (\textit{all } n = 1, 2, \ldots) \tag{4.2.2}$$

if there is a $<$*-monotone operator* $\boldsymbol{T}$ *on* $P_{\mathcal{M}}$ *for which*

$$\boldsymbol{T}_1 < \boldsymbol{T} < \boldsymbol{T}_2. \tag{4.2.3}$$

Proof. (4.2.3) implies $\boldsymbol{T}_1P_1 < \boldsymbol{T}P_1$ and $\boldsymbol{T}P_2 < \boldsymbol{T}_2P_2$ by definition, while the $<$-monotonicity of $\boldsymbol{T}$ with (4.2.1) then implies (4.2.2) for $n = 1$. A simple induction step based on

$$P_1^{n+1} = \boldsymbol{T}_1P_1^n < \boldsymbol{T}P_1^n < \boldsymbol{T}P_2^n < \boldsymbol{T}_2P_2^n = P_2^{n+1}$$

completes the proof.

Remark. The analogue of Theorem 4.2.1 holds in general for operators on partially ordered spaces.

We now consider transition operators of homogeneous Markov chains $\{X_n\}$ with the state space $(E, \mathcal{M})$. The transition operators are described by their transition functions $p(x, B)$, namely

$$p(x, B) = \mathsf{P}(X_{n+1} \in B \mid X_n = x); \quad x \in E,\, B \in \mathcal{M}$$

and, in the case of real-valued processes, by their transition d.f.s $p(x, y)$,

$$p(x, y) = \mathsf{P}(X_{n+1} < y \mid X_n = x) \qquad (x, y \in E \subseteqq \mathbb{R}).$$

Let $\prec$ be a partial ordering on $P_{\mathcal{M}}$ defined as at (1.10.1), where the f are elements of the set $\mathscr{R}_{\prec}$ of functionals having the following property: If for a functional f the inequality (1.10.1) holds for all $P_1, P_2 \in P_{\mathcal{M}}$ with $P_1 \prec P_2$, then $f \in \mathscr{R}_{\prec}$. (Recall that when $\prec$ is one of $\leqq_d$, $\leqq_c$ and $\leqq_{cv}$, the sets $\mathfrak{R}_d(E)$, $\mathfrak{R}_c(E)$ and $\mathfrak{R}_{cv}(E)$ respectively have this property.) In what follows, there are given conditions on the transition functions ensuring the monotonicity or comparability of transition operators.

Theorem 4.2.2. *A transition operator $\boldsymbol{T}$ on $P_{\mathcal{M}}$ is $\prec$-monotone iff for all $f \in \mathscr{R}_{\prec}$ the functionals $f_{\boldsymbol{T}}$ defined by*

$$f_{\boldsymbol{T}}(x) = \int_E f(t)\, p(x, \mathrm{d}t) \qquad (x \in E)$$

are finite and elements of $\mathscr{R}_{\prec}$.

Proof. Take any $P_1, P_2 \in P_{\mathcal{M}}$ and any $f \in \mathscr{R}_{\prec}$, and assume that the integrals $\int_E f(t)\, \boldsymbol{T}P_k(\mathrm{d}t)$ exist for $k = 1, 2$. Then if $f_{\boldsymbol{T}} \in \mathscr{R}_{\prec}$, it follows from

$$\int_E f(t)\, \boldsymbol{T}P_1(\mathrm{d}t) = \int_E f_{\boldsymbol{T}}(t)\, P_1(\mathrm{d}t) \leqq \int_E f_{\boldsymbol{T}}(t)\, P_2(\mathrm{d}t) = \int_E f(t)\, \boldsymbol{T}P_2(\mathrm{d}t)$$

that $\boldsymbol{T}P_1 \prec \boldsymbol{T}P_2$. Conversely, it follows from these same relations that if $\boldsymbol{T}P_1 \prec \boldsymbol{T}P_2$ whenever $P_1 \prec P_2$, then $f_{\boldsymbol{T}} \in \mathscr{R}_{\prec}$.

Theorem 4.2.3. *The transition operators $\boldsymbol{T}_1$ and $\boldsymbol{T}_2$ satisfy $\boldsymbol{T}_1 \prec \boldsymbol{T}_2$ iff their transition functions p_1 and p_2 satisfy*

$$p_1(x, .) \prec p_2(x, .) \qquad (\textit{all } x \in E). \tag{4.2.4}$$

Proof. For any $P \in P_{\mathcal{M}}$, take any $f \in \mathscr{R}_{\prec}$ for which for $k = 1, 2$ the integrals $\int_E f(x)\, \boldsymbol{T}_k P(\mathrm{d}x)$ exist. The property (4.2.4) is equivalent to the functions defined for all $x \in E$ by

$$f_{\boldsymbol{T}_k}(x) = \int_E f(t)\, p_k(x, \mathrm{d}t) \qquad (k = 1, 2)$$

satisfying

$$f_{\boldsymbol{T}_1}(x) \leqq f_{\boldsymbol{T}_2}(x) \qquad (x \in E). \tag{4.2.4'}$$

Thus, writing

$$\int_E f(x)\, \boldsymbol{T}_k P(\mathrm{d}x) = \int_E f_{\boldsymbol{T}_k}(x)\, P(\mathrm{d}x)$$

and using (4.2.4′) shows that then $\boldsymbol{T}_1 \prec \boldsymbol{T}_2$, while the converse is proved by taking P to be a Dirac distribution at x and deducing (4.2.4′).

By using Theorems 2.2.2a and 2.2.2b it can be shown that when P_1, P_2 satisfy (4.2.1), (4.2.3) is a necessary condition for (4.2.2) to hold in the case of either $\leqq_d$ with arbitrary E, or $\leqq_c$ and $\leqq_{cv}$ with $E = \mathbb{R}$.

In the case of real-valued processes it is readily shown by using the classes $\mathcal{F}_d$, $\mathcal{F}_c$ and $\mathcal{F}_{cv}$ defined below (1.2.2), (1.3.6) and (1.4.5) respectively that the transition d.f.s must satisfy the following conditions.

Monotonicity: For all y,

$$\leqq_d\colon p(x_1, y) \leqq p(x_2, y) \quad \text{for all} \quad x_1, x_2 \quad \text{with} \quad x_1 \geqq x_2;$$

$$\leqq_c\colon P_y^{(c)}(x) = \int_y^\infty \big(1 - p(x, t)\big)\, dt \quad \text{is nondecreasing and convex in } x;$$

$$\leqq_{cv}\colon P_y^{(cv)}(x) = \int_{-\infty}^y p(x, t)\, dt \quad \text{is nondecreasing and concave in } x.$$

Comparability: For all x and y,

$$\leqq_d\colon p_1(x, y) \geqq p_2(x, y);$$

$$\leqq_c\colon \int_y^\infty \big(1 - p_1(x, t)\big)\, dt \leqq \int_y^\infty \big(1 - p_2(x, t)\big)\, dt;$$

$$\leqq_{cv}\colon \int_{-\infty}^y p_1(x, t)\, dt \geqq \int_{-\infty}^y p_2(x, t)\, dt.$$

4.2.2 Comparability and monotonicity conditions

Using the results of Section 4.2.1 we are now able to formulate the following theorems concerning the monotonicity and comparability of homogeneous Markov processes in discrete and continuous time. These theorems are important tools for proving the internal and/or external monotonicity properties of stochastic models.

Theorem 4.2.4a (monotonicity, discrete time). *A homogeneous Markov chain* $\{X_n\}$ *with the transition function* p *is nondecreasing (nonincreasing) with respect to* $\prec$ *if*

$$X_1 \prec X_2 \; (X_2 \prec X_1) \tag{4.2.5a}$$

and if p *is* $\prec$*-monotone.*

Under these conditions, if the distributions P_n of the X_n converge suitably towards a stationary distribution P, so that for all n, either $P_n \prec P_{n+1} \prec P$ or $P \prec P_{n+1} \prec P_n$, $P_1, P_2, \ldots$ are then bounds (in the sense of the partial ordering $\prec$) for P. For practical purposes, the accuracy of such a bound will depend on the suitability or otherwise of the choice of P_1. This method (Rossberg and Siegel (1974b) called it the "method of integral inequalities") is used in Section 5.3 below for the queueing model GI/GI/1.

In numerical mathematics there are more efficient methods which use monotonicity properties of operators (see Collatz (1966) and Bohl (1974)) which may be useful also for Markov chains.

Theorem 4.2.4b (monotonicity, continuous time). *A homogeneous Markov process $\{X_t\}$ with the transition functions p^t, namely*

$$p^t(x, B) = \mathsf{P}(X_t \in B \mid X_0 = x)$$

is nondecreasing (nonincreasing) with respect to $\prec$ if positive numbers τ and ϑ exist such that

$$X_0 \prec X_t (X_t \prec X_0) \; (0 \leqq t \leqq \tau) \tag{4.2.5b}$$

and the p^t are $\prec$-monotone for $0 < t < \vartheta$.

Theorem 4.2.5a (comparability, discrete time). *Two homogeneous Markov chains $\{X_n\}$ and $\{Y_n\}$ with the transition functions p_1 and p_2 satisfy*

$$\{X_n\} \prec \{Y_n\} \tag{4.2.6a}$$

if

$$X_1 \prec Y_1 \tag{4.2.7a}$$

and if a $\prec$-monotone transition function p exists such that

$$p_1(x, \cdot) \prec p(x, \cdot) \prec p_2(x, \cdot) \quad (all\ x \in E). \tag{4.2.8a}$$

For real-valued processes with the stationary initial distributions P_1 and P_2,

$$P_1 \prec P_2 \tag{4.2.9}$$

if there exists a $\prec$-monotone p satisfying (4.2.8a) and initial distributions $P_{1,k}$ $(k = 1, 2)$ for which $P_{1,1} \prec P_{1,2}$, $P_{n,k}$ converges weakly to P_k $(n \to \infty)$, and, if $\prec$ is $\leqq_c$ or $\leqq_{cv}$, also

$$\lim_{n\to\infty} \int_0^\infty \overline{P}_{n,k}(t)\,dt = \int_0^\infty \overline{P}_k(t)\,dt \quad or \quad \lim_{n\to\infty} \int_{-\infty}^0 P_{n,k}(t)\,dt = \int_{-\infty}^0 P_k(t)\,dt$$

respectively, where $P_k(t) = P_k\big((-\infty, t]\big)$, $\overline{P}_k(t) = 1 - P_k(t)$ ect.

When $\prec$ stands for $\leqq_d$, the condition (4.2.8a) is equivalent to

$$x_1 \leqq x_2 \text{ implies } p_1(x_1, \cdot) \leqq_d p_2(x_2, \cdot) \tag{4.2.10}$$

for all comparable x_1, x_2. Perhaps a little more convenient is the equivalent condition

$$p_1(x, \cdot) \leqq_d p_2(x + a, \cdot) \quad \text{for all } x \text{ and all } a \geqq 0 \tag{4.2.11}$$

where the p_k are the transition functions for the $\boldsymbol{T}_k$ $(k = 1, 2)$, for we can interpret $p(x, B) \equiv \inf_{a\geqq 0} p_2(x + a, B)$ as the transition function of an operator $\boldsymbol{T}$, and it satisfies (4.2.3) and is monotone.

Theorem 4.2.5b (comparability, continuous time). *Two homogeneous Markov processes* $\{X_t\}$ *and* $\{Y_t\}$ *with the transition functions* $p_1{}^t$ *and* $p_2{}^t$ *satisfy*

$$\{X_t\} \prec \{Y_t\}. \tag{4.2.6b}$$

if

$$X_0 \prec Y_0 \tag{4.2.7b}$$

and if a family $\{p^t\}$ *of* $\prec$*-monotone transition functions exists such that*

$$p_1{}^t(x, \cdot) \prec p^t(x, \cdot) \prec p_2{}^t(x, \cdot) \qquad (\text{all } x, \text{ all } t > 0). \tag{4.2.8b}$$

For the comparability of stationary initial distributions it is enough to have the same conditions as suffice in the case of discrete time chains. In the case of real-valued processes the sufficient conditions on the transition functions for monotonicity (4.2.6) to hold (and for them to hold for $\leqq_d$ for a Polish state space E), are also necessary when they hold for all initial conditions (4.2.5) or (4.2.7) respectively; this can be shown by means of Theorems 2.2.2a and 2.2.2b.

For the relation $\leqq_d$ it is possible to prove the strict monotonicity and the strict and complete comparability of the processes, thereby extending Theorems 4.2.5a and 4.2.5b in an essential way.

Theorem 4.2.6. *A homogeneous Markov process* $\{X_n\}$ *or* $\{X_t\}$ *is strictly nondecreasing with respect to* $\leqq_d$ *if the conditions of Theorem* 4.2.4a *or* 4.2.4b *are true.*

The proof of Theorem 4.2.6 is similar to that of the following theorem.

Theorem 4.2.7. *The homogeneous Markov processes* $\{X_n\}$ *and* $\{Y_n\}$ *or* $\{X_t\}$ *and* $\{Y_t\}$ *are strictly comparable with respect to* $\leqq_d$ *if the conditions of Theorem* 4.2.5a *or* 4.2.5b *are true for* $\leqq_d$.

Proof (discrete time). We show

$$\{X_1, \ldots, X_n\} \leqq_d \{Y_1, \ldots, Y_n\}. \tag{4.2.12}$$

Obviously, (4.2.12) is true for $n = 1$. We assume that (4.2.12) is true for some $m - 1 \geqq 1$ and show the validity for m. Let f be an arbitrary element of $\mathfrak{K}_d(E^m)$ for which the following integrals exist. Then we have

$$\int_{E^m} f(x_1, \ldots, x_m)\, P_k{}^{(m)}(\mathrm{d}x_1, \ldots, \mathrm{d}x_m)$$

$$= \int_{E^{m-1}} \int_E f(x_1, \ldots, x_m)\, p_k(x_{m-1}, \mathrm{d}x_m)\, P_k{}^{(m-1)}(\mathrm{d}x_1, \ldots, \mathrm{d}x_{m-1})$$ where $P_1{}^{(n)}$, $P_2{}^{(n)}$ are the distributions of $\{X_1, \ldots, X_n\}$ and $\{Y_1, \ldots, Y_n\}$ for $n = 1, 2, \ldots$, let $f_k{}^{(m)}$ denote the functionals

$$f_k{}^{(m)}(x_1, \ldots, x_{m-1}) = \int_E f(x_1, \ldots, x_m)\, p_k(x_{m-1}, \mathrm{d}x_m) \qquad (k = 1, 2),$$

and set $f^{(m)}(x_1, \ldots, x_{m-1}) = \int\limits_E f(x_1, \ldots, x_m)\, p(x_{m-1}, \mathrm{d}x_m)$. Because of the assumption on p we have $f^{(m)} \in \mathfrak{R}_d(E^{m-1})$, and because of (4.2.8a)

$$f_1^{(m)}(x_1, \ldots, x_{m-1}) \leqq f^{(m)}(x_1, \ldots, x_{m-1}) \leqq f_2^{(m)}(x_1, \ldots, x_{m-1})$$

for all x_j ($j = 1, 2, \ldots$). Then from the induction assumption there follows

$$\int\limits_{E^m} f(x_1, \ldots, x_m)\, P_1^{(m)}(\mathrm{d}x_1, \ldots, \mathrm{d}x_m) \leqq \int\limits_{E^m} f(x_1, \ldots, x_m)\, P_2^{(m)}(\mathrm{d}x_1, \ldots, dx_m),$$

i.e. (4.2.12) for $n = m$. Consequently, we have (4.2.12) for all n. The complete comparability follows from Theorem 4.1.2.

4.2.3 Homogeneous Markov processes with discrete state space

It is often hard to verify the conditions of Theorems 4.2.4b and 4.2.5b because they relate not merely one transition function or d.f. but families and because the transition functions of Markov processes with continuous time are only seldom known. It is therefore desirable to have conditions on the infinitesimal operators of the processes which ensure monotonicity and comparability. For the case of a discrete state space $E = \{0, 1, \ldots\}$ such conditions, i.e. conditions on the transition rates, are known, as we now describe.

Let $p_{rs}(t)$ denote the transition probabilities,

$$p_{rs}(t) = \mathsf{P}(X_{\tau+t} = s \mid X_\tau = r) \qquad (0 \leqq t, \tau < \infty;\, r, s = 0, 1, \ldots)$$

where $p_{rr}(0) = 1$ for all r and $p_{rs}(0) = 0$ for all $r \neq s$. The transition rates $q_{..}$ are defined by

$$q_{rr} = \lim_{t \to 0} \big(p_{rr}(t) - 1\big)/t \tag{4.2.13}$$

and

$$q_{rs} = \lim_{t \to 0} p_{rs}(t)/t. \tag{4.2.14}$$

Assume that the process is conservative, i.e.

$$-q_{rr} = \sum_{s \neq r} q_{rs} < \infty \qquad \text{(all } r\text{)}. \tag{4.2.15}$$

Then the transition probabilities $p_{rs}(t)$ satisfy the Kolmogorov backward equations (see e.g. Chung (1967)), a system of linear differential equations for the $p_{rs}(t)$ in which the q_{rs} are coefficients. We assume furthermore, that the solution of the system with the initial conditions $p_{rr}(0) = 1$, $p_{rs}(0) = 0$ ($r \neq s$) is unique. We have

Theorem 4.2.8. *Let $\{X_t\}$ and $\{Y_t\}$ be homogeneous Markov processes on the state space $E = \{0, 1, \ldots\}$ with transition rates $q_{rs}^{(1)}$ and $q_{rs}^{(2)}$ respectively. From $X_0 \prec Y_0$ there follows*

$$\{X_t\} \prec \{Y_t\} \tag{4.2.6b}$$

and, additionally, for $\leqq_d$

$$\{X_t\} \text{ str-}\leqq_d \{X_t\} \tag{4.2.6c}$$

as well as complete comparability, if
(1⁰) *when* $\prec$ *denotes* $\leqq_d$,

$$\sum_{u \geqq v} q_{ru}^{(1)} \leqq \sum_{u \geqq v} q_{su}^{(2)} \quad (\text{all } r \leqq s \text{ and all } v \leqq r \text{ and all } v > s); \tag{4.2.16}$$

(2⁰) *when* $\prec$ *denotes* $\leqq_c$, *writing* $f_k(u \mid r) = \sum_{s>u} (s - u)\, q_{rs}^{(k)} \quad (k = 1, 2)$,

$$f_1(u \mid r) \leqq f_2(u \mid s) \quad (\text{all } r \leqq s \leqq u \text{ and all } r = s > u) \tag{4.2.17a}$$

and

$$f_1(u \mid v) \leqq [(v - r)\, f_2(u \mid s) + (s - v)\, f_2(u \mid r)]/(s - r) \ (\text{all } u \leqq r < v < s \text{ if } u < v, \text{ and all } r < v < s \leqq u \text{ if } u > v); \tag{4.2.17b}$$

(3⁰) *when* $\prec$ *denotes* $\leqq_{cv}$, *writing* $g_k(u \mid r) = \sum_{s<u} (u - s)\, q_{rs}^{(k)} \quad (k = 1, 2)$,

$$g_1(u \mid r) \leqq g_2(u \mid s) \qquad (\text{all } u \leqq r \leqq s \text{ and all } u > r = s) \tag{4.2.18a}$$

and

$$g_2(u \mid v) \geqq [(v - r)\, g_1(u \mid s) + (s - v)\, g_1(u \mid r)]/(s - r)$$
$$(\text{all } u \leqq r < v < s \text{ if } u < v, \text{ and all } r < v < s \leqq u \text{ if } u > v); \tag{4.2.18b}$$

a further condition in (2⁰) *and* (3⁰) *being that*

$$\sup_{r \in E} (-q_{rr}^{(k)}) < \infty \qquad (k = 1, 2). \tag{4.2.19}$$

Proof. The proof of the continuous time result (4.2.6b) is reduced to checking that corresponding embedded discrete time chains are similarly ordered. We start by recalling that any continuous time Markov chain $\{X_t\}$ on E with uniformly bounded transition rates as at (4.2.19) has its matrix of transition probabilities $\big(p_{rs}(t)\big)$ expressible, with $Q = (q_{rs})$, as

$$\begin{aligned}\big(p_{rs}(t)\big) &= \exp(Qt) - e^{-\lambda t} \exp\big((I + Q/\lambda)\, \lambda t\big) \\ &= e^{-\lambda t} \sum_{n=0}^{\infty} \big((\lambda t)^n\, (n!)\, (p_{rs}^{(n)})\big)\end{aligned}$$

where the $p_{rs}^{(n)}$ are the n-step transition probabilities for a discrete time Markov chain, $\{X_{(n)}\}$ say, with $(p_{rs}^{(1)}) \equiv I + Q/\lambda$, i.e.

$$p_{rs}^{(1)} = \begin{cases} q_{rs}/\lambda & (s \neq r), \\ 1 - |q_{rr}|/\lambda & (s = r), \end{cases}$$

and λ is any finite quantity $\geqq \sup_{r \in E} |q_{rr}|$. Further, sample paths of X_t can be constructed by starting from sample paths of $X_{(n)}$ for which $X_{(0)}$ and X_0

have the same distribution, and identifying X_t with $X_{(n)}$ for those t for which $t_n < t \leqq t_{n+1}$, $t_0 \equiv 0$ and $\{t_n\colon n = 1, 2, \ldots\}$ being the successive epochs of an independent Poisson process on $(0, \infty)$ with intensity λ.

It is then trivially true that if the discrete time chains $\{X_{(n)}\}$ and $\{Y_{(n)}\}$ constructed as above satisfy $\{X_{(n)}\} \prec \{Y_{(n)}\}$, the Poisson mixtures of their distributions, i.e. $\{X_t\}$ and $\{Y_t\}$, satisfy $\{X_t\} \prec \{Y_t\}$. By Theorem 4.2.1, $\{X_{(n)}\} \leqq_d \{Y_{(n)}\}$ if $X_0 \leqq_d Y_0$ and for all v and all $r \leqq s$,

$$\mathsf{P}(X_{(1)} \geqq v \mid X_{(0)} = r) \leqq \mathsf{P}(Y_{(1)} \geqq v \mid Y_{(0)} = s). \qquad (4.2.16')$$

The expression on the left-hand side here equals

$$\sum_{u \geqq v} p_{ru}^{(1)} = \begin{cases} \sum\limits_{u \geqq v} q_{ru}^{(1)}/\lambda & (v > r), \\ 1 + \sum\limits_{u \geqq v} q_{ru}^{(1)}/\lambda & (v \leqq r). \end{cases}$$

Similar relations hold for the right-hand side of (4.2.16′), and then the conditions (4.2.16) follow on observing that, since λ can be taken arbitrarily large, (4.2.16′) can always be satisfied for $r < v \leqq s$.

Referring to Theorem 4.2.5a, if

$$\begin{aligned} &\inf\big[(s - v)\,\mathsf{E}\big((Y_{(1)} - u)_+ \mid Y_{(0)} = r\big) \\ &+ (v - r)\,\mathsf{E}\big((Y_{(1)} - u)_+ \mid Y_{(0)} = s\big)\big]/(s - r)\,\mathsf{E}\big((X_{(1)} - u)_+ \mid X_{(0)} = v\big) \end{aligned} \qquad (4.2.17')$$

where according to the value of u and v the infimum is taken over either

$$u \leqq r \leqq v \leqq s,\ r < s \quad \text{or} \quad r \leqq v \leqq s \leqq u,\ r < s,$$

and if $\mathsf{E}\big((X_{(1)} - u)_+ \mid X_{(0)} = v\big)$ is increasing in v, then a $\leqq_c$-monotone transition function will exist satisfying (4.2.8a). Expressing (4.2.17′) in terms of the intensity rates yields (4.2.17). (4.2.18) is established by a similar argument.

In order to remove the constraint (4.2.19) in the case of $\leqq_d$, consider in place of the Markov chain $\{X_t\}$ the process $\{X_t{}^a\}$ on state space $\{0, \ldots, a\}$ and transition rates $\{q_{rs}^a\}$ defined by

$$q_{rs}^a = \begin{cases} q_{rs} & \text{if} \quad r, s < a, \\ \sum\limits_{u > a} q_{ru} & \text{if} \quad r < a,\ s = a, \\ 0 & \text{if} \quad r = a \geqq s, \end{cases}$$

so that a is an absorbing state for $\{X_t\}$. For every fixed finite t, the transition probabilities $p_{rs}^a(t) \to p_{rs}(t)$ $(a \to \infty)$, and consequently, for $0 \leqq t \leqq T < \infty$ where T is any fixed finite quantity, $\{X_t{}^a\} \to_d \{X_t\}$ $(a \to \infty)$. Using a similar construction for $\{Y_t\}$, the conditions (4.2.16) for $0 \leqq r \leqq s < a$ suffice to ensure that $\{X_t{}^a\} \leqq_d \{Y_t{}^a\}$ because (4.2.19) holds for finite a. Letting $a \to \infty$ the conclusion still holds, but without needing (4.2.19) to be satisfied.

The proof of (4.2.6c) is similar to that of Theorem 4.2.7.

We remark that if (4.2.6b) holds for all comparable initial distributions, then the conditions (4.2.16) to (4.2.18) are necessarily satisfied (see KIRSTEIN (1976)).

For $q_{rs}^{(1)} = q_{rs}^{(2)}$ Theorem 4.2.8 yields monotonicity conditions for the operators $\boldsymbol{T}_t$ defined by the transition matrices $\big(p_{rs}(t)\big)$. In particular, for $\leqq_d$ we have

Theorem 4.2.9. *All transition operators $\boldsymbol{T}_t$ of a homogeneous Markov process on $\{0, 1, \ldots\}$ with transition rates q_{rs} are monotone iff*

$$\sum_{u \geqq v} q_{ru} \leqq \sum_{u \geqq v} q_{su} \tag{4.2.20}$$

for all r, s with $r \leqq s$ and all $v \leqq r$ and $v > s$.

For homogeneous birth and death processes, the transition rates are

$$q_{rs} = \begin{cases} \lambda_r & (s = r + 1), \\ \mu_r & (s = r - 1), \\ -(\lambda_r + \mu_r) & (s = r), \\ 0 & (\text{otherwise}). \end{cases}$$

Assume $\mu_0 = 0$ and, if the state space is finite so that $E = \{0, \ldots, n\}$ say, $\lambda_n = 0$. The following result follows easily from Theorem 4.2.9.

Proposition 4.2.10. (1°) *Every homogeneous birth and death process satisfies* (4.2.20), *so that its transition operator is $\leqq_d$-monotone.*

(2°) *For homogeneous birth and death processes $\{X_t\}$ and $\{Y_t\}$ with the transition rates $\lambda_r^{(1)}$, $\lambda_r^{(2)}$ and $\mu_r^{(1)}$, $\mu_r^{(2)}$, $X_0 \leqq_d Y_0$ implies $\{X_t\} \leqq_d \{Y_t\}$ and $\{X_t\}$* str-$\leqq_d$ *$\{X_t\}$ if $\mu_r^{(1)} \geqq \mu_r^{(2)}$ and $\lambda_r^{(1)} \leqq \lambda_r^{(2)}$ $(r = 0, 1, \ldots)$, and the processes are then completely comparable.*

Notes and bibliographic comments to Chapter 4

4.1

Most of the existing literature concerning monotonicity and comparability properties deals with homogeneous Markov processes; for a survey of papers on comparability and monotonicity see the notes to Section 4.2. The step from homogeneous to non-homogeneous Markov processes or to conditions for other processes in terms of conditional probabilities causes no mathematical difficulties (see KAMAE, KRENGEL and O'BRIEN (1977) and KAMAE and KRENGEL (1978)). The strict comparability of stochastic processes was used by PLEDGER and PROSCHAN (1973), who were interested in relations of the form (4.1.6) in order to prove bounds for sojourn times in some state sets of reliability systems. They considered only continuous f; FRANKEN and KIRSTEIN (1977) proved Theorem 4.1.1 in the form presented. FRANKEN and KIRSTEIN (1977) also obtained Theorem 4.2.6 which enabled them to establish monotonicity properties for first passage times of homogeneous

Markov processes, properties which Kalmykov (1969) had previously proved by more complicated means. Strict monotonicity with respect to $\leqq_d$ appears to have been used first by Barlow and Proschan (1976). The complete comparability of stochastic processes was studied in Kamae, Krengel and O'Brien (1977) from which paper Theorem 4.1.2 and its proof are taken. Chong (1975) considered for martingales the problem of comparison of r.v.s X_{ν_1} and X_{ν_2} with comparable ν_1, ν_2, where the ν_k are r.v.s with values in the parameter set of the process. The problem may be of interest for other processes.

4.2

Comparability properties of homogeneous Markov processes were first studied by Kalmykov (1962) for $\leqq_d$ and $E = \mathbb{R}$. He used the condition (4.2.11). Daley (1968) also considered real-valued processes in the case of $\leqq_d$ and worked with monotone operators and proved Theorems 4.2.4a and 4.2.5a under the assumption that one of the transition operators is monotone. He also showed (his theorem 4) that for stationary, real-valued, homogeneous Markov chains $\{X_n\}$, the covariances $\operatorname{cov}\big(f(X_n), f(X_{n+m})\big)$ are nonincreasing in m if the transition operators are monotone with respect to $\leqq_d$ and if f is monotonic. Bergmann and Stoyan (1978) gave another proof of this result while Blomqvist (1970) also obtained Daley's result by another method for the model GI/GI/1. Bergmann (1978) showed how it is possible to obtain comparison theorems for covariances and mixed moments of discrete time stochastic processes. Finally, Daley and Tong (1980) proved an assertion in Daley (1968) that a stochastically monotone Markov chain is *associated*, i.e., for any finite subset $X_{n_1}, \ldots, X_{n_k}$ and every pair of functions f and g non-decreasing in each of their arguments such that the expectations concerned are well defined,

$$\operatorname{cov}\big(f(X_{n_1}, \ldots, X_{n_k}), g(X_{n_1}, \ldots, X_{n_k})\big) \geqq 0.$$

The general notion of association has been introduced by Esary, Proschan and Walkup (1967), and it is useful for establishing bounds and inequalities in reliability and statistics. Ahmed, Leon and Proschan (1981) give some criteria for two random vectors X and Y to be associated, which are closely related to monotonicity of transition operators.

The sufficiency of (4.2.3) in the case $\leqq_d$ for the comparison of real-valued Markov processes was shown first by Gaede (1973). Derman and Ignall (1975) gave a similar but not so general condition for Markov chains. Tong (1980b) has given conditions which ensure that $P_1 \leqq_c P_2$ implies $\boldsymbol{T}P_1 \leqq_d \boldsymbol{T}P_2$ and used them to obtain comparison theorems for mixtures of distributions.

Monotone and comparable operators for $\leqq_c$ and $\leqq_{cv}$ and general state spaces were considered by Stoyan (1972a, 1972b). The comparability conditions of Kalmykov (1962) are generalized in Franken and Stoyan (1975). The strict comparability of real-valued homogeneous Markov chains with respect to $\leqq_d$ was first proved by Stoyan (1972b), while Franken and Stoyan (1975) considered general state spaces E, and Franken and Kirstein (1977) proved the equivalence of comparability and strict comparability with respect to $\leqq_d$.

Further important references are Kamae, Krengel and O'Brien (1977) and Kamae and Krengel (1978). O'Brien (1975a) generalized the results of Daley and Kalmykov in another direction: given conditions for $\{X_n\} \leqq_d \{Y_n\}$, he proved sample path relations of the form

$$\mathsf{P}\big(X_n \leqq T_n(Y_0, \ldots, Y_n, X_0);\ n = 1, 2, \ldots\big) = 1$$

for certain mappings T_n: $\mathbb{R}^{n+2} \to \mathbb{R}^n$.

A statement analogous to Theorem 4.2.7 is not true for $\leqq_c$ or $\leqq_{cv}$: Stoyan (1972b) contains a counter-example.

Another variant of comparability of homogeneous Markov chains was studied in Daley (1968) and Spataru (1974). They used a complete ordering of the state space and defined the comparability of distributions via families of sets whose relation with this ordering is the same (i.e., "increasing sets") as the intervals $(-\infty, x)$ with the ordering of real numbers.

A further comparability definition for two-dimensional Markov processes has been used by Gittins (1971) in order to obtain bounds for a GI/GI/1 queue with periodic interruptions (low-tide and high-tide in a harbour) corresponding to the possible epochs for customers to enter and depart from the queue. The first comparison theorem for Markov processes with continuous time was proved by Gaede (1965). He used results from the theory of differential equations to obtain a result for inhomogeneous birth and death processes analogous to Theorem 4.2.8. The results in 4.2.3 are from Kirstein (1976). Comparability properties for other Markov processes can be proved similarly, as for example, for the virtual waiting times in M/GI/1, while Keilson and Kester (1977) showed that comparison theorems for pure jump processes can sometimes be proved given conditions in terms of transition rates. Anderson (1972), Kirstein (1978) and O'Brien (1980) considered diffusion processes. Monotonicity and comparison theorems can be proved for Galton-Watson processes (see e.g. Kirstein (1976) and Stoyan (1974b)). Keilson and Kester (1977) and van Doorn (1980) studied monotonicity properties of birth and death processes.

Open problems.

1. Can the monotonicity properties of Daley for the covariances $\operatorname{cov}\big(f(X_n), f(X_{n+m})\big)$ be generalized to homogeneous Markov chains $\{X_n\}$ on a general state space $(E, \mathscr{M})$ with functionals $f\colon E \to \mathbb{R}$ having suitably-defined monotonicity properties? Consider for example the waiting time vectors in GI/GI/s: these are described by a homogeneous Markov chain with values in $\mathbb{R}^s$ and a $\leqq_d$-monotone transition operator. Also, can these monotonicity properties of the covariances be proved for processes with a $\leqq_c$-monotone transition operator? Under what conditions is it possible to establish bounds for the covariances of Markov chains having comparable transition operators?

2. Do the conditions (4.2.17) and (4.2.18) remain sufficient if (4.2.19) is not true? Can similar statements be established for Markov processes with general state spaces? (Some work in this direction has been done: see KIRSTEIN (1978) and SIEGMUND (1976).)

3. For what functionals f does a statement analogous to Theorem 4.2.6 remain true for $\leqq_c$ or $\leqq_{cv}$? (Such statements have been proved for functionals f of the form

$$f(x_1, \ldots, x_n) = \sum_{m=1}^{n} f_m(x_m) \quad \text{for } f_m \in \mathfrak{R}_c(E) \text{ or } \mathfrak{R}_{cv}(E):$$

see STOYAN (1972b) and FRANKEN and STOYAN (1975).)

Supplementary material added in proof

The comparison of alternating renewal processes is discussed in SONDERMAN (1980) and CHIANG and NIU (1981). WHITT (1981d) studied monotonicity properties of the operator T: $F \to F_R$ where for any d.f. F on $\mathbb{R}_+$ with finite mean, F_R is the forward recurrence time d.f. defined below (1.6.10) on p. 19. NOEBELS (1981, see p. 37) used $\leqq_d$ for the proof of existence of stationary distributions of Markov processes and also considered semi-Markov processes.

CHIANG, D. T., & NIU, S.-C. (1981) Comparing alternating renewal processes. *Naval Res. Log. Quart.* **28**, 207–212.

Notice also the book

VAN DOORN, E. A. (1981) Stochastic monotonicity and queueing applications of birth-death-processes. *Lect. Notes in Statistics* **4**. Springer-Verlag, Berlin–Heidelberg–New York.

5. Monotonicity properties and bounds for queueing systems (I): The single server systems GI/GI/1 and G/G/1

5.0 Basic results for GI/GI/1 and G/G/1

GI/GI/1 denotes the following single-server queueing system. Customers arrive for service at the epochs t_n given by

$$t_0 = 0, \quad t_{n+1} = \alpha_1 + \cdots + \alpha_{n+1} = t_n + \alpha_{n+1} \qquad (n = 0, 1, \ldots).$$

The customer arriving at t_n is called the nth arrival or customer, the time α_n between the $(n-1)^{\text{th}}$ and n^{th} arrivals is the nth interarrival time. The nth customer requires service for a time β_n. $\{\alpha_n\}$ and $\{\beta_n\}$ are independent sequences of r.v.s which are i.i.d., and their d.f.s and parameters are denoted thus:

$$A(t) = \mathsf{P}(\alpha_n \leqq t) = \text{d.f. of } \alpha_n, \quad m_A = \mathsf{E}\alpha_n, \quad \sigma_A^2 = \text{var } \alpha_n;$$

$$B(t) = \mathsf{P}(\beta_n \leqq t) = \text{d.f. of } \beta_n, \quad m_B = \mathsf{E}\beta_n, \quad \sigma_B^2 = \text{var } \beta_n;$$

$$K(t) = \mathsf{P}(U_n \leqq t) = \text{d.f. of } U_n \equiv \beta_n - \alpha_n;$$

$$p \equiv P(\alpha_n > \beta_n) = \int_0^\infty \bar{A}(t) \, \mathrm{d}B(t) = K(0);$$

$$\varrho = m_B/m_A = \lambda m_B = \text{(relative) traffic intensity.}$$

Customers are served in order of arrival ("first-come first-served" discipline, i.e., FCFS), and there are infinitely many waiting places in the system.

If the interarrival times may be interdependent but the service times β_n are still i.i.d. and independent of the interarrival times, the model is denoted by G/GI/1; if the service-times may also be interdependent we use the symbol G/G/1. The model GI/GI/1, and to a lesser extent G/G/1, has been extensively studied in queueing theory. Amongst the many books we mention those of Gnedenko and Kovalenko (1966) for a broad survey of the subject, Cohen (1969) for GI/GI/1 in particular, and Borovkov (1976a) which discusses G/G/1 at some length. Below we recall only those results which are needed for our discussion.

The waiting time w_n of the nth customer satisfies the recursion formula,

$$w_{n+1} = \max(0, w_n + U_n) \equiv (w_n + U_n)_+ \qquad (n = 0, 1, \ldots). \tag{5.0.1}$$

Assume w_0 to be a r.v. independent of all α_n and β_n and having the d.f. W_0. Then corresponding to (5.0.1) we have for the d.f.s W_n of the w_n,

$$W_{n+1}(t) = \int_{0-}^\infty K(t - x) \, \mathrm{d}W_n(x) = \int_{-\infty}^\infty W_n(t - x) \, \mathrm{d}K(x) \ (t \geqq 0). \tag{5.0.2}$$

LINDLEY (1952) proved that in the case $\varrho < 1$, and this assumptions is made in the following, for any initial d.f. W_0 the d.f.s W_n converge weakly towards a unique d.f. W, the stationary waiting time d.f., which is the unique d.f. solution of the integral equation

$$W(t) = \int_{0-}^{\infty} K(t - x)\, \mathrm{d}W(x) = \int_{-\infty}^{\infty} W(t - x)\, \mathrm{d}K(x) \quad (t \geqq 0). \quad (5.0.3)$$

When $m_A < \infty$, the nth moment of W is finite iff the $(n + 1)^{\text{th}}$ moment of B is finite (see KIEFER and WOLFOWITZ (1955)). We denote the first moment of W, the mean stationary waiting time, by

$$m_W = \int_0^{\infty} t\, \mathrm{d}W(t),$$

and the quantity $\overline{W}(0+)$ by p_W. When the function

$$w(t) = \int_{0-}^{\infty} K(t - x)\, \mathrm{d}W(x) \qquad (-\infty < t < \infty)$$

is continuous at 0, p_W equals the probability that a customer has to wait when $\{w_n\}$ is a stationary sequence of r.v.s (see e.g. MARSHALL (1968a)).

$$W^*(\theta) = \int_{0-}^{\infty} \mathrm{e}^{-\theta t}\, \mathrm{d}W(t) = \exp\left[\sum_{n=1}^{\infty} n^{-1} \int_0^{\infty} (1 - \mathrm{e}^{-\theta x})\, \mathrm{d}K^{*n}(x)\right] \quad (5.0.4)$$

where[1] K^{*n} is the n-fold convolution power of K, and hence that

$$m_W = \sum_{n=1}^{\infty} n^{-1} \int_0^{\infty} x\, \mathrm{d}K^{*n}(x). \quad (5.0.4')$$

If $w_0 = 0$, then the mean $m_{W,n}$ of the waiting time w_n is given by

$$m_{W,n} = \mathsf{E}w_n = \sum_{k=1}^{n} k^{-1} \int_0^{\infty} x\, \mathrm{d}K^{*k}(x). \quad (5.0.4'')$$

The rth order cumulant s_r of the stationary waiting time is expressible as

$$s_r = \sum_{k=1}^{\infty} k^{-1} \int_0^{\infty} x^r\, \mathrm{d}K^{*k}(x) \quad (5.0.5)$$

(see BOROVKOV (1970) and BERGMANN (1974)); $s_1 = m_W$ and $s_2 = \sigma_W{}^2$ = variance of stationary waiting time.

---

[1] The asterisk is used to denote both Laplace-Stieltjes transform (L.-S.T.) and convolution: the meaning should always be clear from the context.

MARSHALL (1968a) gave the following formulae for p_W and m_W:

$$1 - p_W = (m_A - m_B)/m_L, \tag{5.0.6}$$

$$m_W = \frac{(m_A - m_B)^2 + \sigma_A^2 + \sigma_B^2}{2(m_A - m_B)} - \frac{m_L^2 + \sigma_L^2}{2m_L}; \tag{5.0.7}$$

here, m_L and σ_L^2 denote the mean and the variance of the idle period of the server for the system in its stationary state, i.e. of that time elapsing between the end of one service and the beginning of the next service when no customer is in the system.

Proof[1] (MARSHALL (1968a)). We use (5.0.1) which with $X_n = \max\left(0, -(w_n + U_n)\right)$ takes the form

$$w_{n+1} - X_n = w_n + U_n. \tag{5.0.8}$$

Assume in the following that the system is in its stationary state, so that w_n and w_{n+1} have the same d.f. W. Then if $X_n > 0$, it is an idle period r.v. Taking expectations in (5.0.8) and assuming $\mathsf{E}w_n = m_W$ is finite we obtain

$$\mathsf{E}X_n = -\mathsf{E}U_n = m_A - m_B, \tag{5.0.8'}$$

Since $\mathsf{E}X_n = \mathsf{E}(X_n; X_n > 0) = \mathsf{E}(X_n \mid X_n > 0)\, \mathsf{P}(X_n > 0) = m_L(1 - p_W)$, (5.0.6) now follows.

Squaring both sides of (5.0.8) and using $w_{n+1}X_n = 0$, we obtain

$$w_{n+1}^2 + X_n^2 = w_n^2 + 2w_nU_n + U_n^2.$$

Assume $\mathsf{E}w_n^2 < \infty$ and take expectations. Then, using the independence of w_n and U_n and
$\mathsf{E}X_n^2 = (1 - p_W)$ (second moment of idle period) $= (1 - p_W)\,(m_L^2 + \sigma_L^2)$, there follows the relation

$$(1 - p_W)\,(m_L^2 + \sigma_L^2) = 2\mathsf{E}U_n\mathsf{E}w_n + \mathsf{E}U_n^2$$

which with (5.0.6) yields (5.0.7).

KÖLLERSTRÖM (1976) proved that the stationary waiting time d.f. W is of type NWU for arbitrary K. DALEY and TRENGOVE (1977) showed that this property is a consequence of representing W as the sum of a geometrically distributed number of i.i.d. positive r.v.s.

For systems of the type M/GI/1, i.e. for systems with Poisson input, so that the α_n are exponentially distributed with parameter $\lambda = m_A^{-1}$, we have

$$W(t) = (1 - \varrho) \sum_{n=0}^{\infty} \varrho^n \left(m_B^{-1} \int_0^t \overline{B}(x)\, dx\right)^{*n} \quad (0 \leq t < \infty) \tag{5.0.9}$$

---

[1] In this proof, due essentially to KINGMAN (1962), the assumption of the finiteness of $\mathsf{E}w_n^2$ is made, though it can be obviated as in KINGMAN (1970) or ROSSBERG (1970a). However the proof given above can also be used in G/G/1.

(see e.g. COHEN (1969), p. 255), and for m_W the Pollaczek-Khinchin formula

$$m_W = \frac{\lambda(\sigma_B^2 + m_B^2)}{2(1-\varrho)} = \mathsf{E}\beta^2/2\mathsf{E}(-U) \tag{5.0.10}$$

where the r.v. β has B as its d.f., and $p_W = \varrho$.

For systems of the type GI/M/1, where the β_n are exponentially distributed with parameter $\mu = m_B^{-1}$, we have (see e.g. COHEN (1969), p. 230)

$$W(t) = 1 - \delta \exp\big(-\mu(1-\delta)\, t\big) \qquad (0 \leqq t < \infty), \tag{5.0.11}$$

$$m_W = \delta m_B/(1-\delta) \tag{5.0.12}$$

and $p_W = \delta$, where δ is the unique root in $0 < z < 1$ of the equation

$$z = \int_0^\infty \exp\big(-\mu(1-z)\, t\big)\, \mathrm{d}A(t). \tag{5.0.13}$$

A quantity similar to p_W is the probability P_{occ} that the system in its stationary state is not empty. Provided the d.f. A is of the non-lattice type,

$$P_{occ} = \lim_{t\to\infty} \mathsf{P}(\nu_t \geqq 1) = \lim_{t\to\infty} t^{-1} \int_0^t \mathsf{P}(\nu_u \geqq 1)\, \mathrm{d}u = \varrho \tag{5.0.14}$$

where ν_t denotes the number of customers in the system at time t; for the last equality, which holds for any A and for G/G/1, see FRANKEN, KÖNIG, ARNDT and SCHMIDT (1981). Observe that in the case of Poisson input, the two probabilities p_W and P_{occ} are equal.

Using the stationary waiting time d.f. W the stationary distribution of the number of waiting customers can be determined (e.g. FINCH (1959)). Let P_k^0 be the probability that an arriving customer in a stationary system finds at least k customers already waiting and let

$$P_k = \lim_{t\to\infty} \mathsf{P}\big(\nu_W(t) \geqq k\big) = \lim_{t\to\infty} \mathsf{P}(\nu_t \geqq k+1) \qquad (k = 1, 2, \ldots)$$

where $\nu_W(t) = (\nu_t - 1)_+$ denotes the number of customers *waiting* at time t. Then

$$P_k^0 = \int_0^\infty A^{*k}(t)\, \mathrm{d}W(t), \tag{5.0.15}$$

$$P_k = \int_0^\infty (A^{*(k-1)} * A_R)(t)\, \mathrm{d}W(t) \tag{5.0.16}$$

where

$$A_R(t) = \lambda \int_0^t \bar{A}(u)\, \mathrm{d}u, \qquad \lambda = 1/m_A.$$

Proof. Assume the system is in its stationary state at time t and number all $\nu_W(t)$ customers there waiting in order of their arrival. The most recent arrival occurred at time $t - \alpha_R'$ where α_R' denotes a backward recurrence time r.v. of the renewal input process, and so has d.f. A_R. When $\nu_W(t) > 1$, the other inter-arrival times between numbered customers then in the system are $\alpha_1, \ldots, \alpha_{\nu_W(t)-1}$ say. Since these customers are *waiting*, it follows that the waiting time of the first numbered customer, w say, exceeds $\alpha_1 + \cdots + \alpha_{\nu_W(t)-1} + \alpha_R'$, and since all these r.v.s are independent with d.f.s W, A and A_R, that for integers $k \geqq 1$,

$$\mathsf{P}(\nu_W(t) \geqq k) = \mathsf{P}(w > \alpha_1 + \cdots + \alpha_{k-1} + \alpha_R')$$

$$= \int_0^\infty (A^{*(k-1)} * A_R)(t)\, \mathrm{d}W(t) \tag{5.0.17}$$

as asserted at (5.0.16). (5.0.15) is proved similarly with α_k replacing α_R'.

From (5.0.15) and (5.0.16) we obtain for the mean queue lengths $L_W = \sum_{k=1}^{\infty} P_k$ and $L_W{}^0 = \sum_{k=1}^{\infty} P_k{}^0$ the formulae

$$L_W = m_W/m_A = \lambda m_W, \tag{5.0.18}$$

and

$$L_W{}^0 = \int_0^\infty H_A(t)\, \mathrm{d}W(t) \tag{5.0.19}$$

where for $t \geqq 0$, $H_A(t) = \sum_{k=1}^{\infty} A^{*k}(t)$ is the renewal function for A and we have used

$$\sum_{k=0}^{\infty} (A^{*k} * A_R)(t) = \lambda t.$$

(5.0.18) is a particular case of the more general relation known as Little's formula "$L = \lambda W$" (see the notes to Section 6.3).

The distribution of the number ν_t of all customers in the system can be determined analogously by using the d.f. of the delay time $\equiv$ waiting time + service time.

In the case of dependent service and interarrival times, i.e. for G/G/1, we have the following statements. If the sequence of interarrival and service times $\{(\alpha_n, \beta_n)\}$ $(n = 0, \pm 1, \pm 2, \ldots)$ is metrically transitive (for example, weakly dependent sequences or the sequence of the intervals between successive points of an ergodic random point process form metrically transitive sequences), and if

$$\varrho \equiv m_B/m_A = \mathsf{E}\beta_0/\mathsf{E}\alpha_0 < 1$$

then there exists a stationary sequence $\{w_n\}$ of waiting times satisfying

$$w_{n+1} = (w_n + \beta_n - \alpha_n)_+ \quad (n = 0, \pm 1, \pm 2, \ldots). \tag{5.0.20}$$

This has been proved by LOYNES (1962); see also BOROVKOV (1976a), SCHASSBERGER (1973) and FRANKEN, KÖNIG, ARNDT and SCHMIDT (1981). The relation (5.0.6) is also true for G/G/1, while (5.0.7) takes the more general form

$$m_W = \frac{2\,\mathrm{cov}\,(w_0, U_0)}{2\mathsf{E}(-U_0)} - \frac{m_L{}^2 + \sigma_L{}^2}{2m_L} \quad \text{where} \quad U_0 \equiv \beta_0 - \alpha_0. \qquad (5.0.21)$$

The proof of (5.0.21) is practically the same as that of (5.0.7) except that w_n and U_n need not be independent so we must take account of

$$\mathsf{E}w_nU_n = \mathsf{E}w_n\mathsf{E}U_n + \mathrm{cov}\,(w_n, U_n) = \mathsf{E}U_0\mathsf{E}w_0 + \mathrm{cov}\,(w_0, U_0).$$

Finally we note that (5.0.14) and (5.0.18) are also true for G/G/1 as proved in FRANKEN (1976) by means of the theory of stationary point processes (see also FRANKEN, KÖNIG, ARNDT, and SCHMIDT (1981)).

5.1 Internal monotonicity properties of waiting times in GI/GI/1

The transition operator of the homogeneous Markov chain $\{w_n\}$ of waiting times given by (5.0.1) is monotone with respect to $\leqq_d$ and $\leqq_c$, as can be shown easily using Theorems 2.2.8 and 4.2.4a. Consequently, we have

Theorem 5.1.1. *If the initial distribution W_0 of the process $\{w_n\}$ $(n = 0, 1, \ldots)$ satisfies*

$$W_0 \prec W_1, \quad \prec \textit{ denoting one of } \leqq_d, \leqq_c, \geqq_d, \geqq_c, \qquad (5.1.1)$$

then for all $n = 0, 1, \ldots$

$$W_n \prec W_{n+1} \qquad (5.1.2)$$

and if the stationary waiting time d.f. W exists,

$$W_n \prec W \quad (\textit{all } n = 0, 1, \ldots) \qquad (5.1.3)$$

where for $\leqq_c$ it must be assumed that $\mathsf{E}W < \infty$.

Similar statements for the queue lengths are true only for $\leqq_d$ in general (see Theorem 6.1.2), though they also hold for $\leqq_c$ in the particular cases M/GI/1 and GI/M/1 as we now show. Let $\nu_n{}^*$ be the number of customers in an M/GI/1 queueing system immediately after the departure of the nth customer. Then

$$\nu_{n+1}^* = (\nu_n{}^* - 1 + X_n)_+ \qquad (5.1.4)$$

where X_n denotes the number of customers arriving during the service of the nth customer. Because of the properties of the Poisson process the X_n are i.i.d., so $\{\nu_n{}^*\}$ satisfies a recurrence relation of the same structure as $\{w_n\}$, and hence there is an analogue of Theorem 5.1.1 for it.

Let ν_n^0 be the number of customers in a GI/M/1 system immediately before the arrival of the nth customer. Then

$$\nu_{n+1}^0 = (\nu_n^0 + 1 + \eta_n)_+ \qquad (n = 0, 1, \ldots) \tag{5.1.5}$$

where η_n denotes the number of customers which can be served during the nth interarrival time. Again because of the properties of the exponential distributions, the η_n are i.i.d., and so an analogue of Theorem 5.1.1 exists for $\{\nu_n{}^0\}$.

If the interarrival and service times are dependent with $\{(\alpha_n, \beta_n)\}$ metrically transitive, we can appeal to Theorem 2.2.9 and prove that when $w_0 = 0$,

$$w_n \leqq_d w_{n+1} \qquad (n = 0, 1, \ldots)$$

i.e. (5.1.2) holds for $\leqq_d$, for putting $X_n = \beta_n - \alpha_n$, $\phi(x, y) = (x + y)_+$ and $Z_n = w_n$, the assumptions of Theorem 2.2.9 are then satisfied with $w_0 = 0$.

5.2 External monotonicity properties of GI/GI/1 and G/G/1

Because of the simple form of the recursion formula (5.0.1)

$$w_{n+1} = (w_n + U_n)_+ \qquad (n = 0, 1, \ldots),$$

it is particularly easy to use the mapping method to study the influence of the d.f. K of difference U_n on the d.f.s W_n of waiting times w_n and the stationary waiting time d.f. W. Since the mapping $\phi\colon \mathbb{R}^2 \to \mathbb{R}$ defined by

$$\phi(x, y) = (x + y)_+,$$

is convex and nondecreasing as a function of (x, y), appealing to Theorem 2.2.6 shows that the w_n are convex and nondecreasing functions of U_n and w_0,

$$w_{n+1} = \Psi_n(w_0, U_0, \ldots, U_n) \qquad (n = 0, 1, \ldots)$$

say. Thus we have (5.2.3) below.

Theorem 5.2.1. *Let Σ_1 and Σ_2 be* GI/GI/1 *queues with the d.f.s K_1 and K_2 and the waiting time d.f.s $W_{n,1}$ and $W_{n,2}$ of nth customers. Then if $\prec$ denotes $\leqq_d$ or $\leqq_c$,*

$$K_1 \prec K_2 \tag{5.2.1}$$

and

$$W_{0,1} \prec W_{0,2} \tag{5.2.2}$$

imply that for all $n = 0, 1, \ldots$,

$$W_{n,1} \prec W_{n,2}; \tag{5.2.3}$$

if the corresponding stationary waiting time d.f.s W_1 and W_2 exist, and, additionally when $\prec$ denotes $\leqq_c$, they have finite means $m_{W,1}$ and $m_{W,2}$, then

(5.2.1) *is sufficient to ensure that*

$$W_1 < W_2, \tag{5.2.4}$$

and

$$m_{W,1} \leqq m_{W,2}. \tag{5.2.5}$$

Proof. Putting $W_{0,1} = W_{0,2} = \Theta_0$, we have for $k = 1, 2$ that $W_{0,k} \leqq_d W_{1,k}$, hence $W_{n,k} \leqq_d W_{n+1,k}$ and, since W_k exists, $W_{n,k}$ converges weakly to W_k as $n \to \infty$. Again, from the monotone convergence of $W_{n,k}$, the corresponding means converge monotonically. Since for these $\{W_{n,k}\}$ $W_{0,1} = W_{0,2}$ we also have (5.2.3) holding, and hence the limits satisfy (5.2.4) and (5.2.5) (cf. Propositions 1.2.3 and 1.3.2).

Remarks (1⁰). Sufficient conditions for (5.2.1) for $\leqq_c$ are that

$$A_2 \leqq_{cv} A_1 \quad \text{and} \quad B_1 \leqq_c B_2. \tag{5.2.1'}$$

Also, recall from (1.4.7) that, in the case of equal means of A_1 and A_2, $A_1 \leqq_c A_2$ and $A_2 \leqq_{cv} A_1$ are equivalent.

(2⁰). Because the functions Ψ_n are nondecreasing and convex in the vectors $(w_0, U_0, \ldots, U_n)$, (5.2.3) remains true for the case of dependent esrvice and interarrival times if for all $n = 0, 1, \ldots$

$$\{w_{0,1}, U_{0,1}, \ldots, U_{n,1}\} \leqq_d (\text{or } \leqq_c) \{w_{0,2}, U_{0,2}, \ldots, U_{n,2}\}. \tag{5.2.1''}$$

If, additionally, the sequences $\{(\alpha_{n,k}, \beta_{n,k})\}$ $(k = 1, 2)$ are metrically transitive and if stationary waiting time d.f.s W_1 and W_2 exist then (5.2.4) holds, as can be proved by starting from (5.2.3) and using the internal monotonicity property (5.1.6). For $\leqq_c$ we also assume that the mean stationary waiting times are finite.

(3⁰). Theorem 5.2.1 yields the intuitively clear statement that "lengthening" service or "shortening" interarrival times (or both), leads to "longer" waiting times. The extremality statement, known in the 1960's (see Rogozin (1966) and Rossberg (1968)), that in the case of constant service or interarrival times the mean stationary waiting times are shorter than for all other d.f.s with the same means, can be embedded in a monotonicity statement for $\leqq_c$. To see this, recall from Section 1.9 that constant service times are $\leqq_c$-minimal and constant interarrival times are $\leqq_{cv}$-maximal.

The comparability of all moments of the stationary waiting times in Σ_1 and Σ_2 follows from Theorem 5.2.1 when (5.2.1) is true for $\leqq_d$ or $\leqq_c$. The following theorem shows that the cumulants are comparable under weaker conditions (see Definition 1.7.1 for $\leqq_c^r$).

Theorem 5.2.2. *Let $s_{n,1}$ and $s_{n,2}$ be the nth order cumulants of the stationary waiting times in Σ_1 and Σ_2, and for some $r \geqq 1$ suppose that $K_1 \leqq_c^r K_2$. Then*

$$s_{n,1} \leqq s_{n,2} \textit{ for integers } n \geqq r. \tag{5.2.6}$$

Proof. Because of the convolution property (C) of $\leqq_c{}^r$, $K_1{}^{*k} \leqq_c{}^r K_2{}^{*k}$ for all $k = 1, 2, \ldots$, and so for all $n \geqq r$,

$$\int_0^\infty x^n \, \mathrm{d}K_1{}^{*k}(x) \leqq \int_0^\infty x^n \, \mathrm{d}K_2{}^{*k}(x).$$

(5.2.6) now follows from (5.0.5).

Theorem 5.2.2 yields extremal properties of cumulants. In the set of all d.f.s of interarrival (or service) times with fixed means and variances, two point d.f.s yield extremal values of the cumulants for $n \geqq 2$, in particular, of the variance $\sigma_W{}^2$ of the stationary waiting time (see Bergmann, Daley, Rolski, and Stoyan (1979)). Bergmann (1979a) proved monotonicity properties of covariances of waiting times with respect to $\leqq_d$. Also, by using the mapping method, statements about queue lengths analogous to (5.2.3) and (5.2.4) can be proved for $\leqq_d$. In the stationary case, the proof can be accomplished by using the formulae (5.0.15) and (5.0.16). For the M/GI/1 system, use of the recursion formulae (5.1.4) and (5.1.5) enables the case $\leqq_c$ to be treated also; Rolski (1976a) shows how this extension is possible for queues with group arrivals or service. However, the formulae (5.0.9) and (5.0.11) enable us to prove the sharper results below.

Theorem 5.2.3. (a) *For* M/GI/1 *queues with arrival rates* λ_k *and service time d.f.s* B_k $(k = 1, 2)$ *with the same mean* m_B, *suppose that* $\lambda_1 \leqq \lambda_2$, $B_1 \leqq_c B_2$, *and set* $\varrho_k = \lambda_k m_B$ *with* $\varrho_k < 1$. *Then the corresponding stationary waiting time d.f.s* W_k *satisfy*

$$W_1 \leqq_d W_2. \tag{5.2.7}$$

(b) *For* GI/M/1 *queues with mean service times* $1/\mu_k$ *and interarrival time d.f.s* A_k *with means* $m_{A,k}$ $(k = 1, 2)$, *assume that* $\mu_1 \geqq \mu_2$, $A_2 \leqq_L A_1$, *and* $\varrho_k \equiv 1/\mu_k m_{A,k} < 1$. *Then the corresponding stationary waiting time d.f.s* W_k *again satisfy* (5.2.7).

Proof. Since B_1 and B_2 have the same means, $B_1 \leqq_c B_2$ is equivalent to

$$\int_0^t \overline{B}_1(x) \, \mathrm{d}x \geqq \int_0^t \overline{B}_2(x) \, \mathrm{d}x \qquad (\text{all } t \geqq 0).$$

Let sequences of i.i.d. r.v.s $\{X_{n,k}\}$ correspond to the d.f.s $(1/m_B) \int_0^t \overline{B}_k(x) \, \mathrm{d}x$ $(k = 1, 2)$, so that, by (5.0.9), r.v.s w_k with the stationary waiting time d.f.s $W_k(\cdot)$ are expressible as

$$w_k = \sum_{n=1}^{N(k)} X_{n,k}$$

where $N(k)$ is geometrically distributed on $\{0, 1, \ldots\}$ with parameter ϱ_k. Since $\mathsf{P}(N(k) \geqq j) = (\varrho_k)^j$, $N(1) \leqq_d N(2)$, and then (5.2.7) follows by appealing to Proposition 2.2.5.

For GI/M/1 queues, $A_2 \leqq_L A_1$ and $\mu_1 \geqq \mu_2$ means by definition that

$$\int_0^\infty \exp\left(-\mu_2(1-z)\,t\right) \mathrm{d}A_2(t) \geqq \int_0^\infty \exp\left(-\mu_1(1-z)\,t\right) \mathrm{d}A_1(t)$$

for all real $z < 1$, and so for the corresponding roots δ_k of (5.0.13) we have $\delta_1 \leqq \delta_2$. (5.2.7) now follows from (5.0.11).

By using (5.0.15) and (5.0.16), comparability with respect to $\leqq_d$ for queue lengths may be obtained under the assumptions of Theorem 5.2.3, or indeed whenever $A_1 \geqq_d A_2$ and $W_1 \leqq_d W_2$ (for (5.0.15)), and $(A_1)_R \geqq_d (A_2)_R$ as well for (5.0.16).

5.3 Bounds obtained from internal monotonicity properties of GI/GI/1

For stationary GI/GI/1 systems Theorem 5.1.1 enables us to obtain bounds for W as follows:

(i) choose a suitable initial d.f. W_0 with the property (5.1.1);

(ii) then obtain better bounds (or, approximations) $W_1, W_2, \ldots$ simply by iteration.

We give here two examples, one for each of the relations $\leqq_c$ and $\leqq_d$.

Proposition 5.3.1. *The mean stationary waiting time m_W satisfies*

$$m_W \geqq c \quad \textit{where} \quad c = \int_{-c}^\infty \bar{K}(t)\,\mathrm{d}t. \tag{5.3.1}$$

Proof. Whenever $W_0 \leqq_c W_1 \leqq_c \ldots$, we have by (1.3.14) that $\mathsf{E}w_0 \leqq \mathsf{E}w = m_W$. Certainly, if $W_0 = \Theta_0$, $W_0 \leqq_c W_1$. Now the function

$$x - \int_{-x}^\infty \bar{K}(t)\,\mathrm{d}t = \int_{-x}^0 K(t)\,\mathrm{d}t - \int_0^\infty \bar{K}(t)\,\mathrm{d}t \tag{5.3.2}$$

is continuous and monotonic in $x \geqq 0$, negative at $x = 0$ and $\to \mathsf{E}(\alpha - \beta) = -\mathsf{E}U > 0$ as $x \to \infty$. Consequently, the equation in c at (5.3.1) defines c uniquely. Also, whenever $W_0 = \Theta_x$ for $x > 0$, $W_1(t) = K(t - x)$ for $t > 0$ so that

$$\mathsf{E}(w_1 - u)_+ = \int_u^\infty \bar{K}(t-x)\,\mathrm{d}t = \int_{-(x-u)}^\infty \bar{K}(t)\,\mathrm{d}t \geqq (x-u)_+ \quad (\text{all}\, u \geqq 0) \tag{5.3.3}$$

provided that $x \leqq c$. Since $(x - u)_+ = \mathsf{E}(w_0 - u)_+$, $0 \leqq x \leqq c$ is thus sufficient for $\Theta_x = W_0 \leqq_c W_1$.

Proposition 5.3.2. *If there exists $\gamma > 0$ for which $\int_0^\infty \exp(\gamma x)\,\mathrm{d}K(x) < \infty$, then the stationary waiting time d.f. W in* GI/GI/1 *satisfies*

$$1 - a_1 \exp(-\theta t) \leqq W(t) \leqq 1 - a_2 \exp(-\theta t) \qquad (\textit{all } t \geqq 0) \tag{5.3.4}$$

where θ is a real number with

$$\int_{-\infty}^{\infty} \exp(\theta x)\, \mathrm{d}K(x) \leqq 1 \tag{5.3.5}$$

and

$$a_1 \equiv \sup_{t>0} f(t), \qquad a_2 \equiv \inf_{t>0} f(t) \qquad (0 \leqq a_1, a_2 \leqq 1), \tag{5.3.6}$$

$$f(t) \equiv \bar{K}(t) \Big/ \int_t^{\infty} \exp\big(\theta(y-t)\big)\, \mathrm{d}K(y) \qquad (0 \leqq t < \infty). \tag{5.3.7}$$

Also, m_W and p_W satisfy

$$a_2/\theta \leqq m_W \leqq a_1/\theta, \tag{5.3.8}$$

$$a_2 \leqq p_W \leqq a_1. \tag{5.3.9}$$

Proof. We shall prove the left hand inequality at (5.3.4), as the proof of the other is analogous, and shall establish in fact that $\overline{W}(t) \leqq a_1 \mathrm{e}^{-\theta t}$ $(t \geqq 0)$.

First note that since

$$0 > \mathsf{E}U = -(\mathrm{d}/\mathrm{d}\theta) \int_{-\infty}^{\infty} \mathrm{e}^{\theta x}\, \mathrm{d}K(x)|_{\theta=0},$$

the existence of γ as in the theorem implies that there do exist some $\theta > 0$ such that (5.3.5) is satisfied. Suppose then that

$$\overline{W}_0(t) = a\, \mathrm{e}^{-\theta t} \qquad (t \geqq 0) \tag{5.3.10}$$

for some constants a and θ to be determined. Substituting this $\overline{W}_0(t)$ in (5.0.2) leads to

$$\overline{W}_1(t) = \bar{K}(t) + a\, \mathrm{e}^{-\theta t} \int_{-\infty}^{t} \mathrm{e}^{\theta x}\, \mathrm{d}K(x).$$

Now let θ satisfy (5.3.5) and set $a = a_1$ as at (5.3.6); then it follows that

$$\overline{W}_0(t) \geqq \overline{W}_1(t), \tag{5.3.11}$$

i.e., $W_0 \geqq_d W_1$ as required. (5.3.8) and (5.3.9) follow immediately from (5.3.4).

Define

$$\theta(K) = \sup\left\{\theta\colon \int_{-\infty}^{\infty} \mathrm{e}^{\theta x}\, \mathrm{d}K(x) \leqq 1\right\}.$$

Then it is worth noting that $\theta(K)$ is monotonic in K in the sense that

$$K_1 \leqq_c K_2 \ \text{ implies } \ \theta(K_1) \geqq \theta(K_2) \tag{5.3.12}$$

(and, recall that $K_1 \leqq_d K_2$ implies that $K_1 \leqq_c K_2$ so (5.3.12) holds then as well). If in particular A is of the type NBUE and $B \leqq_c \operatorname{Exp}(1/m_B)$, then

$$W(t) \geqq 1 - \varrho \exp\left(-(m_B^{-1} - m_A^{-1})\, t\right) \qquad (t \geqq 0) \tag{5.3.13}$$

as shown by MIYAZAWA (1976b) and STOYAN (1977b) using firstly, that for A of type NBUE $W \leqq_d W_v$ where W_v denotes the stationary virtual waiting time d.f., and secondly Takacs's identity

$$W_v(t) = 1 - \varrho + \varrho(W * B_R)(t) \qquad (t \geqq 0) \tag{5.3.14}$$

from which we have $W \leqq_d W_v$ for W_v in M/GI/1; (5.3.13) then follows on using (5.2.7).

BOROVKOV (1976a) used (5.3.4) to establish the following bound for the rate of convergence towards the stationary waiting time d.f.s in GI/GI/1.

Theorem 5.3.3. *Let W_n be the waiting time d.f. of the nth customer when $W_0 = \Theta_0$, and let W be the stationary waiting time d.f. corresponding to the d.f. K. For all $\gamma > 0$ for which*

$$e(\gamma) = \int_0^\infty \exp(\gamma t)\, \mathrm{d}K(t) \leqq 1,$$

$$0 \leqq W_n(t) - W(t) \leqq \exp(-\gamma t)\,[e(\gamma)]^n \quad (t \geqq 0,\ n = 0, 1, \ldots). \tag{5.3.15}$$

Proof. As is well-known, the waiting time w_n of the nth customer has the same d.f. as M_n,

$$M_n = \max(0, U_1, \ldots, U_1 + \cdots + U_n), \tag{5.3.16}$$

and W can be interpreted as the d.f. of the r.v.

$$M = \sup_{n \geqq 0} Y_n, \qquad Y_n = \sum_{k=1}^{n} U_k.$$

Consequently, we have

$$\begin{aligned} W_n(t) - W(t) &= \mathsf{P}(M > t) - \mathsf{P}(M_n > t) \\ &= \mathsf{P}(M_n \leqq t, M > t) \leqq \int_0^t \mathsf{P}(M > t - x)\, \mathrm{d}\mathsf{P}(Y_n \leqq x) \\ &\leqq \int_0^t \mathrm{e}^{-\gamma(t-x)}\, \mathrm{d}\mathsf{P}(Y_n \leqq x) \leqq \mathrm{e}^{-\gamma t}\big(e(\gamma)\big)^n. \end{aligned}$$

5.4 Bounds obtained from external monotonicity properties of GI/GI/1

For a given GI/GI/1 queue Σ we can easily obtain bounds for the stationary waiting time d.f. W by using systems Σ_1 and Σ_2 for which

$$K_1 \prec K \prec K_2 \text{ where } \prec \text{ denotes } \leqq_c \text{ or } \leqq_d$$

and appealing to Theorems 5.2.1 and 5.2.2. The relation $\leqq_c$ is of especial interest, because with it we can find bounds for quantities in Σ from systems with the same traffic intensity ϱ. In particular, we obtain practically useful bounds for m_W by using equations (5.0.10) and (5.0.12) for M/GI/1 and GI/M/1, as follows.

(a) If

$$A \leqq_c \text{Exp}\,(1/m_A), \tag{5.4.1}$$

then

$$m_W \leqq \mathsf{E}\beta^2/2m_A(1-\varrho) = (\sigma_B{}^2 + m_B{}^2)/2m_A(1-\varrho). \tag{5.4.2}$$

Criteria for (5.4.1) are given in Sections 1.5 and 1.6. In particular, by Proposition 1.6.2, (5.4.1) is true if A is NBUE, in which case since we have also the lower bound at (5.6.8) below (note also (5.6.5)), m_W is bounded quite closely.

(b) If

$$B \leqq_c \text{Exp}\,(1/m_B), \tag{5.4.3}$$

then

$$m_W \leqq \delta m_B/(1-\delta) \tag{5.4.4}$$

where δ is the root in $0 < \delta < 1$ of the equation

$$\delta = \int_0^\infty \exp\left(-(1-\delta)\,t/m_B\right) \mathrm{d}A(t).$$

In many cases this bound is better than (5.4.2). The values of m_W in E_k/M/1 given in Table 5.1 provide bounds for those E_k/GI/1 queues satisfying (5.4.3).

Table 5.1. Stationary mean waiting time m_W/m_B in E_k/M/1 queues

k	ϱ								
	0.1	0.2	0.3	0.4	0.5	0.6	0.7	0.8	0.9
1	0.111	0.250	0.429	0.667	1.000	1.500	2.333	4.000	9.000
2	0.030	0.104	0.217	0.380	0.618	0.984	1.601	2.844	6.589
3	0.013	0.063	0.152	0.289	0.494	0.813	1.358	2.459	5.785
4	0.007	0.045	0.122	0.244	0.433	0.729	1.237	2.267	5.383
5	0.004	0.035	0.104	0.218	0.396	0.679	1.164	2.152	5.142
6	0.003	0.029	0.093	0.201	0.372	0.645	1.116	2.067	4.981
7	0.002	0.025	0.085	0.189	0.355	0.621	1.082	2.021	4.867
8	0.001	0.022	0.079	0.180	0.342	0.604	1.056	1.980	4.781
9	0.001	0.020	0.075	0.173	0.333	0.590	1.036	1.948	4.714
10	0.001	0.019	0.071	0.168	0.325	0.579	1.020	1.922	4.660
∞	0.000	0.007	0.043	0.120	0.255	0.480	0.876	1.693	4.179

A general lower bound for m_W is obtained by using the extremal property of constant interarrival times, i.e., that $\Theta_{m_A} \geqq_c A$.

Proposition 5.4.1. *For any* GI/GI/1 *system with* $\varrho < 1$,

$$m_W \geqq \frac{\sigma_B{}^2}{2m_A(1-\varrho)} - {}^1/_2 m_B, \tag{5.4.5}$$

while if $b \equiv \sup\{t: B(t) = 0\} > 0$, *then*

$$m_W \geqq \frac{\sigma_B{}^2}{2m_A(1-\varrho)} - {}^1/_2(m_B - b). \tag{5.4.6}$$

Proof. When the interarrival d.f. is IFR, the bound (5.6.5) holds, and in particular, since Θ_{m_A} is IFR, this bound applies to D/GI/1 queues. Thus m_W is bounded below by the mean waiting time in D/GI/1 with interarrival time d.f. Θ_{m_A} and service time d.f. B, which in turn is bounded as at (5.4.5).

The bound (5.4.6) is proved by applying (5.4.5) to a D/GI/1 queue with the service time d.f. $B(t + b)$ and the interarrival times $m_A - b$. (For this queue the U_n have the same d.f. as in the D/GI/1 queue above and thus the waiting time d.f. is not changed.)

5.5 Bounds for the mean busy and idle periods

Those time intervals which start at service completion epochs where the server becomes idle because no further customers are in the system, and conclude with the next arrival, are called *idle periods.* Similarly, the time intervals between two consecutive idle periods are called *busy periods.* In the queue GI/GI/1 the respective busy and idle periods form two sequences of i.i.d. r.v.s, but any given busy period and the idle period immediately following it are in general dependent.

Denoting the corresponding means by m_L (idle) and m_F (busy), the probability P_{occ} that the server is busy in its stationary state can be written in the form

$$P_{occ} = m_F/(m_L + m_F), \tag{5.5.1}$$

and on account of (5.0.14),

$$m_L = (1-\varrho)\, m_F/\varrho. \tag{5.5.2}$$

Consequently, a bound for m_F yields also a bound for m_L: this is of interest because it usually seems to be simpler to bound m_F than m_L. For example we have the obvious relation

$$m_F \geqq m_B \tag{5.5.3}$$

and, consequently,

$$m_L \geqq (1-\varrho)\, m_B/\varrho = (1-\varrho)\, m_A = m_A - m_B. \tag{5.5.4}$$

A somewhat better bound is obtained by using the internal monotonicity of GI/GI/1. Starting from the initial d.f. $W_0 = \Theta_0$ we can obtain

$$1 - p_W \leqq p = \int_0^\infty \bar{A}(t)\,\mathrm{d}B(t) = \mathsf{P}(U < 0),$$

and by (5.0.6) and (5.5.2) we then have

$$m_F \geqq m_B/\mathsf{P}(U < 0). \tag{5.5.5}$$

Furthermore, we have the following

Proposition 5.5.1. *If the interarrival time d.f. A is of type NBU (NWU), then*

$$m_F \geqq (\leqq)\, m_B \Big/ \left(1 - \int_0^\infty H_A(t)\,\mathrm{d}B(t)\right) \tag{5.5.6}$$

where $H_A = \sum_{n=1}^{\infty} A^{*n}$ *is the renewal function of A.*

Proof. Suppose a busy period is initiated by a customer whose service time is β and that during this time there are ν arrivals. If $\nu = 0$ then the duration Z of the busy period equals β. Otherwise, we may write

$$Z = \beta + Z_1' + \cdots + Z_\nu'$$

by regarding each Z_j' as a quasi busy period in which the first customer served is one of the ν arrivals and all others (if any) served in Z_j' arrive during the period. Thus, Z_j' in general differs in distribution from Z by having the first arrival epoch occur at a time α_j^R after its start, α_j^R being a residual interarrival time.

Now modify the system by replacing α_j^R by an interarrival time α_j^0 that is independent of all other interarrival times, leading to a possibly different busy period Z_j that is now distributed like Z. Then if $\alpha_j^R \leqq_d \alpha_j^0$, it follows from a sample path argument that $Z_j < Z_j'$, and hence that

$$\beta + \sum_{j=1}^{\nu} Z_j \leqq_d Z. \tag{5.5.7}$$

Since $\leqq_d$ has the property (E) and A being NBU ensure $\alpha_j^R \leqq_d \alpha_j^0$ ($\sim A$), it then follows from Wald's identity that

$$m_F = \mathsf{E}Z \geqq m_B + m_F \mathsf{E}\nu$$

whence, using

$$\mathsf{E}\nu = \mathsf{E}H_A(\beta) = \int_0^\infty H_A(t)\,\mathrm{d}B(t),$$

the inequality (5.5.6) follows.

When A is NBUE, $\lambda t - 1 \leqq H_A(t) \leqq \lambda t$ where $\lambda = 1/m_A$, so then (5.5.6) implies that $m_F \geqq m_B/(2 - \varrho)$ which, being smaller than m_B, is a bound of no interest. When A is NWUE, $\lambda t \leqq H_A(t) \leqq \lambda t + \lambda^2 \operatorname{var} \alpha$, so then $m_F \leqq m_B/(\varrho + \lambda^2 \operatorname{var} \alpha)$.

A counterpart to Proposition 5.5.1 for idle periods is the following

Proposition 5.5.2. *If the interarrival time d.f. A is of type NBU (NWU) then the idle period d.f. L satisfies*

$$L \leqq_d (\geqq_d) A, \tag{5.5.8}$$

while if A is only NBUE (NWUE), then

$$m_L \leqq (\geqq) m_A. \tag{5.5.9}$$

Proof. The idle period is a "residual interarrival time", and its d.f. L has the form

$$L(t) = \int_0^\infty \mathsf{P}(\alpha_1 \leqq t + x \mid \alpha_1 > x) \, \mathrm{d}F(x) \tag{5.5.10}$$

where F is the d.f. of the delay time, defined as the sum of a stationary waiting time r.v. and an independent service time. When A is NBU (NWU)

$$\mathsf{P}(\alpha_1 > t + x \mid \alpha_1 > x) \leqq (\geqq) \mathsf{P}(\alpha_1 > t) \qquad (\text{all } t, x \geqq 0)$$

which when combined with (5.5.10) yields (5.5.8). Similarly (5.5.9) follows on using the interpretation (1.6.9) of the property NBUE (NWUE).

Loulou (1978) used random walk theory and Berry-Esseen bounds to establish the crude bound

$$m_F \leqq c_1 m_B\big(\sigma_U/(-m_U)\big) \exp\big((c_2 \mathsf{E} \, |U|^3/\sigma_U^2 - c_3 m_U)/\sigma_U\big) \tag{5.5.11}$$

where $m_U = \mathsf{E}U = m_B - m_A$, $\sigma_U = (\sigma_A^2 + \sigma_B^2)^{1/2}$ and c_1, c_2 and c_3 are constants which can be taken as $c_1 = 0.9$, $c_2 = 5.4$, and $c_3 = 0.8$. Since $m_F = m_B/(1 - p_W)$, (5.5.11) can be regarded as giving a bound on $\mathsf{P}(w = 0)$. In particular, if for a sequence of systems $m_U \to 0$ while σ_U^2 and $\mathsf{E} \, |U|^3$ converge to finite limits, (5.5.11) shows that $m_F = O(-1/m_U)$ rather than $O(1/m_U^2)$.

5.6 Bounds for the mean waiting time m_W and the probability of waiting via idle period bounds

Marshall (1968a) gave bounds for m_W and p_W by bounding the last term in the identity

$$m_W = \frac{(m_A - m_B)^2 + \sigma_A^2 + \sigma_B^2}{2(m_A - m_B)} - \frac{m_L^2 + \sigma_L^2}{2m_L}. \tag{5.0.7}$$

Upper bounds for m_W follow from using the Cauchy-Schwarz inequality

$$(m_L^2 + \sigma_L^2)/2m_L \geqq m_L/2 \tag{5.6.1}$$

in conjunction with lower bounds for m_L such as in section 5.5. Thus from (5.5.4) we have

$$m_W \leqq (\sigma_A^2 + \sigma_B^2)/2m_A(1 - \varrho), \tag{5.6.2}$$

and from (5.5.5) we have

$$m_W \leqq \frac{\sigma_A^2 + \sigma_B^2}{2m_A(1 - \varrho)} - \frac{(1 - p)(1 - \varrho)\, m_A}{2p} \tag{5.6.3}$$

where $p = \int_0^\infty \bar{A}(t)\, \mathrm{d}B(x)$. With the aid of Theorem 1.6.1 we can obtain better bounds for $(m_L^2 + \sigma_L^2)/2m_L$ than by (5.6.1). Denoting the idle period d.f. by L as before, if the interarrival time d.f. is DMRL (IMRL) then

$$\frac{m_L^2 + \sigma_L^2}{2m_L} \leqq (\geqq) \frac{m_A^2 + \sigma_A^2}{2m_A} \tag{5.6.4}$$

from which it follows that

$$m_W \geqq (\leqq) \frac{\sigma_A^2 + \sigma_B^2}{2m_A(1 - \varrho)} - {}^1/_2 m_A(\varrho + \sigma_A^2/m_A^2) \quad \text{if } A \text{ is DMRL (IMRL).} \tag{5.6.5}$$

Further bounds can be given for interarrival d.f.s of type γ-MRLA (γ-MRLB). As in Proposition 5.5.2, we then have

$$m_L \leqq (\geqq)\, \gamma, \tag{5.6.6}$$

while Marshall showed (cf. Proposition 1.6.3) that then

$$(m_L^2 + \sigma_L^2)/2m_L \leqq (\geqq)\, \gamma. \tag{5.6.7}$$

(5.6.7) and (5.0.7) give for γ-MRLA (γ-MRLB) interarrival times

$$m_W \geqq (\leqq) \frac{\sigma_A^2 + \sigma_B^2}{2m_A(1 - \varrho)} - {}^1/_2 m_A(1 + \varrho) - (\gamma - m_A), \tag{5.6.8}$$

the last term vanishing when A is NBUE (NWUE) since then $\gamma = m_A$. In the case of systems of the type E_k/GI/1, A is then NBUE and γ-MRLB with $\gamma = m_A/k$, so

$$0 \leqq m_W - \frac{\sigma_A^2 + \sigma_B^2}{2m_A(1 - \varrho)} - {}^1/_2 m_A(1 + \varrho) \leqq m_A(1 - k^{-1}). \tag{5.6.9}$$

(see Fainberg (1976) and Fainberg and Stoyan (1978)). Kreinin (1980) sharpened (5.6.9) to

$$m_W \leqq \frac{m_B^2/k + \sigma_B^2}{2m_A(1 - \varrho)} = \frac{\varrho^2 \sigma_A^2 + \sigma_B^2}{2m_A(1 - \varrho)}, \tag{5.6.10}$$

equality holding for M/GI/1.

Bounds for p_W are obtained by starting from (5.0.6) and bounding m_L by (5.5.9) or (5.6.6). Thus, if A is of type NBUE (NWUE)

$$p_W \leqq (\geqq)\, \varrho, \tag{5.6.11}$$

while for Erlang interarrival times, i.e. $A = \mathrm{Erl}\,(1/km_A,\, k)$ as for (5.6.9), we have

$$1 - (1 - \varrho)\, k \leqq p_W \leqq \varrho. \tag{5.6.12}$$

5.7 Other methods for bounding m_W

A wide variety of methods has been used to derive bounds for m_W in GI/GI/1; Daley and Trengove (1977) identified at least seven methods used to give at least twenty different bounds, excluding the restriction to classes of d.f.s like IFR or DMRL or NBUE as in the previous section, and there have been further notes published since then.

It was shown in Daley (1977) that in the case of the second moment of the tail of a r.v., the Cauchy-Schwarz inequality, which is basically what is used in establishing Kingman's (1962) bound at (5.6.2), can be strengthened to the inequality

$$\mathsf{E}(Y - Z)_+^2 \geqq \left(\mathsf{E}\,Y^2/(\mathsf{E}\,Y)^2\right)\left(\mathsf{E}(Y - Z)_+\right)^2 \tag{5.7.1}$$

where Y and Z are independent non-negative r.v.s with finite second moments. Then in place of the argument leading to (5.6.2) we can write either

$$\begin{aligned} 2m_A(1 - \varrho)\, m_W &= \mathsf{E}U^2 - \mathsf{E}(\alpha - \beta - w)_+^2 \\ &\leqq \mathsf{E}U^2 - \left(\mathsf{E}(\alpha - \beta)_+^2/(\mathsf{E}(\alpha - \beta)_+)^2\right)\mathsf{E}(\alpha - \beta - w)_+\right)^2 \\ &= \mathrm{var}\, U - (-\mathsf{E}U/\mathsf{E}U_-)^2\, \mathrm{var}\, U_- \end{aligned} \tag{5.7.2}$$

or else the weaker bound which does not however involve U but only α and β,

$$\begin{aligned} 2m_A(1 - \varrho)\, m_W &\leqq \mathsf{E}U^2 - \left(\mathsf{E}\alpha^2/(\mathsf{E}\alpha)^2\right)\left(\mathsf{E}(\alpha - \beta - w)_+\right)^2 \\ &= \mathrm{var}\, U - (-\mathsf{E}U/\mathsf{E}\alpha)^2\, \mathrm{var}\, \alpha \\ &= \sigma_B^2 + \varrho(2 - \varrho)\, \sigma_A^2. \end{aligned} \tag{5.7.3}$$

The inequality at (5.7.1) is proved by showing that $\mathsf{E}(Y - z)_+^2/\left(\mathsf{E}(Y - z)_+\right)^2$ is monotonic non-decreasing in z, and then a conditioning argument yields (5.7.1).

Marshall (1968a) used a convexity argument to establish the lower bound at (5.3.1). Calo and Schwarz (1977) and Daley (1976c) also used a convexity argument to show that

$$m_W \leqq \eta \tag{5.7.4}$$

where η is the positive root of

$$\eta^2 = \mathsf{E}(U + \eta)_+^2. \tag{5.7.5}$$

To see this, it is not difficult to check that

$$g(x) \equiv x^2 - \mathsf{E}(U + x)_+^2$$

is convex in x with $g(0) < 0$ and $g(x) \to \infty$ for $x \to \infty$. Then by the recurrence relation (5.0.1), and Jensen's inequality,

$$0 = \mathsf{E}w^2 - \mathsf{E}(U + w)_+^2 = \mathsf{E}g(w) \geqq g(m_W),$$

which leads to (5.7.4).

Karpelevich and Kreinin (1980) proved an inequality for queues with finitely divisible interarrival and service times, that is, when α and β can be expressed as

$$\alpha = \alpha_1' + \cdots + \alpha_m', \qquad \beta = \beta_1' + \cdots + \beta_m'$$

for some integer $m \geqq 2$ and i.i.d. r.v.s $\alpha_1', \ldots, \alpha_m'$ and $\beta_1', \ldots, \beta_m'$. Then the mean stationary waiting time $m_W' = \mathsf{E}w'$ in a system with generic r.v.s α' and β' is related to that of a system with α and β by

$$m_W \leqq m_W', \tag{5.7.6}$$

because w' is distributed like

$$\max(0, \alpha_1' - \beta_1', \quad \alpha_1' + \alpha_2' - \beta_1' - \beta_2', \ldots)$$

$$\geqq \max(0, \alpha_1 - \beta_1, \alpha_1 + \alpha_2 - \beta_1 - \beta_2, \ldots)$$

and the right hand side is distributed like w.

The same authors (Karpelevich and Kreinin (1976)) also established (5.6.5) and (5.6.8) when A has IFR, though the weaker DMRL property suffices as noted above, as well as the result (with the same assumption, but DMRL sufficing — see Daley and Trengove (1977) or Proposition 1.6.4 above) that

$$2m_A(1 - \varrho)\, m_W \leqq \sigma_A^2 + \sigma_B^2 - \mathrm{var}\,(\alpha - \xi)_+ \tag{5.7.7}$$

where ξ is real and positive and satisfies

$$\mathsf{E}(\alpha - \xi)_+ = (1 - \varrho)\, \mathsf{E}\alpha. \tag{5.7.8}$$

5.8 Heuristic approximations for GI/GI/1

There are many bounds and approximations for m_W, p_W and other characteristics obtained by more or less heuristic methods, especially in connection with heavy traffic and diffusion approximations. Kingman (1964) showed that (5.6.2) is a heavy traffic approximation, and (cf. also Kingman

(1965)) in KLEINROCK (1976) it is shown that diffusion approximation arguments give the same result. Also using a diffusion approximation and a further heuristic step KOBAYASHI (1974) obtained for the stationary state probabilities $p_k = \lim_{t\to\infty} \mathsf{P}(\nu_t = k)$

$$p_k \simeq \begin{cases} 1 - \varrho & (k = 0) \\ \varrho(1 - \varrho)\, \hat{\varrho}^{k-1} & (k = 1, 2, \dots) \end{cases} \tag{5.8.1}$$

where

$$\hat{\varrho} = \exp\left(-2(1 - \varrho)/(\varrho\sigma_A^2/m_A^2 + \sigma_B^2/m_B^2)\right);$$

see KLEINROCK (1976) for further discussion.

SUNAGA, KONDO and BISNOS (1978) gave diffusion approximations for the mean number of customers for small and large ϱ. Their results can also be used for G/G/1 queues.

Other formulae can be obtained by direct manipulation from existing formulae, as for example, starting from M/GI/1. Thus. MARCHAL (1976) (see also MARCHAL and HARRIS (1976)) proposed

$$m_W \simeq \frac{1 + \sigma_B^2/m_B^2}{1/\varrho^2 + \sigma_B^2/m_B^2} \times \frac{\sigma_A^2 + \sigma_B^2}{2m_A(1 - \varrho)}, \tag{5.8.2}$$

and KRÄMER and LANGENBACH-BELZ (1976) suggested

$$m_W \simeq \frac{\varrho m_B}{2(1 - \varrho)} \left(\frac{\sigma_A^2}{m_A^2} + \frac{\sigma_B^2}{m_B^2}\right) \times \begin{cases} \exp\left\{-\dfrac{2(1 - \varrho)}{3\varrho} \cdot \dfrac{(1 - \sigma_A^2/m_A^2)^2}{\sigma_A^2/m_A^2 + \sigma_B^2/m_B^2}\right\} & (\sigma_A^2 \leqq m_A^2), \\ \exp\left\{-(1 - \varrho) \cdot \dfrac{\sigma_A^2/m_A^2 - 1}{\sigma_A^2/m_A^2 + 4\sigma_B^2/m_B^2}\right\} & (\sigma_A^2 > m_A^2), \end{cases}$$

$$p_W \simeq \varrho + \varrho(1 - \varrho)\,(\sigma_A^2/m_A^2 - 1)\, A'$$

where

$$A' = \begin{cases} \dfrac{1 + \sigma_A^2/m_A^2 + \varrho\sigma_B^2/m_B^2}{1 + \varrho(\sigma_B^2/m_B^2 - 1) + \varrho^2(4\sigma_A^2/m_A^2 + \sigma_B^2/m_B^2)} & (\sigma_A^2 \leqq m_A^2), \\ \dfrac{4\varrho}{\sigma_A^2/m_A^2 + \varrho^2(4\sigma_A^2/m_A^2 + \sigma_B^2/m_B^2)} & (\sigma_A^2 > m_A^2). \end{cases}$$

FINDLY, FREDERICKS (1980) discussed heuristic approximations for W, m_W and p_W with $W(t) \approx 1 - c\, \mathrm{e}^{-at}$.

5.9 Bounds in the case of non-renewal input

Using methods similar to those used for independent interarrival and service times, we can obtain bounds in some cases where the interarrival and service times may be stochastically dependent. For example, external monotonicity properties with respect to $\leqq_c$ can be used, or bounds can be obtained using (5.0.21), which reduces to (5.0.7) in the case of independence.

In the following we assume that the sequence $\{(\alpha_n, \beta_n)\}$ is metrically transitive and that the mean stationary waiting time is finite.

If the service times are mutually independent and independent of the nterarrival times, then as analogues of (5.4.5) and (5.4.6) we have

$$m_W \geqq \sigma_B^2/2m_A(1-\varrho) - m_B/2 \tag{5.9.1}$$

and

$$m_W \geqq \sigma_B^2/2m_A(1-\varrho) - (m_B - b)/2 \tag{5.9.2}$$

for any b for which $B(b) = 0$. The proof of these formulae is the same as in the case of a renewal input, for it uses the fact that m_W is greater than the mean stationary waiting time in D/GI/1. If the input process is $\leqq_c$-thinner than a Poisson process with the same intensity (see Section 1.10.2) we obtain, analogously to (5.4.2),

$$m_W \leqq (\sigma_B^2 + m_B^2)/2m_A(1-\varrho). \tag{5.9.3}$$

From (5.1.6) there follow the very rough bounds for G/G/1,

$$p_W \geqq 1 - p = \bar{K}(0), \tag{5.9.4}$$

$$m_W \geqq \int_0^\infty t \, \mathrm{d}K(t), \tag{5.9.5}$$

where K is the d.f. of $U_0 = \beta_0 - \alpha_0$ (recall that $\{U_n\}$ is stationary). For the mean busy period m_F in a stationary G/G/1 system,

$$m_F \geqq m_B \quad \text{and} \quad m_F \geqq m_B/p$$

hold, because under our assumptions the formulae (5.5.1) and (5.5.2) are valid. With them we can obtain bounds for m_W in G/G/1, much as in the independent case, starting from (5.0.21). Thus, if cov $(w_0, U_0) \leqq 0$, then

$$m_W \leqq -(\operatorname{var} U_0)/2\mathsf{E}U_0 = \left(\sigma_A^2 + \sigma_B^2 - 2\operatorname{cov}(\alpha_0, \beta_0)\right)/2m_A(1-\varrho) \tag{5.9.6}$$

and

$$m_W \leqq (\operatorname{var} U_0)/2\mathsf{E}(-U_0) - (1-p)(1-\varrho)\, m_A/2p. \tag{5.9.7}$$

These two inequalities reduce to (5.6.2) and (5.6.3) for G/GI/1.

Granot, Granot and Lemoine (1975) considered a special G/GI/1 queue with the following non-renewal input. The nth customer is scheduled to arrive at time point nm_A: however, he may be early or late, and actually appears at $t = nm_A + \Delta_n$, where the Δ_n are i.i.d. with $\mathsf{E}\Delta_n = 0$ and variance σ_Δ^2 and $\mathsf{P}(|\Delta_n| < d) = 1$ for some $d > 0$. For this system

$$m_W \leqq (\sigma_B^2 + 2\sigma_\Delta^2)/2m_A(1-\varrho). \tag{5.9.8}$$

Further, if $\mathsf{E} \exp(\gamma\beta_n) < \infty$ for some $\gamma > 0$, then there is $\theta > 0$ with

$$W(x) \geqq 1 - \exp\left(-\theta(x - d)\right) \quad (x \geqq d). \tag{5.9.9}$$

KINGMAN (1962) discussed the heavy traffic behaviour of such a system and suggested $m_W \approx \sigma_B{}^2/2(m_A - m_B)$ as a heavy traffic approximation.

Notes and bibliographic comments to Chapter 5

5.1

The internal monotonicity of GI/GI/1 for $\leqq_d$, at least for $W_0 = \Theta_0$, has been known since LINDLEY (1952). It plays a role in the proof of existence of the stationary waiting time d.f. In the general form given here, Theorem 5.1.1 was proved by DALEY (1968) for $\leqq_d$ (see also KINGMAN (1970)), and by STOYAN (1972a) for $\leqq_c$.

LOYNES (1962) found the monotonicity property (5.1.6) and used it to prove the existence of stationary waiting time d.f.s in G/G/1.

Problem. Is it possible to prove that in G/G/1, internal monotonicity as at (5.1.6) holds for initial states other than $w_0 = 0$, as well as for $\leqq_c$?

5.2

External monotonicity was first proved by GAEDE (1965) and by DALEY and MORAN (1968). The external nature of systems with constant service or interarrival times was found by ROGOZIN (1966), who used the equation (5.0.4′) for m_W and Jensen's inequality. ROSSBERG (1968, 1970b) worked similarly and, furthermore, considered other characteristics such as p_W and delay times, proving that m_W has a strict minimum for $A = \Theta_{m_A}$. Theorem 5.2.1 for $\leqq_c$ was proved, first by STOYAN and STOYAN (1969) (in order to generalize the results of ROGOZIN and ROSSBERG), and shortly after by BOROVKOV (1970). JANSSEN (1976) studied the monotonicity properties for $\leqq_d$ of a special G/G/1 queue (semi-Markovian interarrival and service times). ROSS (1978) and ROLSKI (1980) compared m_W in M/GI/1 with m_W in a GI/GI/1 queue with "interrupted Poisson input" (see Section 7.5) and doubly stochastic Poisson input. BOROVKOV (1970) found the formula (5.0.5) for the cumulants of the stationary waiting time r.v. in GI/GI/1 and proved the monotonicity property given in Theorem 5.2.2 for $\leqq_c$. BERGMANN (1974) obtained similar results and proved Theorem 5.2.2 in the form presented here.

ROLSKI (1972) proved part (b) of Theorem 5.2.3. Using (5.0.9) we can show that for an M/GI/1 system for which $B \leqq_c \operatorname{Exp}(1/m_B)$, we have $W \leqq_c W'$, where $W'(t) = 1 - \varrho \exp(-\varrho t/m_W)$. VOLODIN and KOLIN (1972) suggested W' as an approximation for W. The mapping method be used to establish external monotonicity properties of virtual waiting times in GI/GI/1 with respect to $\leqq_d$ and $\leqq_c$ (see also Section 6.2). For its stationary d.f. W_v, this can be done directly via (5.3.14); see also HARRISON (1977).

Open problems. 1. Under what conditions are stationary and nonstationary queue length d.f.s comparable with respect to $\leqq_c$? In particular, are stationary distributions comparable when (5.2.1) holds for $\leqq_c$?

2. For GI/M/1 and M/GI/1, $K_1 \leqq_c K_2$ implies $W_1 \leqq_d W_2$: for what other systems can this hold? (See also WHITT (1983a).)

3. What conditions imply $W_1 \leqq_L W_2$?

4. For a given class of GI/GI/1 queues prescribed by fixing the first and second moments of α_n and β_n, DALEY and TRENGOVE (1977) contains conjectures concerning sup m_W and inf m_W within the class.

5.3

MARSHALL (1968a) found the bound in Proposition 5.3.1; the proof above comes from STOYAN and STOYAN (1974). They also gave upper bounds for W and m_W using internal monotonicity with respect to $\leqq_c$, using as initial d.f. the stationary waiting time d.f. in M/M/1, and assuming (5.4.1) and (5.4.3).

KINGMAN (1970) deduced (5.3.4) by using internal monotonicity. ROSSBERG and SIEGEL (1974a, b) demonstrated the wider applicability of his method in applying it to GI/GI/1 queues in which for example the busy period starts with a warm-up period.

ROSS (1974) showed that the left hand of (5.3.4), as had KINGMAN (1964) for the right hand side, could be established from a martingale inequality. (The sequence $\{Z_n\} = \left\{\exp\left(\theta \sum_{j=1}^{n} U_j\right)\right\}$ is a martingale if $\mathsf{E} \exp(\theta U_1) = 1$; this is used with the inequality and $W(t) = \mathsf{P}(\sum_{j=1}^{n} U_j > t$ for some $n = 1, 2, \ldots)$ to furnish a proof.) TAN (1979) has refined this method further. BERGMANN and STOYAN (1976) gave (5.3.12), which can be proved as $W_1 \leqq_d W_2$ in Theorem 5.2.3. They also gave simpler quantities that approximate the coefficients a_1 and a_2, and conjectured the inequality (5.3.13). MIYAZAWA (1976a) earlier proved an analogous relation for the queue length. DELBROUCK (1975) also studied exponential bounds, and considered relations between (general) W and stationary waiting time d.f.s in M/GI/1 in the case of concave service time d.f.s.

KÖLLERSTRÖM (1976) gave a real variable derivation identifying the largest real θ satisfying (5.3.5) and then (5.3.4), by first demonstrating analytically that W is NWU. Identifying W with the distribution of the maximum of a random walk with negative mean step-length starting at the origin, this NWU property is an easy consequence of the regenerative nature of the finite set of ascending ladder heights of the walk (ASMUSSEN, 1980, private communication; see also DALEY and TRENGROVE (1977)). (5.3.4) can be used to give geometric bounds on P_k^0 and P_k at (5.0.15) and (5.0.16), and those representations can be used with the NWU property of W to deduce that $\{P_k^0\}$ and $\{P_k\}$ also are NWU (ASMUSSEN and DALEY, unpublished observation).

KÖLLERSTRÖM (1978) showed that

$$\exp(-t/m_W) - \delta \leqq 1 - W(t) \leqq \exp\left[-t/m_W + \delta\big(\exp(t/m_W) - 1\big)\right]$$

for suitable δ with $\delta \to 0$ as $\varrho \to 1$.

STOYAN (1977b) used bounds for NWU d.f.s (see KORZENIOWSKI and OPAWSKI (1976)) to conclude that

$$\overline{W}(t) \leqq \frac{c}{c+t} \quad (t \geqq 0, c \geqq m_W).$$

BARTFAI (1973) found a bound for the convergence similar to (5.3.15); furthermore, he proved theorems on "long" waiting times, for example, on $\lim_{n\to\infty} \sup_{1\leqq j\leqq n} w_j$. For further discussion of (5.3.15) and related inequalities see KOTZUREK (1979), who showed that in (5.3.15) a factor $a \leqq 1$ in the right hand can be introduced and n can be replaced by $n + 1$. For further bounds on the rate of convergence, see HEATHCOTE (1967) and VERAVERBEKE and TEUGELS (1975). Interesting convergence properties of mean waiting times are proved by MORI (1976).

Open problems. 1. Do there exist other initial d.f.s for which useful bounds for W can be obtained by using internal monotonicity? (See also KINGMAN (1970)).

2. Can useful bounds for queue lengths be obtained using internal monotonicity?

5.4

GAEDE (1965) and DALEY and MORAN (1968), having proved external monotonicity, suggested its use for bounding, but they worked with $\leqq_d$ and their bounds were relatively rough. The bounds for m_W in this section mostly come from STOYAN (1974a) and STOYAN and STOYAN (1974). MORI (1975) and KARPELEVICH and KREININ (1976) proved (5.4.2) for NBUE interarrival times, and MORI (1975) found (5.4.5). (5.4.2) can be proved under weaker assumptions (NBUE defined with the help of $\leqq_{cv}$, i.e. $A_R \leqq_{cv} A$; see BERGMANN (1979b)).

KINGMAN's (1965) remarks on bounds for GI/GI/1 by means of normal d.f.s have been used in STOYAN (1972e). If K is smaller with respect to $\leqq_c$ than the normal d.f. with the parameters $m < 0$ and σ^2, then

$$m_W \leqq \sum_{n=1}^{\infty} \sigma n^{-1/2}[\phi(z_n) + z_n \Phi(z_n)] \quad (z_n \equiv n^{-1/2} m/\sigma;\ n = 1, 2, \ldots)$$

where ϕ denotes the density and Φ the d.f. of a standard normal d.f. The sum on the right hand side is approximately equal to $\sigma\big(1/(2\alpha) - c + O(\alpha)\big)$ where $\alpha = -m/\sigma$ and $c = 0.58$ (see KINGMAN (1965)).

Just as for m_W in (5.4.2) and (5.4.4), the variance $\sigma_W{}^2$ of the stationary waiting time can be bounded by using Theorem 5.2.2 and formulae for $\sigma_W{}^2$ in M/GI/1 and GI/M/1. See also BERGMANN, DALEY, ROLSKI, and STOYAN (1979), KINGMAN (1962), MORI (1975), and FAINBERG (1979).

In principle, bounds for W can be improved by using internal monotonicity.

5.6

KINGMAN (1962) proved (5.6.2). The other bounds in Section 5.6 are mainly results of MARSHALL (1968a). Similar methods yield bounds for GI/GI/1 queues with group arrivals (see MARSHALL (1968b)); for cost characteristics in controlled GI/GI/1 queues, see HEYMAN and MARSHALL (1968). STOYAN and STOYAN (1976) proved (5.6.3). FAINBERG (1976) proved (5.6.10) by means of the Erlang phase method (in the formulation of SCHASSBERGER (1973)). It was proved first for queues whose service time d.f.s are mixtures of Erlang d.f.s, and then FAINBERG generalized this inequality using robustness properties of GI/GI/1. This idea may well be fruitful in other cases also.

SIEGEL (1974) also gave bounds for W and m_W, making use of the properties IFR ect. The starting point of his investigations is the relation that stems from ladder properties of random walks, (see e.g. FELLER (1966))

$$W(t) = \sum_{n=0}^{\infty} p_n \hat{H}^{*n}(t), \qquad p_n = (1 - p_W)\, p_W{}^n \qquad (n = 0, 1, \ldots)$$

where $\hat{H}(t) = H(t)/H(\infty)$ and on the set that $\nu = \min \left\{n : \sum_{j=1}^{n} U_j \geqq 0\right\} < \infty$, $H(t) = \mathsf{P}\left(\sum_{j=1}^{\nu} U_j \leqq t, \nu < \infty\right)$, so $H(\infty) = \mathsf{P}(\nu < \infty)$. He gave conditions under which $\hat{H} \leqq_d B$. If K_+ is of type NBUE (NWUE), where K_+ is the d.f. $K_+(t) = \mathsf{P}(U_1 < t \mid U_1 \geqq 0)$ with the mean m_{K_+}, he was able to show $m_W \leqq (\geqq)\, p_W m_{K_+}$. This condition is fulfilled if B has IFR (DFR) and A has DFR (IFR).

5.9

Most bounds presently known for G/G/1 are unsatisfactory and crude. This situation may undergo some change in the future as a result of the general tendency of queueing theory to give more attention to models with dependence.

The idea of using (5.0.21) for bounding G/G/1 queues comes from BRUMELLE (1971a). SAKASEGAWA and YAMAZAKI (1977) have given inequalities and approximations for waiting times in tandem queues by reducing them to single-server queues equivalent in some sense.

SHIMSHAK (1979) compared some approximations for single server series systems using mean waiting times from simulations.

NIU (1981) (or else see NIU (1977)) showed a monotonicity property for a tandem queueing system consisting of r single server queues in series as follows. Let $\beta_n{}^j$ be the service time of the nth customer at server j, and $Z_n{}^j$ the epoch at which it leaves that server (equivalently, when it enters the $(j + 1)^{\text{th}}$ system), so that

$$Z_n{}^0 = \sum_{k=1}^{n} \alpha_k, \quad Z_n{}^j = \max\,(Z_n{}^{j-1}, Z_{n-1}^j) + \beta_n{}^j.$$

From these relations, comparability with respect to $\leqq_d$ or $\leqq_c$ between the interarrival and/or service times of two such systems Σ_1 and Σ_2, implies the comparability of Z_n^j, and of the means of the total waiting time W_n of the nth customer. For the same interarrival time distribution in Σ_1 and Σ_2 the waiting times W_n themselves are comparable with respect to $\leqq_d$ or $\leqq_c$ respectively. He also proved an inequality of the form (5.6.2) for the mean waiting time at the second server in GI/D/1/GI/1, see Niu (1980a).

Open problems. 1. Is it possible to give practicable criteria for the comparison with respect to $\leqq_c$ of non-renewal point processes with Poisson processes of equal intensity? In particular, for what output processes of queueing systems is this possible? (For output processes see Daley (1976b) and Disney and Cherry (1974)).

In Stoyan and Stoyan (1976) it was conjectured that M/GI/1 output is majorized by the Poisson process of equal intensity. Even qualitative properties of output processes are of interest, because their distributions are very hard to determine in most cases. In Stoyan and Stoyan (1976) some simple monotonicity properties with respect to $\leqq_d$ are given: for comparable interarrival times in GI/GI/1 the output processes are comparable (see also Whitt (1981a)).

2. When is cov $(w_0, U_0) \leqq 0$ true? (See around (5.9.6).)

3. Can the inequalities of Suzuki and Maruta (1970) for queues in series be derived rigorously?

Supplementary material added in proof

The inequality (5.6.10) is true not merely for E_k/GI/1 but for a wider class of d.f.s including all the gamma distributions that interpolate $\{E_k\}$ in the sense of $\leqq_c$ between the limits $k = 1$ and ∞ (corresponding to M/GI/1 and D/GI/1): see Vainshtein and Kreinin (1981). Heyman (1982) and Rolski (1983) are further references concerning comparisons of systems with doubly stochastic Poisson input.

Daley and Whitt (1983) have surveyed the results and methods concerning the d.f.s A and B for α and β with prescribed first two moments yielding inf m_W and sup m_W. All results are consistent with sup m_W being attained, for given d.f. A (B), if B (A) is the two-point distribution with mass at zero, or sup m_W for given A equals the limit as $\varepsilon \to 0$ of m_W in the class of systems for which $B = D^{(\varepsilon)}$, i.e.

$$\mathsf{P}(\beta = m_B - \varepsilon\sigma_B) = 1 - \mathsf{P}(\beta = m_B + \sigma_B/\varepsilon) = 1/(1 + \varepsilon^2).$$

Daley and Rolski's (1983) study of m_W in light traffic is also consistent with the two-point distribution mentioned above for α yielding sup m_W. In unpublished work Daley has shown that, with $D^{(\varepsilon)}$ as above,

$$\lim_{\varepsilon \to 0} m_W(\text{GI/D}^{(\varepsilon)}/1) = m_W(\text{GI/D/1}) + \lim_{\varepsilon \to 0} m_W(\text{D/D}^{(\varepsilon)}/1).$$

About bounds for tandem queues see also Yamazaki (1981). Flipo (1981) proved comparison theorems for various queue disciplines in G/G/1. Ott has shown, via discrete G/D/1-queue, that (5.4.5) is tight.

6. Monotonicity properties and bounds for queueing systems (II): the many-server system GI/GI/s

6.0 Basic facts

The queue GI/GI/s is a generalisation of the model GI/GI/1 and differs from this only by the existence of s similar servers instead of one server. On their arrival, customers are assigned with equal probability to any empty server if such exist, or else, all servers being busy, the customers wait in a common queue from which they go to service in order of their arrival (FCFS queue discipline) as soon as any server becomes free. The characteristics of the model are denoted as in the case of the model GI/GI/1 in Section 5.0.

GI/GI/s queues are, of course, much more complicated than GI/GI/1 queues. For example, in studying waiting times, s-dimensional vectors must be considered, and consequently there is no such (relatively) simple integral equation for the waiting time d.f. as that of LINDLEY at (5.0.3). For further information on results for GI/GI/s, we would refer to GNEDENKO and KOVALENKO (1966) and BOROVKOV (1976a) and the original papers of KIEFER and WOLFOWITZ (1955, 1956).

Following KIEFER and WOLFOWITZ (1955) the waiting time process in a GI/GI/s system can be studied as an s-dimensional vector process $\{\boldsymbol{w}'(t)\} = \{(w_1'(t), \ldots, w_s'(t))\}$ in which the kth component equals the sum of the residual service time of the customer (if any) currently in service at time t with the kth server and of the service times of those customers (if any) waiting in the queue and subsequently served by the kth server, while if this sum is empty $w_k'(t) = 0$. Rather than $\boldsymbol{w}'(t)$, it proves more convenient to rearrange its components in ascending order to yield the vector

$$\boldsymbol{w}(t) = (w_1(t), \ldots, w_s(t))$$

where $0 \leqq w_1(t) \leqq \cdots \leqq w_s(t)$. We then use $\boldsymbol{w}_n$ to denote the vector $\boldsymbol{w}(t_n - 0)$ at the epoch t_n of arrival of the nth customer,

$$\boldsymbol{w}_n = (w_1(t_n - 0), \ldots, w_s(t_n - 0)) = (w_{1,n}, \ldots, w_{s,n}) \quad \text{say},$$

yielding $w_{1,n}$ as the waiting time of the nth customer. Corresponding to (5.0.1) we then have

$$\boldsymbol{w}_{n+1} = (\boldsymbol{R}(\boldsymbol{w}_n + \beta_n \boldsymbol{e}) - \alpha_n \boldsymbol{i})_+ \qquad (n = 1, 2, \ldots) \tag{6.0.1}$$

where $\boldsymbol{e}$ and $\boldsymbol{i}$ denote the s-vectors $(1, 0, \ldots, 0)$ and$(1, 1, \ldots, 1)$ respectively, and for any s-vector $\boldsymbol{x}$, $(\boldsymbol{x})_+$ denotes $(\max(0, x_1), \ldots, \max(0, x_s))$ and

$\boldsymbol{R}(\boldsymbol{x})$ consists of the components of $\boldsymbol{x}$ rearranged in ascending order. Finally, let W_n denote the d.f. of $w_{1,n}$ and $\boldsymbol{W}_n$ that of $\boldsymbol{w}_n$.

Kiefer and Wolfowitz (1955) showed that when

$$sm_A > m_B, \text{ i.e. } \varrho/s < 1 \quad \text{where } \varrho = m_B/m_A, \tag{6.0.2}$$

there exists a uniquely determined stationary d.f. $\boldsymbol{W}$ for the Markov chain with sample paths satisfying (6.0.1), and for any initial d.f. $\boldsymbol{W}_0$,

$$\boldsymbol{W}_n \text{ converges weakly to } \boldsymbol{W} \quad (n \to \infty). \tag{6.0.3}$$

Since for given $\boldsymbol{W}_0$, $W_n(y) = \boldsymbol{W}_n(y, \infty, \ldots, \infty)$, it follows from (6.0.3) that

$$W_n \text{ converges weakly to } W \quad (n \to \infty). \tag{6.0.4}$$

Kiefer and Wolfowitz (1956) also showed that the mean m_W of W is finite if $\mathsf{E}\beta^2 < \infty$, i.e., if the service time d.f. has a finite second moment.

For both GI/GI/s and the more general dependent system G/G/s (see Franken, König, Arndt and Schmidt (1981)), the mean number n_{occ} of busy servers for the system in its stationary state satisfies

$$n_{occ} = \mathsf{E} \# \{k : w_k'(t) > 0\} = \varrho, \tag{6.0.5}$$

if $\boldsymbol{w}'(t)$ is such a stationary process. Little's formula relating m_W and the mean queue length L_W holds, namely

$$L_W = \lambda m_W, \quad \lambda \equiv 1/m_A \tag{6.0.6}$$

(cf. (6.3.4) and bibliographic on p. 125).

However, practical formulae for m_W or W, or even $p_W = \mathsf{P}(w_1 > 0)$, are known only in a few special cases, as for example M/M/s (see e.g. section 1.2.7 of Gnedenko and Kovalenko (1966)), M/D/s (see e.g. section 1.7 of Prabhu (1965)), and GI/M/s (see e.g. section 29 of Borovkov (1976a)), for which the formulae are as follows.

(a) M/M/s (see also Table 6.2, p. 120):

$$m_W = m_B p_W/(s - \varrho), \tag{6.0.7}$$

$$p_W = \varrho^s P_0/(s-1)!\,(s - \varrho), \tag{6.0.8}$$

$$1/P_0 = \sum_{k=0}^{s} \varrho^k/k! + \varrho^{s+1}/s!(s - \varrho). \tag{6.0.9}$$

(b) M/D/s (see also Table 6.3, p. 120):

$$m_W = m_B \sum_{j=1}^{\infty} e^{-j\varrho} \sum_{k=js+1}^{\infty} (k - js)\,(j\varrho)^{k-1}/k! \quad (\varrho = m_B/m_A). \tag{6.0.10}$$

(c) GI/M/s (see also Table 6.4, p. 121 for D/M/s):

$$m_W = C m_B/s(1 - \delta)^2, \tag{6.0.11}$$

δ being the root in $0 < \delta < 1$, with $A^*(\theta) = \int_0^\infty e^{-\theta t}\, dA(t)$, of

$$\delta = A^*\big((1 - \delta)\, s/m_B\big), \tag{6.0.12}$$

$$\frac{1}{C} = \frac{1}{1-\delta} + \sum_{k=1}^{s} \frac{\binom{s}{k}}{C_k(1-a_k)} \cdot \frac{s(1-a_k) - k}{s(1-\delta) - k},$$

$$C_k = \prod_{j=1}^{k} a_j/(1 - a_j), \quad a_j = A^*(j/m_B),$$

and the stationary waiting time d.f. W is then

$$W(y) = 1 - C(1-\delta)^{-1} \exp\big(-s(1-\delta)\, y/m_B\big) \quad (y \geqq 0). \tag{6.0.13}$$

In the case of constant service times, i.e. of systems of the type GI/D/s, it is enough to consider single-server queues of the type GI/D/1 for the study of waiting times, because when the customers are assigned to the servers *cyclically*, i.e., the $(js + k)^{\text{th}}$ customer goes to the kth server for all j irrespective of the state of the system, the waiting times are no different from those under the FCFS discipline. With the cyclic discipline, GI/GI/s functions like s staggered GI/GI/1 systems with the interarrival d.f. A^{*s}.

Finally, if the stationary waiting time d.f. W is known, the queue length d.f. can be determined via (5.0.15) and (5.0.16); if $\boldsymbol{W}$ is known the d.f. of the number of customers in the system can be determined as well.

6.1 Internal monotonicity properties of GI/GI/s

In general the queue GI/GI/s has similar internal monotonicity properties to those of GI/GI/1 only with respect to $\leqq_d$. Similar statements are true for $\leqq_c$ only in particular cases, e.g., GI/D/s systems with cyclic assignment of customers. This is a consequence of the mapping

$$\phi\colon \mathbb{R}_+^{s+2} \to \mathbb{R}_+^{s} \text{ given by}$$

$$\phi(\boldsymbol{w}, \alpha, \beta) = \big(\boldsymbol{R}(\boldsymbol{w} + \beta \boldsymbol{e}) - \alpha \boldsymbol{i}\big)_+, \tag{6.1.1}$$

being monotonic in $\boldsymbol{w}$, α and β (nondecreasing, nondecreasing, and nonincreasing respectively), but not convex. From Theorem 2.2.8 there follows

Theorem 6.1.1. *If the initial d.f.* $\boldsymbol{W}_0$ *of the process* $\{\boldsymbol{W}_n\}$ *satisfies*

$$\boldsymbol{W}_0 \leqq_d \boldsymbol{W}_1, \tag{6.1.2}$$

then for all $n = 0, 1, \ldots,$

$$\boldsymbol{W}_n \leqq_d \boldsymbol{W}_{n+1} \tag{6.1.3}$$

and, if the stationary waiting time d.f. $\boldsymbol{W}$ *exists,*

$$\boldsymbol{W}_n \leqq_d \boldsymbol{W}. \tag{6.1.4}$$

If $W_0 \geqq_d W_1$, *then the stationary waiting time d.f.* W *exists, and* (6.1.3) *and* (6.1.4) *hold with the reverse inequality.*

The last part of the theorem is a consequence of $\boldsymbol{w}_n$ decreasing in distribution and being a.s. bounded below by zero.

For the number of customers in the system there exists a similar statement, proved by using a particular Markov chain as follows. Let ν_n denote the number of customers in the system at $t_n - 0$, let $X_{1,n}, \ldots, X_{s,n}$ denote the residual service times of customers being served at $t_n - 0$, specified in increasing order so that $X_{1,n} \leqq \cdots \leqq X_{s,n}$ and $X_{s-\nu_n,n} = 0 < X_{s-\nu_n+1,n}$ if $\nu_n \leqq s$, and when $\nu_n > s$, let $Y_{1,n}, \ldots, Y_{(\nu_n-s),n}$ denote the service times of the $\nu_n - s$ customers waiting at $t_n - 0$, specified in order of their arrival; if $\nu_n \leqq s$ there are no such Y r.v.s. Then the sequence $\{Z_n\}$ for which

$$Z_n = \{\nu_n, X_{1,n}, \ldots, X_{s,n}, Y_{1,n}, \ldots, Y_{(\nu_n-s),n}\}$$

is a homogeneous Markov chain on state space $(E, \mathscr{M})$ where E is the set of all finite sequences $\boldsymbol{x} = \{x_0, \ldots, x_j\}$ where $j = \max(s, x_0)$, $x_0 \in \{0, 1, \ldots\}$, $x_i \geqq 0$ $(i = 1, \ldots, j)$, and $\mathscr{M}$ the σ-algebra generated by sets M of the form

$$M = \begin{cases} (i, \{0\}^{s-i} \times B) & (i = 0, \ldots, s-1), \\ (i, B) & (i = s, s+1, \ldots), \end{cases}$$

where B is any Borel subset of $\mathbb{R}_+{}^i$. Z_{n+1} is then given by

$$Z_{n+1} = \Psi(Z_n, \alpha_{n+1}, \beta_n)$$

where the mapping $\Psi\colon E \times \mathbb{R}_+ \times \mathbb{R}_+ \to E$ is measurable and has the monotonicity property for $\boldsymbol{x}_k \in E$, $a_k, b_k \in \mathbb{R}_+$ $(k = 1, 2)$, $\boldsymbol{x}_1 \leqq_E \boldsymbol{x}_2$, $a_1 \geqq a_2$, $b_1 \leqq b_2$ implies $\Psi(\boldsymbol{x}_1; a_1, b_1) \leqq_E \Psi(\boldsymbol{x}_2; a_2, b_2)$ when as the partial ordering $\leqq_E$ on E we use $\boldsymbol{x} \leqq_E \boldsymbol{y}$ to mean that (i) $x_0 \leqq y_0$, and (ii) for each $i \in I_x \equiv \{1, \ldots, \max(s, x_0)\}$ there exists $j(i) \in I_y \equiv \{1, \ldots, \max(s, y_0)\}$ such that $x_i \leqq y_{j(i)}$ and if $i_1 \neq i_2$, $j(i_1) \neq j(i_2)$, and also if $s < i_1 < i_2$, then $s < j(i_1) < j(i_2)$ while if $i \leqq s$ then $j(i) \leqq s$. The lengthy proof is elementary and omitted here, being given in Stoyan (1973). This monotonicity of Ψ means that $\{Z_n\}$ satisfies the assumptions of Theorem 2.2.8 for $\leqq_d$ generated by $\leqq_E$ on the set of d.f.s on $\mathscr{M}$, and hence

Theorem 6.1.2. *If the initial d.f.* P_0 *of the process* $\{Z_n\}$ *satisfies*

$$P_0 \leqq_d P_1, \tag{6.1.5}$$

then for all $n = 0, 1, \ldots$,

$$P_n \leqq_d P_{n+1}, \tag{6.1.6}$$

and if a stationary distribution P *exists, then*

$$P_n \leqq_d P. \tag{6.1.7}$$

If $P_0 \geqq_d P_1$, *then a stationary distribution* P *exists and* (6.1.6) *and* (6.1.7) *hold with the reverse inequality.*

For example, (6.1.5) is satisfied when $Z_0 = (0; 0, 0)$. Of course, inequalities for the number of customers follow from (6.1.6) and (6.1.7).

6.2 External monotonicity properties of GI/GI/s

In general, just as for internal monotonicity, external monotonicity properties of GI/GI/s can be proved only with respect to $\leqq_d$. Examples can be given to show that an analogous relation to (6.2.1) with $\leqq_d$ replaced by $\leqq_c$, is not true in general (see WOLFF (1977a, b) and WHITT (1980b)).

Theorem 6.2.1. *For $k = 1, 2$, let Σ_k be GI/GI/s queues with the interarrival and service time d.f.s A_k and B_k, and let $\boldsymbol{W}_{n,k}$ be the corresponding d.f.s of the waiting time vectors at $t_n - 0$. Then*

$$\boldsymbol{W}_{0,1} \leqq_d \boldsymbol{W}_{0,2},\ A_1 \geqq_d A_2,\ \text{and}\ B_1 \leqq_d B_2\ \text{imply}\ \boldsymbol{W}_{n,1} \leqq_d \boldsymbol{W}_{n,2} \tag{6.2.1}$$

for $n = 1, 2, \ldots$, while for the stationary d.f.s $\boldsymbol{W}_1$ and $\boldsymbol{W}_2$ (assuming they exist),

$$A_1 \geqq_d A_2\ \text{and}\ B_1 \leqq_d B_2\ \text{imply}\ \boldsymbol{W}_1 \leqq_d \boldsymbol{W}_2. \tag{6.2.2}$$

Analogous inequalities for the waiting time d.f.s follow from (6.2.1) and (6.2.2). Theorem 6.2.1 follows easily from Theorem 2.2.7 on observing that the mapping ϕ defined by (6.1.1) is nondecreasing in $\boldsymbol{w}_n$ and β_n and nonincreasing in α_n.

D.f.s of the number of customers in the system or in the queue at arrival instants have analogous monotonicity properties with respect to $\leqq_d$. Similarly, just as for G/G/1, we also have external monotonicity properties with respect to $\leqq_d$ for G/G/s (i.e. allowing for the interarrival and service times to be dependent). Because of the relation between GI/D/s and GI/D/1 noted at the end of Section 6.0, we have the following result for stationary waiting time d.f.s W in GI/D/s.

Theorem 6.2.2. *For $k = 1, 2$ let Σ_k be GI/D/s queues with interarrival time d.f.s A_k and service times $m_{B,k}$. Then*

$$A_1 \geqq_d A_2\ \text{and}\ m_{B,1} \leqq m_{B,2}\ \text{imply}\ W_1 \leqq_d W_2, \tag{6.2.3'}$$

$$A_2 \leqq_{cv} A_1\ \text{and}\ m_{B,1} \leqq m_{B,2}\ \text{imply}\ W_1 \leqq_c W_2. \tag{6.2.3''}$$

Perhaps surprisingly, external monotonicity properties cannot be proved in general for such quantities as the number of customers in the system at time t ($t < \infty$) or the virtual waiting time, as shown in the following example (cf. JACOBS and SCHACH (1972)).

Example 6.2(a). For $k = 1, 2$ consider the GI/GI/2 queues Σ_k with interarrival times $\alpha_{n,k}$ and service times $\beta_{n,k}$, and denote by $w_{t,k}$ and $\nu_{t,k}$ the virtual waiting time and the queue lengths in Σ_k at time t, respectively. If Σ_1 and Σ_2 are empty at $t = 0$ and

$$\mathsf{P}(\alpha_{n,1} = 2) = \mathsf{P}(\alpha_{n,1} = 6) = 0.5 = \mathsf{P}(\alpha_{n,2} = 2) = \mathsf{P}(\alpha_{n,2} = 8),$$

$$\beta_{n,1} = \beta_{n,2} = 3 \quad (n = 0, 1, \ldots),$$

then

$$\mathsf{P}(w_{9.5,2} = 0) = .375 > \mathsf{P}(w_{9.5,1} = 0) = .25 = \mathsf{P}(w_{10.5,1} = 0) > \mathsf{P}(w_{10.5,2} = 0),$$

and

$$\mathsf{P}(\nu_{8.5,2} = 0) = .875 > \mathsf{P}(\nu_{8.5,1} = 0) = .75, \mathsf{P}(\nu_{11.5,1} = 0) = .8125$$
$$> \mathsf{P}(\nu_{11.5,2} = 0) = .6875.$$

However, with the additional assumption that Σ_1 and Σ_2 have the same interarrival distribution time, we can then prove monotonicity properties for the virtual waiting time and number of customers in the system at time t.

Theorem 6.2.3. *For $k = 1, 2$ let Σ_k be initially empty* G/GI/s *queues with the same input distribution and service time d.f.s B_k and denote by $\nu_{t,k}$ and $w_{t,k}$ the virtual waiting time and number of customers in Σ_k at time t respectively. Then*

$$B_1 \leqq_d B_2 \tag{6.2.4}$$

implies that for all $t \geqq 0$,

$$w_{t,1} \leqq_d w_{t,2} \tag{6.2.5}$$

and

$$\nu_{t,1} \leqq_d \nu_{t,2}. \tag{6.2.6}$$

Proof. We can assume that Σ_1 and Σ_2 have the same sequence $\{\alpha_n\}$ of interarrival times. For given $t > 0$, suppose that

$$t_n = \sum_{i=1}^{n} \alpha_i \leqq t < \sum_{i=1}^{n+1} \alpha_i = t_{n+1}.$$

By Theorem 6.2.1 the waiting times $w_{n,k}$ of the customers arriving at t_n in Σ_k satisfy $w_{n,1} \leqq_d w_{n,2}$. (6.2.5) now follows from the relation $w_{t,k} = \big(w_{n,k} - (t - t_n)\big)_+$. (6.2.6) is proved by a similar reduction (cf. Theorem 6.1.2).

Whitt (1980b) considered further characteristics of two GI/GI/s systems Σ' and Σ'' say. By means of the mapping method it is easily shown that $A' \leqq_{cv} A''$ and $B' \leqq_c B''$ imply

$$L_n' = \sum_{i=1}^{s} w'_{i,n} \leqq_c \sum_{i=1}^{s} w''_{i,n} = L_n'', \tag{6.2.7}$$

$$C_n' = w'_{s,n} \leqq_c w''_{s,n} = C_n'', \tag{6.2.8}$$

and, more generally, for non-decreasing convex functions $f(.)$, mapping $\mathbb{R}^s$ into $\mathbb{R}$, that are invariant under permutation of their arguments,

$$D_n' \equiv f(w_n') \leqq_c f(w_n'') = D_n''. \tag{6.2.9}$$

For example, writing $L_n = \Phi_n(\alpha_1, \ldots, \alpha_n, \beta_1, \ldots, \beta_n)$, we have for all arguments fixed except β_k say, $L_n = \max(c, \beta_k - d)$ for some c and d, and this is non-decreasing and convex; L_n also has these properties for $-\alpha_k$ (and, recall that $\alpha'' \leqq_{cv} \alpha'$ implies $-\alpha' \leqq_c -\alpha''$).

Inequalities analogous to (6.2.7), (6.2.8) and (6.2.9) are also true in the stationary case, and can be used to bound certain characteristics. For example, since $w_{1,n} \leqq s^{-1} \sum_{i=1}^{s} w_{i,n}$, the stationary mean waiting time m_W in GI/GI/s systems for which $A \leqq_c \text{Exp}(1/m_A)$ and $B \leqq_c \text{Exp}(1/m_B)$, as holds when both A and B are NBUE, satisfies

$$m_W \leqq (m_B/s)\, M_s \tag{6.2.10}$$

where M_s is the mean number of customers in the stationary M/M/s system,

$$M_s = \varrho/s + \lambda m_W(\mathrm{M/M}/s). \tag{6.2.11}$$

(6.2.10) is a relatively good heavy traffic bound for m_W; combining (6.2.11) yields the representation

$$m_W \leqq \varrho m_B/s^2 + \varrho m_W(\mathrm{M/M}/s)/s,$$

from which equality in the case of an M/M/1 system can be verified.

6.3 Bounds and approximations for many-server queues GI/GI/s

The present section surveys some methods for obtaining bounds and approximations for the mean stationary waiting time m_W and the stationary probability of wait p_W in GI/GI/s queues. Consequently several results are stated without proof; rather, the methodical aspect plays the more important role.

An important tool for bounds and approximations and bounds in many server queues is the use of *system approximations*: we construct systems whose system quantities yield approximations for the unknown quantities of the original system. Often these approximations can be simplified by using monotonicity properties. For constructing approximating systems for GI/GI/s there are, amongst others, the following possibilities:

(a) variation of the number of servers;
(b) variation of the queue discipline;
(c) modification of the normal service procedure;
(d) approximation by many-server systems with different interarrival or service time d.f.s.

Throughout this Section 6.3, we denote the GI/GI/s queue under study by Σ and its interarrival and service time d.f.s by $A(.)$ and $B(.)$ respectively.

6.3.1 *Variation of the number of servers*

6.3.1.1 Use of single-server queues ($s \to 1$)

In this section we approximate Σ by GI/GI/1 queues having the same traffic intensity. Of particular interest are the systems

Σ_1: interarrival times $\alpha_1, \alpha_2, \ldots$ are as in Σ, service times are $\beta_1/s, \beta_2/s, \ldots$; and

Σ_1': interarrival times are $s\alpha_1, s\alpha_2, \ldots$, service times $\beta_1, \beta_2, \ldots$ are as in Σ.

It is readily seen that Σ_1 and Σ_1' differ only in the time scale in the latter being s times that of Σ_1.

The systems Σ_1 is in fact the more interesting system because it corresponds to an s-server system in which all servers are occupied simultaneously with the one customer, so his service needs are satisfied s times as fast as in Σ. Denote by $V_1(t)$ and $V(t)$ the workload at time t in Σ_1 and Σ respectively, i.e., the sum of service times of all waiting customers and of all residual service times of customers currently being served. If $V_1(0) = V(0)$ then it is not difficult to verify that, when Σ and Σ_1 are both governed by the same sequence $\{(\alpha_n, \beta_n)\}$,

$$V_1(t) \leqq V(t) \text{ a.s.} \tag{6.3.1}$$

and hence that

$$\mathsf{E} \lim_{t\to\infty} t^{-1} \int_0^t V_1(x)\, dx \leqq \mathsf{E} \lim_{t\to\infty} t^{-1} \int_0^t V(x)\, dx. \tag{6.3.2}$$

Brumelle (1971a) showed how a lower bound for m_W can be found from (6.3.2): the proof for the following result may be found in Brumelle (1971b) or Franken (1976).

Proposition 6.3.1. *Let the sequence $\{(\alpha_n, \beta_n)\}$ of the stationary queueing system* G/G/s *be metrically transitive, and assume that $m_B < s m_A$ and*

$$\mathsf{E}(w_{1,n} + \beta_n)^2 < \infty, \qquad \mathsf{E} w_{s,n}^2 < \infty \tag{6.3.3}$$

where $w_{1,n}$ denotes the waiting time of the nth customer and $w_{s,n}$ is the maximal component of the vector $\boldsymbol{w}_n$. Then

$$\mathsf{E} \lim_{t\to\infty} t^{-1} \int_0^t V(x)\, dx = \lambda\big(\mathsf{E}(\beta_n w_{1,n}) + {}^1\!/_2 \mathsf{E}\beta_n^2\big), \quad \lambda = 1/m_A \tag{6.3.4}$$

which for a GI/GI/s *system yields*

$$\mathsf{E} \lim_{t\to\infty} t^{-1} \int_0^t V(x)\, dx = \lambda\big(m_B m_W + {}^1\!/_2(m_B^2 + \sigma_B^2)\big). \tag{6.3.4'}$$

From (6.3.2) and (6.3.4') we obtain

Proposition 6.3.2. *The stationary mean waiting time m_W in GI/GI/s satisfies*

$$m_W \geqq m_W^{(1)} - {}^1/_2(1 - s^{-1})\,(m_B^2 + \sigma_B^2)/m_B \tag{6.3.5}$$

where $m_W^{(1)}$ is the mean stationary waiting time in Σ_1, i.e. the GI/GI/1 queue with interarrival time d.f. $A(t)$ and service time d.f. $B(st)$.

In (6.3.5), $m_W^{(1)}$ can be replaced by a lower bound for $\mathsf{E}w$ in GI/GI/1 from Chapter 5. It should also be noted that MORI (1975) showed that when A is NBUE and $s - 1 > \varrho \equiv m_B/m_A$, the factor $1 - s^{-1}$ can be replaced by ϱ/s.

BRUMELLE (1971a) showed how Proposition 6.3.2 can be used to give a partial solution to the problem of giving conditions under which there should be an inequality analogous to (6.3.2) but for the mean stationary delay times, which in Σ equals $m_Q \equiv m_W + m_B$ and in Σ_1 equals $m_Q^{(1)} \equiv m_W^{(1)} + m_B/s$. It follows easily from (6.3.5) that

Corollary 6.3.2a. *If $\sigma_B \leqq m_B$ then*

$$m_Q \geqq m_Q^{(1)}. \tag{6.3.6}$$

By using either the mapping method or the sample path method, it can be shown that the departure epochs in Σ_1 precede those in Σ. More precisely, the departure epoch of the nth customer is smaller (in the sense of $\leqq_d$) in Σ_1 than in Σ (see ARJAS and LEHTONEN (1978)).

As complements to (6.3.5) it would appear (see also MORI (1975)) that

$$p_W \leqq p_W^{(1)}, \tag{6.3.7}$$

$$m_W \leqq m_W^{(1)} \tag{6.3.8}$$

where $p_W^{(1)}$ and $m_W^{(1)}$ denote the stationary probability of waiting and mean stationary waiting time in Σ_1 respectively.

Example 6.3.1(a). STOYAN (1974a) showed that for M/GI/s,

$$p_W \leqq \varrho/s = p_W^{(1)}. \tag{6.3.9}$$

Proof. The mean number of occupied servers in a stationary G/G/s queue equals $\varrho = \lambda m_B$, and therefore the probability P_{occ} that all servers are busy satisfies

$$P_{occ} \leqq \varrho/s,$$

the bound here being the greatest probability which the point s can have for a discrete distribution on $0, \ldots, s$ with the mean ϱ. Since $P_{occ} = p_W$ in M/GI/s, (6.3.9) follows. Also, since $p_W^{(1)} = \varrho/s$, we must have (6.3.7) for M/GI/s.

Example 6.3.1(b). The relation (6.3.8) is true for M/M/s and M/D/s.

Example 6.3.1(c). BRUMELLE (1973) and MORI (1975) showed that for GI/M/s and GI/D/s,

$$W \leqq_d W_1 \tag{6.3.10}$$

for the stationary waiting time d.f.s W and W_1 in Σ and Σ_1, from which (6.3.7) and (6.3.8) then follow for these systems.

Example 6.3.1(d). In order to show that when $\sigma_B > m_B$, (6.3.6) need not be satisfied BRUMELLE (1971a) gave an example which can be described as follows. Consider an M/M/2 system with arrival rate $p\lambda$ and service rate $\mu/2$: by (6.0.7) its mean stationary waiting time equals $(2/\mu)\,\varrho^2/(4-\varrho^2)$ where $\varrho = 2p\lambda/\mu$. Now modify it into an M/GI/2 system Σ with arrival rate λ and service d.f. $(1-p)\,\Theta_0 + p\,\mathrm{Exp}\,(\mu/2)$ by superposing an independent Poisson arrival process, at rate $(1-p)\,\lambda$, of customers requiring a.s. zero service times. Such arrivals do not alter the load on the servers at all; their mean waiting time is the same as for M/M/2 arrivals, while the mean stationary delay time in Σ equals

$$m_Q = (2/\mu)\,[\varrho^2/(4-\varrho^2) + p].$$

The M/GI/1 Σ_1 approximating system has Poisson arrivals at rate λ and service d.f. $(1-p)\,\Theta_0 + p\,\mathrm{Exp}\,(\mu)$, so its mean stationary waiting time equals

$$(p\lambda/\mu^2)/(1-p\lambda/\mu) \quad \text{and} \quad m_Q^{(1)} = (2/\mu)/(2-\varrho).$$

It is easily checked that for $0 < p < 1$, $\sigma_B^2 = p(2-p)/\mu^2 > (p/\mu)^2 = m_B^2$, and that (cf. (6.3.6)) $m_Q^{(1)} > m_Q$ if $p < (1+\varrho)/(2+\varrho)$, i.e., if $\lambda/\mu > (2p-1)/2p(1-p)$.

If (6.3.8) is true for M/GI/s, the bound

$$m_W \leqq \lambda(\sigma_B^2 + m_B^2)/2s^2(1-\varrho/s) \tag{6.3.11}$$

given by MAALØE (1973) on the basis of heuristic arguments is also true.

Let $\boldsymbol{w}_n = (w_{1,n}, \ldots, w_{s,n})$ denote the vector of server-loads at arrival epochs as at (6.0.1), and introduce

$$Z_n = \sum_{i=2}^{s} (w_{i,n} - w_{1,n}) = \sum_{i=2}^{s} (w_{i,n} - w_n)$$

since $w_{1,n} = w_n$. KIEFER and WOLFOWITZ (1955) showed that for the waiting times $\{w_n'\}$ in Σ_1', if $w_0' \leqq sw_{1,0} + Z_0$, then

$$w_n' \leqq sw_{1,n} + Z_n \quad (\text{all } n), \tag{6.3.12}$$

so that in the stationary case, we have on taking expectations in (6.3.12) that

$$m_W \geqq (m_W' - \mathsf{E}Z_n)/s. \tag{6.3.12'}$$

In principle therefore, use of a lower bound for m_W' in Σ_1' (and recall that $m_W' = sm_W^{(1)}$) and an upper bound for $\mathsf{E}Z_n$, yields a lower bound for m_W. Bounding $\mathsf{E}Z_n$ is not such a simple matter, except for $s = 2$ where, as SUZUKI and YOSHIDA (1970) noted, we have

Proposition 6.3.3. *The mean stationary waiting time* m_W *in* GI/GI/2 *satisfies*

$$m_W \geqq m_W^{(1)} - (m_B^2 + \sigma_B^2)/4m_B \tag{6.3.13}$$

where $m_W^{(1)}$ *is the mean stationary waiting time in the* GI/GI/1 *queue with the d.f.s* $A(t)$ *and* $B(2t)$.

Proof. (6.3.13) follows from (6.3.12′) and the inequality $\mathsf{E}Z_n \leqq \mathsf{E}\beta^2/2\mathsf{E}\beta$ which follows from

$$Z_{n+1} \leqq |\beta_n - Z_n| \tag{6.3.14}$$

by squaring and taking expectations. This last inequality follows from using $|(x + y)_+ - y_+| \leqq |x|$ in the expression

$$\begin{aligned} Z_{n+1} &= |(w_{1,n} + \beta_n - \alpha_n)_+ - (w_{2,n} - \alpha_n)_+| \\ &= |(\beta_n - Z_n + w_{2,n} - \alpha_n)_+ - (w_{2,n} - \alpha_n)_+|. \end{aligned}$$

Brumelle (1971a) suggested an entirely different approximating G/G/1 system Σ_B. He assumed that in the GI/GI/s queue considered the assignment is uniformly at random when there is more than one idle server. Then, focussing attention on just one particular server and setting $I_n = 1$ if the nth customer goes to this server and $I_n = 0$ otherwise, he shows that

(a) the sequence $\{(\alpha_n, I_n\beta_n)\}$ is strictly stationary with $\mathsf{E}(\alpha_n - I_n\beta_n) > 0$;

(b) $W \leqq_d W_B$, where W_B denotes the stationary waiting time d.f. in the G/G/1 queue Σ_B with input $\{(\alpha_n, I_n\beta_n)\}$ (W_B is assumed to exist);

(c) for fixed n, the r.v.s α_n, I_n and β_n are independent and cov $(w_n, I_n\beta_n - \alpha_n) \leqq 0$, where $\{w_n\}$ is the stationary sequence of waiting times in Σ_B.

Thus, the inequalities (5.9.6) and (5.9.7) can be used for bounding m_W. This single-server approximation to GI/GI/s also applies to G/G/s, and $W \leqq_d W_B$ continues to hold provided $\{(\alpha_n, \beta_n)\}$ is metrically transitive and W_B exists.

Generalizing an inequality for M/GI/s systems in the original German edition Rolski (1979) showed the result of

Proposition 6.3.4. *For a* GI/GI/s *queue* Σ *with* NBUE *interarrival time d.f. with mean* $m_A = 1/\lambda$, $W \leqq_d W_M$ *where* W_M *denotes the stationary waiting time d.f. in an* M/GI/1 *system with the same service time d.f. and arrival rate* λ/s, *i.e.,*

$$W(t) \geqq (1 - \varrho/s) \sum_{k=0}^{\infty} (\varrho/s)^k B_R^{*(k)}(t) \quad (t \geqq 0), \tag{6.3.15}$$

and hence

$$m_W \leqq \lambda(m_B^2 + \sigma_B^2)/2(s - \varrho). \tag{6.3.16}$$

Proof. Let V_j be the d.f. of the jth component of the ordered virtual waiting time vector in Σ (see Miyazawa (1977) and Rolski (1978)). The virtual waiting time d.f. $W_v = V_1$ satisfies

$$W_v(t) = 1 - \varrho/s + (\varrho/s)\,(W * B_R)\,(t) + \varepsilon(t)$$

where

$$\varepsilon(t) = s^{-1} \sum_{j=2}^{s} \left(W_v(t) - V_j(t)\right) \geqq 0,$$

and since $W \leqq_d W_v$ for NBUE interarrival times (MORI (1975) and STOYAN (1977b)) and

$$s^{-1} \sum_{j=1}^{s} V_j(t) = 1 - \varrho/s + (\varrho/s)\,(W * B_R)\,(t),$$

(6.3.15) now follows.

STOYAN and STOYAN (1977) used some knowledge on the form of the marginal d.f.s of the stationary waiting time vector $\boldsymbol{W}$ and inequality (6.3.18) below to give bounds on p_W in M/GI/s, including

$$p_W \leqq \varrho - (s-1) + \sum_{i=0}^{s-2} (s-1-i)\, \varrho^i\, \mathrm{e}^{-\varrho}/i! \qquad (s \geqq 2). \quad (6.3.17)$$

We mention finally, that ARJAS and LEHTONEN (1978) considered approximations for G/GI/s in which the customers are considered in successive groups of s consecutive arrivals. Given the arrival and service times in Σ, these are replaced in the approximating systems $\Sigma_1{}^*(\Sigma_1{}^{**})$ by the first (last) arrival and smallest (largest) service times within each group. The authors were able to show that $\Sigma_1{}^*$ and $\Sigma_1{}^{**}$ are bounding systems with respect to $\leqq_d$ for certain stationary quantities such as departure and waiting times, though the upper bounding system $\Sigma_1{}^{**}$ need not be stable; simulations by these authors indicated that $\Sigma_1{}^*$ is a worse bounding system than Σ_1.

6.3.1.2 Use of systems with infinitely many servers ($s \to \infty$)

Consider as an approximating system the GI/GI/∞ queue Σ_∞ with the same interarrival and service times as in Σ. Supposing that both Σ and Σ_∞ have the same initial state, it is immediately clear that for all $t \geqq 0$ as well as for each arrival epoch the number of customers $\nu_t{}^\infty$ in Σ_∞ is smaller than in Σ, i.e.,

$$\nu_t{}^\infty \leqq \nu_t \text{ a.s.} \qquad (6.3.18)$$

Consequently, by using (6.3.18) and formulae for GI/GI/∞ we can obtain bounds for m_W, p_W and other quantities. In the special case of M/GI/s the relatively simple formulae for M/GI/∞ can be used, namely, the probability P_n that n customers are in the system in its stationary state follows the Poisson distribution

$$P_n = \mathrm{e}^{-\varrho} \varrho^n/n! \qquad (n = 0, 1, \ldots;\ \varrho = \lambda m_B) \qquad (6.3.19)$$

(see e.g. section 33 of BOROVKOV (1976a)). Using this we obtain for the mean stationary queue length L_W in M/GI/s

$$L_W \geqq \sum_{n=s+1}^{\infty} (n-s)\, \mathrm{e}^{-\varrho} \varrho^n/n! = L_W^\infty \qquad (6.3.20)$$

from which by means of Little's equation at (6.0.6) we have

$$m_W \geqq m_A L_W^{\infty}. \tag{6.3.21}$$

For the stationary probability of wait p_W we obtain

$$p_W \geqq \sum_{n=s}^{\infty} \mathrm{e}^{-\varrho} \varrho^n / n! \tag{6.3.22}$$

For the case of time dependent arrival rate $\lambda(t)$ it is known that for $n = 0, 1, \ldots,$

$$P_n(t) = \mathrm{e}^{-\varrho(t)} \big(\varrho(t)\big)^n / n!$$

where

$$\varrho(t) = \int_0^t \lambda(t - x) \big(1 - B(x)\big) \, \mathrm{d}x. \tag{6.3.23}$$

Using $P_n(t)$ in place of P_n leads to formulae analogous to (6.3.20) to (6.3.22) but for time dependent quantities, with ϱ replaced by $\varrho(t)$.

6.3.2 Variation of queue discipline

6.3.2.1 Cyclic service

Kingman (1965, 1970) suggested the following approximating system Σ_c in which customers are assigned cyclically to the servers irrespective of the system state, i.e., for all $n = 0, 1, \ldots,$ and $k = 1, \ldots, s$, the $(ns + k)^{\text{th}}$ customer is served by server k which has its own queue. In this discipline every server works like a GI/GI/1 queue with the service time d.f. B and the interarrival time d.f. A^{*s}. Denoting the stationary waiting time d.f. in Σ_c by W_c and its mean by $m_W{}^{(c)}$, Gittins (1978) proved that

$$m_W \leqq m_W{}^{(c)}, \tag{6.3.24}$$

and Wolff (1977a) showed more generally that

$$W \leqq_c W_c \tag{6.3.25}$$

(see example at the end of this section). Then from the bound (5.7.1) it follows that

$$m_W \leqq \frac{(\varrho/s)\,(2 - \varrho/s)\, \sigma_A{}^2 + \sigma_B{}^2/s}{2m_A(1 - \varrho/s)}. \tag{6.3.26}$$

In connection with the so-called heavy traffic theory of queues, it was conjectured in Kingman (1965) and subsequently proved by Köllerström (1974; see also 1979) that the d.f. W in Σ is approximately negative exponential with mean

$$\frac{\sigma_A{}^2 + \sigma_B{}^2/s^2}{2m_A(1 - \varrho/s)}. \tag{6.3.27}$$

Both this and a plausible but incomplete argument based on SUZUKI and YOSHIDA's (1970) work led DALEY to conjecture that (6.3.27) is an upper bound to m_W, and that a tighter bound (cf. (6.3.26)) is

$$m_W \leqq \frac{(\varrho/s)\,(2 - \varrho/s)\,\sigma_A^2 + \sigma_B^2/s^2}{2m_A(1 - \varrho/s)}. \tag{6.3.28}$$

SUZUKI and YOSHIDA (1970) proved that (6.3.27) is a bound when $\varrho < 1$; similar methods as in their paper lead also to

$$m_W \leqq \frac{\sigma_A^2 + \sigma_B^2/s^2 + (m_B^2 + \sigma_B^2)\max(0, 1 - 1/\varrho)\,(s - 1)/s^2}{2m_A(1 - \varrho/s)} \tag{6.3.29}$$

(see MORI (1975)) and, for GI/E$_k$/s queues,

$$m_W \leqq \frac{\sigma_A^2 + \sigma_B^2/s^2}{2m_A(1 - \varrho/s)} + (1 - s^{-1})\,m_B. \tag{6.3.30}$$

Also, better bounds than (6.3.26) can be obtained for M/GI/s queues. For these it follows from (6.3.24) that m_W is bounded by the mean stationary waiting time in E$_s$/GI/1 queues with the service time d.f. B and parameters $\lambda = m_A^{-1}$ and s of the Erlang interarrival d.f. Upper bounds for m_W in E$_s$/GI/1 then follow for example by (5.4.2), (5.4.4) and (5.6.10), yielding for m_W in M/GI/s

$$m_W \leqq \frac{\lambda(m_B^2 + \sigma_B^2)}{2s - 2\varrho} = \frac{(\varrho/s)\,\varrho\sigma_A^2 + \sigma_B^2/s}{2m_A(1 - \varrho/s)}; \tag{6.3.16'}$$

if

$$B \leqq_c \operatorname{Exp}(1/m_B),$$

then

$$m_W \leqq (\zeta^{-1} - 1)\,m_B \tag{6.3.31}$$

where ζ is the solution in $0 < z < 1$ of $z = \big(1 + (1 - z)/\varrho\big)^{-s}$; and finally by (5.6.10), a stronger bound than (6.3.30),

$$m_W \leqq \frac{(\varrho/s)^2\,\sigma_A^2 + \sigma_B^2/s}{2m_A(1 - \varrho/s)}. \tag{6.3.32}$$

There are analogous formulae for E$_k$/GI/s queues; for example if $B \leqq_c \operatorname{Exp}(1/m_B)$ as required for (6.3.31) then that inequality continues to hold with sk replacing s in the equation defining ζ, while the form of (6.3.32) remains unchanged. See also KREININ (1981).

The inequality at (5.7.4) is another possible starting point for bounding m_W in E$_k$/GI/s (see KARPELEVICH and KREININ (1980)).

The cyclic assignment of customers has the following *extremal property*. Let Σ_r be a system with the same interarrival and service time d.f.s as in Σ and Σ_c, and for which the $(ns + k)^{\text{th}}$ customer is assigned, independently of the state of the system, and independently of other customers, to the server

$$k + j \text{ (or } k + j - s \text{ if } k + j > s\text{) with probability } p_j$$

$$(j, k = 1, \ldots, s;\ n = 0, 1, \ldots).$$

As special cases we have:

$$p_1 = 1,\ p_i = 0 \quad (i = 2, \ldots, s)\text{: cyclic assignment;}$$

$$p_i = s^{-1} \quad (i = 1, \ldots, s)\text{: random assignment.}$$

Every server in Σ_r works like a GI/GI/1 queue with the service time d.f. B and the interarrival time d.f. A_r, where A_r has the form $A_r(t) = \mathsf{P}\left(\sum_{i=1}^{\nu} \alpha_i < t\right)$, the α_i being i.i.d. r.v.s with the d.f. A and ν is a r.v. with mean s which is independent of the α_i. This r.v. ν describes the order of assignment; $\nu = s$ a.s. for cyclic assignment, while for random assignment ν is geometrically distributed on $\{1, 2, \ldots\}$ with the parameter $1 - s^{-1}$,

$$\mathsf{P}(\nu = i) = s^{-1}(1 - s^{-1})^{i-1} \qquad (i = 1, 2, \ldots).$$

We have $s \leqq_c \nu$, and hence by Proposition 2.2.5

$$\sum_{i=1}^{s} \alpha_i \leqq_c \sum_{i=1}^{\nu} \alpha_i,$$

leading by Theorem 5.2.1 to

Proposition 6.3.1. *For the stationary waiting time d.f.s W_r and W_c in Σ_r and Σ_c,*

$$W_c \leqq_c W_r, \tag{6.3.33}$$

so for the mean stationary waiting times,

$$m_W^{(c)} \leqq m_W^{(r)}. \tag{6.3.33'}$$

The inequality (6.3.27) for M/GI/s may be replaced by inequalities of the form

$$m_W \leqq \frac{c\lambda(m_B^2 + \sigma_B^2)}{2s - 2\varrho} = \frac{c(\sigma_B^2/s + (\varrho/s)\, \varrho\sigma_A^2)}{2m_A(1 - \varrho/s)}$$

with $c \leqq 1$. Schmidt (1978) showed that in the case $\varrho = \lambda m_B < 1$, c can be chosen as $c = \varrho(1 - \varrho)$, while Stoyan (1978b) showed that for all $\varrho < s$,

$$m_W \leqq \frac{p_W}{\varrho/s} \cdot \frac{\lambda(m_B^2 + \sigma_B^2)}{2s(1 - \varrho/s)} = \frac{m_B(1 + \sigma_B^2/m_B^2)\, p_W}{2(s - \varrho)}. \tag{6.3.33''}$$

To derive (6.3.33″), consider the system in its stationary state and suppose that an arriving customer finds i customers in the system for some $i \geqq s$. Let $r_1, \ldots, r_s$ be the residual service times of these customers in service and $\beta_{s+1}, \ldots, \beta_i$ the service times of the waiting customers. Then the r_j are identically distributed (but, in general, not independent), and their d.f. is B_R because (a) the stationary residual service time d.f. is B_R and (b) in the case of Poisson input, customer and time distributions are equal (see Cohen (1976), König, Rolski, Schmidt and Stoyan (1978) or Franken, König, Arndt, and Schmidt (1981)). Now by assumption the β_k are i.i.d with d.f. B, and thus the mean waiting time $m_{W,i}$ of our customer is claimed to satisfy

$$m_{W,i} \leqq m_R + ([i/s] - 1)\, m_B$$

where m_R is the mean of B_R, so $m_R = (m_B{}^2 + \sigma_B{}^2)/2m_B$, and $[i/s]$ is the largest integer $\leqq i/s$. The right hand of the last inequality is the mean waiting time of the customer if the queue discipline is changed from FCFS to a cyclic assignment of the $(i - s)$ waiting customers. Clearly, $[i/s] \leqq i/s$ and so taking expectations on $i \geqq s$,

$$m_W \leqq p_W m_R + L_W{}^0 m_B/s,$$

where $L_W{}^0$ is the mean stationary queue length at an arrival instant. In the case of a Poisson input $L_W{}^0$ equals the time-stationary mean queue length L_W, and using $L_W = \lambda m_W$, we obtain $m_W \leqq p_W m_R + \varrho m_W/s$, from which (6.3.33″) follows.

This section concludes with an example showing that, for initially empty GI/GI/2 systems with the same interarrival and service time distributions, one system with the FCFS discipline and the other with the cyclic discipline the waiting time, the sum of the residual work times, and the maximum residual work time, can be larger in the former (apparently more efficient) system for certain customers (c.f. Table 6.1).

Table 6.1. Waiting times in a GI/GI/2 queue with FCFS and cyclic discipline

			FCFS		Cyclic	
n	α_n	β_n	$w_{n,1}$	$w_{n,2}$	$w'_{n,1}$	$w'_{n,2}$
0	1	100	0	0	0	0
1	1	1	0	99	99	0
2	1	100	0	98	98	0
3	197	200	97	99	197	0
4	1	100	0	100	0	3
5		100	99	99	99	2

The basis of Kingman's approximation is the conjecture (cf. (6.3.25b)) that $W \leqq_d W_c$, but Whitt (1981c) gave examples refuting this suggestion. Consider for example a GI/M/2 queue with $m_B = 2$ and $\mathsf{P}(\alpha = 1) = 1 - \mathsf{P}(\alpha = 3000) = 0.99999$. The known formulae for GI/M/s yield

$$W(0) = 0.00604 < W_c(0) = 0.00631.$$

6.3.2.2 A modified preemptive resume LCFS discipline

Practical experience has shown (see LEE and LONGTON (1957), KRAMPE, KUBAT and RUNGE (1974) and NOZAKI and ROSS (1975, 1978)) that the following approximations for M/GI/s are very accurate:

$$m_W \approx {}^1/_2(1 + \sigma_B{}^2/m_B{}^2)\, m_W(\mathrm{M/M}/s), \tag{6.3.34}$$

$$p_W \approx p_W(\mathrm{M/M}/s), \tag{6.3.35}$$

where $m_W(\mathrm{M/M}/s)$ and $p_W(\mathrm{M/M}/s)$ denote the mean stationary waiting time and the stationary probability of waiting in M/M/s with the parameters λ and m_B respectively. Formally, we can interpret (6.3.34) as a generalization of the Pollaczek-Khinchin formula (5.0.10) to which it reduces for $s = 1$. The expressions at (6.3.34) and (6.3.35) can be deduced by the two steps below; they also serve in part to explain their accuracy.

(1°). Consider a queue Σ_L with Poisson input process at rate λ and service time d.f. B as for Σ, but for which the FCFS discipline is replaced by a variant of a work-conserving preemptive resume LCFS discipline which we describe via $\nu(t)$, the number of customers in the system at time t, and (for $\nu(t) \geqq 1$) the $\nu(t)$-vector $\boldsymbol{X}(t) = \big(X_1(t), \ldots, X_{\nu(t)}(t)\big)$ of residual service times, these being specified in order of priority for service so that the customers corresponding to the first $\min\big(s, \nu(t)\big)$ components of $\boldsymbol{X}(t)$ are in service at t. A customer arriving at t is allocated a priority rating at random from amongst the first $\nu_{occ}(t) \equiv \min\big(s, \nu(t-0) + 1\big)$ components of $\boldsymbol{X}(t+0)$. Thus, if an arrival at t has service time β, then for each $k = 1, \ldots, \nu_{occ}(t)$, we have with probability $1/\nu_{occ}(t)$ that $X_k(t+0) = \beta$ and $X_j(t+0) = X_j(t-0)$ $(j = 1, \ldots, k-1)$ $X_{j+1}(t+0) = X_j(t-0)$ $\big(j = k, \ldots, \nu(t-0)\big)$. Omitting terms that are $o(h)$ for $h \downarrow 0$, we can then write down the relations

$$\mathsf{P}\big(\nu(t+h) = 0\big) = (1 - \lambda h)\, \mathsf{P}\big(\nu(t) = 0\big) + \mathsf{P}\big(\nu(t) = 1,\, X_1(t) \leqq h\big)$$

and, for $i = 1, 2, \ldots,$

$$\begin{aligned}
\mathsf{P}\big(\nu(t+h) = i,\ & X_j(t+h) \leqq x_j\ (j = 1, \ldots, i)\big)\\
&= (1 - \lambda h)\, \mathsf{P}\big(\nu(t) = i,\, h < X_j(t) \leqq x_j + h\\
&\qquad (j = 1, \ldots, i(s)),\, X_j(t) \leqq x_j\ (j = s+1, \ldots, i)\big)\\
+ \sum_{k=1}^{(i+1)(s)} \mathsf{P}\big(\nu(t) &= i+1,\, X_k(t) < h,\, h < X_j(t) < x_j + h \quad (j = 1, \ldots, k-1),\\
&\qquad h < X_j(t) < x_{j-1} + h\ (j = k+1, \ldots, (i+1)\,(s)),\\
&\qquad X_j(t) < x_{j-1} \quad (j = s+1, \ldots, i+1)\big)\\
+ \lambda h \sum_{k=1}^{i(s)} \big(i(s)\big)^{-1} B(x_k)\, \mathsf{P}\big(\nu(t) &= i-1,\ h < X_j(t) < x_j + h \quad (j = 1, \ldots, k-1),\\
h < X_j(t) < x_{j+1} + h\ (j = k, \ldots, i(s) - 1),&\ X_j(t) < x_{j+1}\ (j = s, \ldots, i-1)\big),
\end{aligned}$$

where $i(s) = \min(s, i)$, it being understood that undefined r.v.s corresponding to null sets for the index j, occur for $k = 1$ and for $i \leq s$ and are to be omitted. By the usual procedure of letting $h \downarrow 0$ and $t \to \infty$, these relations lead to the following system of equations for the stationary distribution whose existence and uniqueness can be proved using the theory of regenerative phenomena since $\{t: \nu(t) = 0\}$ is a regenerative set for the process (see e.g. GNEDENKO and KOVALENKO (1966)) and in which we write $P_0 = \lim_{t\to\infty} \mathsf{P}(\nu(t) = 0)$ and for $i = 1, 2, \ldots$ and i-vectors $\boldsymbol{x} = (x_1, \ldots, x_i)$ with positive components, $F_i(\boldsymbol{x}) = \lim_{t\to\infty} \mathsf{P}(\nu(t) = i,\ \boldsymbol{X}(t) \leq \boldsymbol{x})$, and $D_j = \partial/\partial x_j$:

$$\lambda P_0 = D_1 F_1(x_1)|_{x_1=0},$$

$$\lambda F_i(\boldsymbol{x}) = \sum_{k=1}^{i(s)} \left(D_k F_i(\boldsymbol{x}) - D_k F_i(\boldsymbol{x})|_{x_k=0}\right)$$

$$+ \sum_{k=1}^{(i+1)(s)} D_{k'} F_{i+1}(x_1, \ldots, x_{k-1}, x_{k'}, x_k, \ldots, x_i)|_{x_{k'}=0}$$

$$+ \left(\lambda/i(s)\right) \sum_{k=1}^{i(s)} B(x_k)\, F_{i-1}(x_1, \ldots, x_{k-1}, x_{k+1}, \ldots, x_i),$$

interpreting $F_0 = P_0$ in the last term when $i = 1$. It can be verified that this system of equations has as solution

$$P_0 = P_0(\mathrm{M/M}/s), \tag{6.3.36}$$

$$F_i(\boldsymbol{x}) = P_i(\mathrm{M/M}/s) \prod_{j=1}^{i} B_R(x_j) \qquad (i = 1, 2, \ldots), \tag{6.3.37}$$

B_R denoting the length-biased sampling distribution for B given by

$$B_R(t) = m_B^{-1} \int_0^t \left(1 - B(u)\right) \mathrm{d}u$$

and $P_i(\mathrm{M/M}/s) = \mathsf{P}(\nu(t) = i \text{ in M/M}/s \text{ system})$, so P_0 is as at (6.0.9).

Using

$$\sum_{i=s}^{\infty} F_i(\infty, \ldots, \infty) = \sum_{i=s}^{\infty} P_i(\mathrm{M/M}/s)$$

as an approximation for p_W leads to (6.3.35).

(2⁰). Now consider an arriving customer in Σ_L in its stationary state but suppose that he is served in the FCFS discipline rather than the modified LCFS, and denote his waiting time by w_{mod}. Direct computation of w_{mod} is complicated although we know from (6.3.37) that, if there are $\nu(t - 0) = i$ customers in the system on his arrival at t say, all the i residual service times are independent and have the same d.f. B_R with mean $m_{B_R} = {}^1/_2(1 + \sigma_B^2/m_B^2)\, m_B$. An approximation to $\mathsf{E} w_{\mathrm{mod}}$ is obtained by supposing these residual service times to be exponentially distributed with mean m_{B_R}, which would imply that

$$\mathsf{E}\left(w_{\text{mod}} \mid \nu(t-0) = i\right) \approx (i + 1 - s)_+ \, (m_{B_R}/s),$$

and hence

$$\mathsf{E} w_{\text{mod}} \approx \sum_{i=s}^{\infty} (i + 1 - s) \, (m_{B_R}/s) \, P_i(\mathrm{M/M}/s)$$

$$= (m_{B_R}/m_B) \, m_W(\mathrm{M/M}/s).$$

leading to (6.3.34). Clearly, in the case where B is exponential, $B = B_R$ and the steps (1^0) and (2^0) lead to exact values for p_W and m_W.

Other state probabilities in M/GI/s, in addition to p_W, may be approximated by those in M/M/s. A further argument which justifies the approximation of M/GI/s by M/M/s is the "convergence" as $s \to \infty$ of the state probabilities of M/GI/s to those of M/GI/∞, this last system having the same insensivity property of Σ_L as at (6.3.37), namely that its state probabilities coincide with those of M/M/∞.

6.3.3 Modification of the service procedure

Again we assume that Σ is of type M/GI/s and consider the following approximating system Σ^*. The input is the same as in Σ. Any arriving customer finding an idle server has the service d.f. B as in Σ, while an arrival finding all servers busy has service d.f. B_z given by

$$B_z(t) = \big(B(z + t) - B(z)\big) / \big(1 - B(z)\big) \tag{6.3.38}$$

where z is the minimum of the residual work times of the s servers as in the Kiefer-Wolfowitz description of M/GI/s, and, of course, the customer goes to that server yielding the minimum (i.e., the FCFS discipline is preserved). As an aside, note that $B(z) < 1$ a.s.

This curious system is of interest, because Jurkevich (1973) was able to obtain exact formulae for its mean stationary waiting time $m_W{}^*$ and the probability of waiting $p_W{}^*$, and also because it is suitable for system approximations in the sense that we can give conditions under which

$$m_W{}^* \leqq (\geqq) \; m_W, \quad p_W{}^* \leqq (\geqq) \, p_W. \tag{6.3.39}$$

If for all z

$$B_z \leqq_d (\geqq_d) \, B \tag{6.3.40}$$

holds (i.e., B is NBU (NWU)), we obtain from the monotonicity of G/G/s that the residual work time vectors in Σ^* are smaller (larger) with respect to $\leqq_d$ than those in Σ, from which (6.3.39) follows. However, unless B is NBU, it is possible that $m_W{}^*$ and $p_W{}^*$ may be larger than m_W and p_W, or even that Σ^* possesses no stationary waiting time d.f. even though $\varrho/s < 1$.

From (6.3.39) we obtain for M/GI/s queues with NBU service times that

$$m_W \geqq m_W^* = (P_0^* \lambda^s/s!) \int_0^\infty (1 + \lambda x) \big(f(x)\big)^s e^{\lambda x} \, dx, \tag{6.3.41}$$

$$p_W \geqq p_W^* = (P_0^* \lambda^s/s!) \left(m_B^s + \lambda \int_0^\infty \big(f(x)\big)^s e^{\lambda x} \, dx \right), \tag{6.3.42}$$

where

$$f(x) = \int_x^\infty \overline{B}(u) \, du \quad \text{and} \quad 1/P_0^* = \sum_{k=0}^{s} (\lambda m_B)^k/k! + (\lambda^{s+1}/s!) \int_0^\infty \big(f(x)\big)^s e^{\lambda x} \, dx.$$

The proof of these formulae is omitted here and can be found in JURKEVICH (1973). It must be noted that he considered systems with limited waiting time; our system Σ^* is the case of unlimited waiting time.

Since B is NBU, $B \leqq_c \text{Exp}\,(1/m_B)$, so $f(x) \leqq m_B \, e^{-x/m_B}$ for $x \geqq 0$. This inequality gives upper bounds on p_W^*, contrary to the assertion in previous editions of this monograph.

6.4 Heuristic bounds and approximations for M/GI/s

Many authors have given approximations for M/GI/s queues. From practical experience it is known that m_W and p_W for M/GI/s lie between the corresponding characteristics of M/D/s and M/M/s if λ and m_B are fixed. Here the variance σ_B^2 of service time plays an important role: for $\sigma_B^2 \approx m_B^2$ the values are similar to those of M/M/s, for $\sigma_B^2 \approx 0$ to those of M/D/s. These observations prompt the approximations, writing $c_B^2 = \sigma_B^2/m_B^2$,

$$m_W \approx (1 - c_B^2) \, m_W(\text{M/D/}s) + c_B^2 m_W(\text{M/M/}s), \tag{6.4.1}$$

$$p_W \approx (1 - c_B^2) \, p_W(\text{M/D/}s) + c_B^2 p_W(\text{M/M/}s) \tag{6.4.2}$$

(see PAGE (1972) and HEIN (1973)). Values of $m_W(\text{M/D/}s), \ldots, p_W(\text{M/M/}s)$ can be found in Tables 6.2 and 6.3.

PAGE (1972) suggested for $\text{E}_k/\text{E}_l/s$ the approximation

$$m_W \approx (1 - c_A^2) \, c_B^2 m_W(\text{D/M/}s) + c_A^2 (1 - c_B^2) \, m_W(\text{M/D/}s) + c_A^2 c_B^2 m_W(\text{M/M/}s) \tag{6.4.3}$$

(note that $c_A^2 = 1/k$, $c_B^2 = 1/l$ for such a system), and showed its accuracy particularly in the case of large ϱ; in Table 6.4 p_W and m_W are given for D/M/s queues. Similarly, SAKASEGAWA (1977) obtained heuristic approximations for GI/GI/s queues. The approximations of COSMETATOS (1974, 1976) for GI/M/s and M/GI/s queues are also of a similar nature.

Another approximation for m_W in M/GI/s has been suggested by BOXMA,

Table 6.2. Waiting probability p_W and mean waiting time m_W/m_B in M/M/s queues

ϱ/s	Number of servers s 1	2	3	4	5	6	7	8	9	10
0.1	**0.100**	**0.018**	**0.004**	**0.001**	**0.000**	**0.000**	**0.000**	**0.000**	**0.000**	**0.000**
	0.111	0.010	0.001	0.000	0.000	0.000	0.000	0.000	0.000	0.000
0.2	**0.200**	**0.067**	**0.025**	**0.010**	**0.004**	**0.002**	**0.001**	0.000	0.000	0.000
	0.250	0.042	0.010	0.003	0.001	0.000	0.000	0.000	0.000	0.000
0.3	**0.300**	**0.138**	**0.070**	**0.037**	**0.020**	**0.011**	**0.006**	**0.004**	**0.002**	**0.001**
	0.429	0.099	0.033	0.013	0.006	0.003	0.001	0.001	0.000	0.000
0.4	**0.400**	**0.229**	**0.141**	**0.091**	**0.060**	**0.040**	**0.027**	**0.018**	**0.013**	**0.009**
	0.667	0.191	0.078	0.038	0.020	0.011	0.006	0.004	0.002	0.002
0.5	**0.500**	**0.333**	**0.237**	**0.174**	**0.130**	**0.099**	**0.076**	**0.059**	**0.046**	**0.036**
	1.000	0.333	0.158	0.087	0.052	0.033	0.022	0.015	0.010	0.007
0.6	**0.600**	**0.450**	**0.355**	**0.287**	**0.236**	**0.197**	**0.165**	**0.140**	**0.119**	**0.101**
	1.500	0.562	0.296	0.179	0.118	0.082	0.059	0.044	0.033	0.025
0.7	**0.700**	**0.577**	**0.492**	**0.429**	**0.378**	**0.336**	**0.301**	**0.271**	**0.245**	**0.222**
	2.333	0.961	0.547	0.357	0.252	0.187	0.143	0.113	0.091	0.074
0.8	**0.800**	**0.711**	**0.647**	**0.597**	**0.554**	**0.512**	**0.486**	**0.458**	**0.432**	**0.409**
	4.000	1.778	1.079	0.746	0.554	0.432	0.347	0.286	0.240	0.205
0.9	**0.900**	**0.853**	**0.817**	**0.788**	**0.763**	**0.740**	**0.720**	**0.702**	**0.685**	**0.669**
	9.000	4.263	2.724	1.969	1.525	1.234	1.029	0.877	0.761	0.669

Table 6.3. Waiting probability p_W and mean waiting time m_W/m_B in M/D/s queues

ϱ/s	Number of servers s 1	2	3	4	5	6	7	8	9	10
0.1	**0.100**	**0.018**	**0.004**	**0.000**	**0.000**	**0.000**	**0.000**	**0.000**	**0.000**	**0.000**
	0.056	0.006	0.001	0.000	0.000	0.000	0.000	0.000	0.000	0.000
0.2	**0.200**	**0.064**	**0.024**	**0.009**	**0.004**	**0.001**	**0.001**	0.000	0.000	0.000
	0.125	0.024	0.007	0.002	0.001	0.000	0.000	0.000	0.000	0.000
0.3	**0.300**	**0.134**	**0.067**	**0.035**	**0.019**	**0.011**	**0.006**	**0.003**	**0.002**	**0.001**
	0.214	0.055	0.020	0.009	0.004	0.002	0.001	0.001	0.000	0.000
0.4	**0.400**	**0.221**	**0.134**	**0.085**	**0.056**	**0.037**	**0.025**	**0.017**	**0.012**	**0.008**
	0.333	0.103	0.045	0.023	0.012	0.007	0.004	0.003	0.002	0.001
0.5	**0.500**	**0.323**	**0.225**	**0.163**	**0.121**	**0.092**	**0.070**	**0.054**	**0.042**	**0.033**
	0.500	0.177	0.087	0.050	0.031	0.020	0.014	0.009	0.007	0.005
0.6	**0.600**	**0.439**	**0.340**	**0.272**	**0.221**	**0.183**	**0.153**	**0.128**	**0.109**	**0.093**
	0.750	0.293	0.158	0.098	0.066	0.047	0.034	0.026	0.020	0.015
0.7	**0.700**	**0.565**	**0.476**	**0.410**	**0.358**	**0.316**	**0.282**	**0.252**	**0.226**	**0.204**
	1.167	0.494	0.276	0.190	0.136	0.102	0.079	0.063	0.051	0.042
0.8	**0.800**	**0.701**	**0.631**	**0.576**	**0.532**	**0.494**	**0.461**	**0.431**	**0.406**	**0.382**
	2.000	0.903	0.554	0.386	0.289	0.226	0.183	0.152	0.128	0.110
0.9	**0.900**	**0.847**	**0.808**	**0.776**	**0.748**	**0.724**	**0.702**	**0.682**	**0.664**	**0.647**
	4.500	2.147	1.378	1.000	0.777	0.630	0.527	0.450	0.391	0.345

Further tables for M/GI/s queues are given in GROENEVELT, VAN HOORN and TIJMS (1982) and HILLIER and YU (1981).

Table 6.4. Waiting probability p_W and mean waiting time m_W/m_B in D/M/s queues

ϱ/s	Number of servers s									
	1	2	3	4	5	6	7	8	9	10
0.1	**0.000**	**0.000**	**0.000**	**0.000**	**0.000**	**0.000**	**0.000**	**0.000**	**0.000**	**0.000**
	0.000	0.000	0.000	0.000	0.000	0.000	0.000	0.000	0.000	0.000
0.2	**0.007**	**0.001**	**0.000**	**0.000**	**0.000**	**0.000**	**0.000**	**0.000**	**0.000**	**0.000**
	0.007	0.003	0.000	0.000	0.000	0.000	0.000	0.000	0.000	0.000
0.3	**0.041**	**0.009**	**0.002**	**0.001**	**0.000**	**0.000**	**0.000**	**0.000**	**0.000**	**0.000**
	0.043	0.005	0.001	0.000	0.000	0.000	0.000	0.000	0.000	0.000
0.4	**0.107**	**0.040**	**0.016**	**0.007**	**0.003**	**0.001**	**0.001**	**0.000**	**0.000**	**0.000**
	0.120	0.022	0.006	0.002	0.001	0.000	0.000	0.000	0.000	0.000
0.5	**0.203**	**0.103**	**0.057**	**0.033**	**0.019**	**0.012**	**0.007**	**0.004**	**0.003**	**0.001**
	0.255	0.065	0.024	0.010	0.005	0.002	0.001	0.001	0.000	0.000
0.6	**0.324**	**0.205**	**0.139**	**0.097**	**0.070**	**0.051**	**0.037**	**0.028**	**0.021**	**0.015**
	0.480	0.152	0.069	0.036	0.021	0.013	0.008	0.005	0.003	0.002
0.7	**0.467**	**0.347**	**0.271**	**0.217**	**0.177**	**0.146**	**0.122**	**0.102**	**0.086**	**0.073**
	0.876	0.326	0.170	0.102	0.066	0.046	0.033	0.024	0.018	0.014
0.8	**0.629**	**0.528**	**0.458**	**0.405**	**0.361**	**0.326**	**0.295**	**0.268**	**0.245**	**0.225**
	1.693	0.711	0.411	0.273	0.195	0.146	0.113	0.090	0.073	0.060
0.9	**0.807**	**0.746**	**0.702**	**0.665**	**0.634**	**0.607**	**0.583**	**0.561**	**0.541**	**0.522**
	4.179	1.933	1.211	0.861	0.657	0.524	0.431	0.363	0.211	0.270

Cohen and Huffels (1979) using the "insensitivity" (in the sense of Section 6.3.2.2) of M/GI/s with the processor sharing discipline.

Hokstad (1978) suggested the approximation

$$m_W \approx p_W(\text{M/M/}s)\, m_B(1 + \sigma_B^2/m_B^2)/2(s - \varrho). \tag{6.4.4}$$

Using arguments of the theory of regenerative processes, Tijms, van Hoorn and Federgruen (1981) obtained heuristic approximations for p_W and other steady-state probabilities.

Finally, Takahashi (1977) has given approximations for m_W and σ_W^2 in M/GI/s, which are more accurate than (6.4.1), namely,

$$m_W \approx m_W^{(T)} \equiv (\mu_B^{(a)}/m_B^a)^{1/(a-1)}\, m_W(\text{M/D/}s) \tag{6.4.5}$$

where

$$\mu_B^{(a)} = \int_0^\infty t^a \,\mathrm{d}B(t)$$

and the power a is determined so that the approximation is exact for M/M/s. Some values for a can be found in Table 6.5, computed by Takahashi. For the waiting time variance σ_W^2 he suggested

$$\sigma_W^2 \approx (\mu_B^{(b)}/m_B^b)^{2/(b-1)} \left(\sigma_W^2(\text{M/D/}s) - m_W^2(\text{M/D/}s)\right) + (m_W^{(T)})^2 \tag{6.4.6}$$

where σ_W^2 (M/D/s) is the variance of the stationary waiting time in M/D/s (see Table 6.6) and b is as in Table 6.7.

Table 6.5. Values of a for TAKAHASHI's approximation (see (6.4.5))

ϱ/s	Number of servers s 1	2	2	4
0.1	2	1.21	0.84	0.64
0.2	2	1.40	1.07	0.86
0.3	2	1.55	1.27	1.08
0.4	2	1.66	1.45	1.29
0.5	2	1.75	1.59	1.47
0.6	2	1.83	1.71	1.62
0.7	2	1.88	1.81	1.75
0.8	2	1.93	1.89	1.85
0.9	2	1.97	1.95	1.93

Table 6.6. Values of $[\sigma_W^2(\mathrm{M/D/}s) - m_W^2(\mathrm{M/D/}s)]/m_B^2$ (cf. (6.4.6))

ϱ/s	Number of servers s 1	2	3	4
0.1	0.037	0.003	0.000	0.000
0.2	0.083	0.013	0.003	0.001
0.3	0.143	0.029	0.008	0.003
0.4	0.222	0.055	0.020	0.009
0.5	0.333	0.094	0.039	0.020
0.6	0.500	0.157	0.073	0.040
0.7	0.778	0.267	0.133	0.078
0.8	1.333	0.492	0.261	0.016
0.9	3.000	1.176	0.659	0.043

Table 6.7. Values of b for TAKAHASHI's M/GI/s approximation (see (6.4.6))

ϱ/s	Number of servers s 1	2	3	4
0.1	3	1.71	1.17	0.88
0.2	3	1.94	1.42	1.12
0.3	3	2.12	1.66	1.36
0.4	3	2.26	1.87	1.60
0.5	3	2.38	2.05	1.82
0.6	3	2.48	2.20	2.01
0.7	3	2.56	2.33	2.18
0.8	3	2.63	2.45	2.33
0.9	3	2.69	2.54	2.45

Approximations for M/GI/s systems with limited queue length (and priorities) have been given by BASHARIN and LYSENKOVA (1972) and NOZAKI and ROSS (1978). For the state probabilities P_n in M/GI/s/r the first authors suggested

$$P_n = \begin{cases} P_0(\varrho/s)^n/n! & (n = 0, \ldots, s), \\ P_s\gamma\phi^{n-s} & (n = s+1, \ldots, s+r-1), \\ P_s\phi^r & (n = s+r) \end{cases}$$

where $\gamma = 2/(1 + \sigma_B^2/m_B^2)$, $\phi = (\varrho/s)^\gamma$, $\varrho = \lambda m_B$, and the probabilities are normalized by $P_0 + \cdots + P_{s+r} = 1$.

Diffusion approximations for G/G/s and GI/GI/s queues have been given in Halachmi and Franta (1977, 1978), Sunaga, Kondo and Bisnos (1978) and Sunaga, Bisnos and Nishida (1982).

Notes and bibliographic comments to Chapter 6

6.1

The internal monotonicity of waiting times of GI/GI/s was used by Kiefer and Wolfowitz (1955), while for the number of customers it was proved by H. Stoyan (1973). Whitt (1981b) uses monotonicity and continuity for proving the existence of limiting distributions in the GI/GI/s queue.

Open problems 1. For what initial distributions $\boldsymbol{w}_0$ different from $\mathsf{P}(\boldsymbol{w}_0 = 0) = 1$ is (6.1.2) satisfied? With their help new bounds could be given, as in Section 5.3.
2. When $s = 1$, W is NWU; is the same true for $s \geqq 2$?

6.2

The external monotonicity properties of GI/GI/s with respect to $\leqq_d$ were proved by Jacobs and Schach (1972) and Stoyan (1972c); see also Kiefer and Wolfowitz (1955). Whitt (1980b) proved (6.2.7), (6.2.8), (6.2.9) in a somewhat different way. Other related work is included in Whitt (1981a) and Sonderman (1979a, b).

Open problem. It is reasonable to conjecture that monotonicity properties for stationary waiting time d.f.s similar to those for GI/GI/1 also exist for classes of GI/GI/s queues. In particular (cf. Theorem 5.2.3) (a) For two M/GI/s systems with equal mean service times and equal arrival rates, does $B_1 \leqq_c B_2$ imply $W_1 \leqq_d W_2$ or at least $W_1 \leqq_c W_2$? Are similar inequalities true for the stationary distribution of queue length? (See also Avis (1977).)

(b) Is a statement analogous to Theorem 5.2.3 (b) true for GI/M/s systems? (The monotonicity of the solution of (6.0.12) can be proved as in Section 5.2 (see Eckberg (1977)); the analysis of C is more difficult.)

6.3

The examples in 6.3 show that complicated systems must often be used in order to obtain bounds for GI/GI/s queues.

Potthoff (1962) gave (6.3.22) but without rigorous proof. The results in 6.3.1.2 are from STOYAN (1974a, 1976a); many similar formulae were obtained by NEWELL (1973) using heuristic methods. Systems with infinitely many servers are used as bounding systems by BLOOMFIELD and COX (1972), and, similarly, loss systems by STOYAN (1973a). VASICEK (1977) proved that, for the stationary waiting time distribution W_d in GI/GI/s with any general queueing discipline of the class that has no idle servers if customers are waiting,

$$\int_0^\infty f(x)\,\mathrm{d}W(x) \leqq \int_0^\infty f(x)\,\mathrm{d}W_d(x) \leqq \int_0^\infty f(x)\,\mathrm{d}W_L(x),$$

where W is the stationary waiting time d.f. under the discipline FCFS and W_L that under the normal LCFS and f is a convex function. From this it follows that $W \leqq_c W_d \leqq_c W_L$; the corresponding means are equal. Similar inequalities are true in the nonstationary case for G/G/s queues.

WOLFF (1977a) and SMITH and WHITT (1981) compared characteristics of pooled queueing systems with characteristics of their components. For example, let $p_v(s, \lambda, \mu)$ be the stationary loss probability in M/M/s/0, then

$$(\lambda_1 + \lambda_2)\, p_v(s_1 + s_2, \lambda_1 + \lambda_2, \mu) \leqq \lambda_1 p_v(s_1, \lambda_1, \mu) + \lambda_2 p_v(s_2, \lambda_2, \mu).$$

Furthermore, when $m_W(s, \lambda, \mu)$ denotes the stationary mean waiting time in M/M/s,

$$m_W(s_1 + s_2, \lambda_1 + \lambda_2, \mu) \leqq [\lambda_1/(\lambda_1 + \lambda_2)]\, m_W(s_1, \lambda_1, \mu) + [\lambda_2/(\lambda_1 + \lambda_2)]\, m_W(s_2, \lambda_2, \mu).$$

Equations (6.3.36) and (6.3.37), which describe the "insensitivity" of M/GI/s under the modified preemptive resume LCFS discipline, were given by STOYAN and STOYAN (1976) and earlier by MACK (1967); the justification behind the approximation in Section 6.3.2.2 was suggested by STOYAN (1976a). That the queue size distribution that follows from (6.3.37) depends on B only through its mean m_B, is an insensitivity property typical of certain queueing systems with Poisson arrivals (cf. the book by FRANKEN, KÖNIG, ARNDT, and SCHMIDT (1981)).

LOULOU (1973, 1974) used a further system approximation in which the idle servers assist the busy ones, so that the s servers are always either all active or all inactive. In this way he obtained upper and lower bounds for the work load $L_n = \sum_{i=1}^{s} w_{i,n}$ in GI/GI/s.

YU (1974) proved for queues with heterogeneous Erlang servers two-sided bounds with respect to $\leqq_d$ for W. He used the stationary waiting time d.f. of the single-server queue Σ_1, using methods by which STIDHAM (1970) had proved optimal properties of certain single server queues. ARJAS and LEHTONEN (1978) obtained similar bounds for service time d.f.s of type NBU and NWU.

AL-KHAYAL and GROSS (1977) give several references for approximations of GI/M/s queues; see also YECHIALI (1977) and COSMETATOS and GODSAVE

(1980). HALFIN and WHITT (1981) and WHITT (1982a) use $\leqq_d$ to get heavy-traffic limits for the steady-state distribution of GI/M/s and GI/GI/∞ systems.

We mention finally, that POLYAK (1968) gave bounds for M/G/s queues with group arrivals.

Equation (6.3.4) is one of many of the type "$L = \lambda W$" relating the mean L of a quantity, stationary in time, to the mean W of the corresponding quantity observed at arrival instants for which the mean interarrival interval equals λ^{-1}. For further discussion see BRUMELLE (1971b), STIDHAM (1974), FRANKEN (1976), and FRANKEN, KÖNIG, ARNDT, and SCHMIDT (1981). In this last reference and MARSHALL and WOLFF (1971), KÖNIG, SCHMIDT and STOYAN (1976), MIYAZAWA (1976a) and KÖNIG and SCHMIDT (1980), inequalities are proved for moments and distributions of queue characteristics defined at arrival instants (like W) and arbitrary time points (like L).

Open problems 1. Under which conditions is (6.3.3) true for G/G/s queues?
2. When do the inequalities $m_W \leqq m_W'$ and $p_W \leqq p_W'$ hold between these quantities for the systems Σ and Σ_1' described at the start of Section 6.3.1.1?
3. Is it possible to give classes of service time d.f.s for which $\leqq$ or $\geqq$ holds in (6.3.34)? Can the error in (6.3.34) be estimated for any classes of service time d.f.s?
4. Consider the following approximating GI/GI/1 system Σ_{rB} in which the input is the same as for Σ but the service times $\{\beta_n\}$ are replaced by $\{I_n\beta_n\}$, where $\{I_n\}$ is a sequence of i.i.d. indicator r.v.s, independent of $\{\alpha_n\}$ and $\{\beta_n\}$, and equal to 1 with probability $1/s$ and otherwise 0. Let $\{w_n^{rB}\}$ denote a sample realization of waiting times in Σ_{rB}, so that

$$w_{n+1} = (w_n + I_n\beta_n - \alpha_n)_+ .$$

Does the stationary d.f. W_{rB} for $\{w_n^{rB}\}$ satisfy $W \leqq_d W_{rB}$? (Compare with Brumelle's approximation on p. 110!) holds in (6.3.34)?
5. Notwithstanding the fact that $W \leqq_d W_c$ is not necessarily true, even in GI/M/2, what are sufficient conditions on A and/or B to ensure that $W \leqq_d W_c$ does hold?

Added in proof. The argument in STOYAN and STOYAN (1977) leading to (6.3.17) is incomplete, but the result is nevertheless true: it can be proved using the fact that in a stationary M/GI/s system, the processes $w_i(t)$ introduced at the beginning of § 6.0 being stationary and identically distributed for $i = 1, \ldots, s$ and having the property at (6.0.5), have

$$\mathsf{P}\big(w_i(t_n - 0) = 0\big) = \mathsf{P}\big(w_i(t) = 0\big) = 1 - \varrho/s .$$

For bounds for E_k/GI/s see also KREININ (1981) and VAINSHTEIN and KREININ (1981). In PAGE (1982), (6.4.3) is compared with SAKASEGAWA's (1977) approximation. On the basis of numerical calculations, HILLIER and YU (1981) suggested approximate formulae for m_W in $E_k/E_l/s$. GREEN (1982) has comments on some approximations to M/GI/s queues.

7. Monotonicity properties and bounds for some other stochastic models

7.0 Introductory remarks

The aim of this chapter is to provide further examples in which monotonicity properties or bounds are derived for various stochastic models that arise in reliability, inventory, branching processes, and renewal theory, and certain sundry queueing applications. What we give is a small selection of many results that are becoming increasingly available through the study of qualitative properties of stochastic models. The final section summarizes in tabular form some references to several other examples.

The bounds and approximations given below are, with few exceptions, those which are based on monotonicity properties or comparison techniques. There are of course many other practically useful approximation methods, while for many classes of models there exist methods that are specifically adapted to the properties of the models, as for example in the repairman problem in queueing (see e.g. PAGE (1972)), loss systems in teletraffic (KOSTEN (1973, 1978), LE GALL (1962), WILKINSON (1956), SHNEPS (1974), BASHARIN, KOKOTUSHKIN and NAUMOV (1979, 1980)), "heavy traffic" situations in queueing and branching processes (COHEN (1973), KINGMAN (1965), WHITT (1974b), FAHADY, QUINE and VERE-JONES (1971), BOROVKOV (1980)), reliability systems (BARLOW and PROSCHAN (1965, 1975), KOZLOV and USHAKOV (1975)), and network planning (GOLENKO (1972)). An interesting survey on approximation methods in queueing is BHAT, SHALABY and FISCHER (1979).

The last decade has seen increasing interest in diffusion approximations, and not only in heavy traffic of queueing applications (see e.g. around p. 62 of KLEINROCK (1976), GELENBE (1975), KOGAN and LITVIN (1979), BOROVKOV (1980), and several references at the end of Chapters 5 and 6), because such approximations usually have the virtue of being specified in terms of the first one or two moments of the constituent components of the model.

7.1 Bounds on the rate of convergence to stationary distributions in some queueing models

For many practical problems it is important to know how fast some time-dependent system characteristics converge from given initial values toward their stationary values. The reason for this is that the more easily determined stationary values may be reasonable as approximations to the time-dependent quantities only under additional assumptions such as suitably large time. In the following we exhibit the use of partial orderings to obtain bounds on the rate of convergence to a stationary distribution or a moment in three such problems.

7.1.1 State probabilities in M/GI/∞

Let $P_n(t)$ denote the probability that at time t there are n customers in an M/GI/∞ queueing system assumed to be empty at $t = 0$, so $P_0(0) = 1$. Recall from (6.3.23) that

$$P_n(t) = \mathrm{e}^{-\varrho(t)}\big(\varrho(t)\big)^n/n!\ (n = 0, 1, \ldots) \text{ where } \varrho(t) = \lambda \int_0^t \overline{B}(x)\,\mathrm{d}x.$$

In the case of finite m_B this Poisson distribution with mean $\varrho(t)$ converges as $t \to \infty$ to the stationary distribution, being Poisson with mean $\varrho = \lambda m_B$, as at (6.3.19):

$$P_n(t) \to P_n = \mathrm{e}^{-\varrho}\varrho^n/n!\ (n = 0, 1, \ldots)\ (t \to \infty).$$

The definition of $\varrho(.)$ shows, by reference to (1.4.4), that for service time distributions B_k $(k = 1, 2)$,

$$\text{if } B_1 \leqq_{cv} B_2 \text{ then } \varrho_1(t) \geqq \varrho_2(t) \quad (\text{all } t \geqq 0). \tag{7.1.1}$$

This inequality underlies the simple proof of

Proposition 7.1.1. *For* $k = 1, 2$, *suppose two* M/GI/∞ *service systems have the same arrival rate* λ *and that their service time d.f.s* B_k *have the same mean* m_B. *Then if either* $B_1 \leqq_{cv} B_2$ *or* $B_2 \leqq_c B_1$, *the respective state probabilities* $\{P_{n,k}(t)\}$ *satisfy*

$$\sum_{n=0}^{j}\big(P_n - P_{n,1}(t)\big) \leqq \sum_{n=0}^{j}\big(P_n - P_{n,2}(t)\big) \quad (\textit{all } t \geqq 0;\, j = 0, 1, \ldots) \tag{7.1.2}$$

where $\lim_{t\to\infty} P_{n,k}(t) = P_n = \mathrm{e}^{-\varrho}\varrho^n/n!$ $(k = 1, 2;\, n = 0, 1, \ldots)$, $\varrho = \lambda m_B$.

Proof. Under the assumptions on the d.f.s B_k, (7.1.1) holds (recall (1.4.5)). Then applying the property (1.2.11) of Poisson distributions to (6.3.23) with the parameter values $\varrho_k(t)$ $(k = 1, 2)$ shows that

$$\sum_{n=0}^{j} P_{n,1}(t) \geqq \sum_{n=0}^{j} P_{n,2}(t) \quad (\text{all } t \geqq 0;\, j = 0, 1, \ldots),$$

from which (7.1.2) follows because $P_{n,1}(\infty) = P_{n,2}(\infty) = P_n$ when $\varrho_1(\infty) = \varrho_2(\infty) = \varrho$.

The interpretation of this result is that, since $B_1 \leqq_{cv} B_2$ and $B_2 \leqq_c B_1$ are equivalent when the means are the same, the rate of convergence of $\{P_{n,1}(t)\}$ to $\{P_n\}$ is slower than for $\{P_{n,2}(t)\}$ when B_1 is smaller than B_2 with respect to $\leqq_{cv}$, i.e., when B_1 has a heavier positive tail than B_2. In the particular case that $B \leqq_c \text{Exp}(1/m_B)$ (equivalently, $\text{Exp}(1/m_B) \leqq_{cv} B$),

$$\sum_{n=0}^{j} \mathrm{e}^{-\varrho_e(t)}\big(\varrho_e(t)\big)^n/n! \geqq \sum_{n=0}^{j} P_n(t) \geqq \sum_{n=0}^{j} \mathrm{e}^{-\varrho_c(t)}\big(\varrho_c(t)\big)^n/n! \qquad (7.1.3)$$

where $\varrho_e(t) = \lambda m_B(1 - \mathrm{e}^{-t/m_B})$, $\varrho_c(t) = \lambda \min(t, m_B)$.

7.1.2 Mean waiting times in GI/GI/1

For $k = 1, 2$, consider queueing systems Σ_k of the type GI/GI/1 with d.f.s K_k for the difference of service and interarrival times. Assume that at $t = 0$, the arrival instant of the first customer, both systems are otherwise empty. For customers $n = 0, 1, \ldots$ denote the mean waiting time in Σ_k by $m_{W,n}^{(k)}$, and denote the corresponding stationary means (assuming they exist) by $m_{W^{(k)}}$.

Proposition 7.1.2. *With the above notation,* $K_1 \leqq_c K_2$ *implies that for all* $n = 0, 1, \ldots,$

$$m_{W^{(1)}} - m_{W,n}^{(1)} \leqq m_{W^{(2)}} - m_{W,n}^{(2)}. \qquad (7.1.4)$$

Proof. By (5.0.4″) we have

$$m_{W^{(k)}} - m_{W,n}^{(k)} = \sum_{r=n+1}^{\infty} r^{-1} \int_0^{\infty} x \,\mathrm{d}K_k^{*r}(x).$$

Now $K_1 \leqq_c K_2$ implies that $K_1^{*r} \leqq_c K_2^{*r}$, and since $f(x) = \max(0, x)$ is nondecreasing and convex, (7.1.4) now follows.

Just as for M/GI/∞ in the last section, the interpretation of this result is that when K_2 has a heavier tail than K_1 (i.e., $K_1 \leqq_c K_2$), as for example when $B_1 \leqq_c B_2$ with $A_1 = A_2$, the rate of convergence of $m_{W,n}^{(2)}$ to its limit is slower than that of $m_{W,n}^{(1)}$.

For numerical illustrations of this rate of convergence, reference may be made to tables for m_W and $m_{W,n}$ in Heathcote and Winer (1969) for M/D/1, D/M/1 and M/M/1, and Table 7.1 for values for M/M/1. For a

Table 7.1. Mean waiting time $m_{W,n}/m_B$ of nth arrival in M/M/1 queues ($m_{W,0} \equiv 0$, $m_{W,1} = \mathsf{E}(\beta - \alpha)_+$)

n	ϱ 0.1	0.2	0.3	0.4	0.5	0.6	0.7	0.8	0.9
1	0.091	0.167	0.231	0.286	0.333	0.375	0.412	0.444	0.474
3	0.110	0.236	0.371	0.506	0.638	0.763	0.881	0.991	1.093
5	0.110	0.247	0.409	0.589	0.781	0.976	1.170	1.358	1.537
7	0.111	0.249	0.421	0.625	0.860	1.112	1.373	1.636	1.894
9	0.111	0.250	0.426	0.645	0.907	1.205	1.526	1.860	2.195
11	0.111	0.250	0.427	0.655	0.937	1.271	1.646	2.046	2.458
13	0.111	0.250	0.428	0.660	0.957	1.321	1.742	2.206	2.692
15	0.111	0.250	0.428	0.663	0.970	1.358	1.821	2.344	2.904
∞	0.111	0.250	0.429	0.667	1.000	1.500	2.333	4.000	9.000

queueing system GI/GI/1 with $B \leqq_c \text{Exp}(1/m_B)$ and $A \leqq_c \text{Exp}(1/m_A) = \text{Exp}(\varrho/m_B)$, we have

$$(m_W - m_{W,n})/m_B \leqq \varrho/(1-\varrho) - (\text{Value in Table 7.1 for } n, \varrho). \quad (7.1.5)$$

7.1.3 Probability of emptiness in M/GI/1

Let $P_0(t)$ be the probability that the queue M/GI/1 is empty at time t assuming the system to be empty at $t = 0$. Then by e.g. p. 77 of TAKACS (1962),

$$P_0(t) = \sum_{n=0}^{\infty} \left(e^{-\lambda t}(\lambda t)^n/n!\right) \int_0^t (1 - x/t)\, dB^{*n}(x), \quad (7.1.6)$$

and for $t \to \infty$ it follows in the case that $\varrho = \lambda m_B < 1$,

$$P_0(t) \to P_0 \equiv 1 - \varrho \quad (t \to \infty). \quad (7.1.7)$$

Observe that, with S_n denoting a r.v. with the d.f. B^{*n} and hence having mean nm_B, the integral in (7.1.6) can be expressed as

$$\begin{aligned} \mathsf{E}(1 - S_n/t)_+ &= 1 - nm_B/t + \mathsf{E}(S_n/t - 1)_+ \\ &= 1 - nm_B/t + t^{-1} \int_t^{\infty} \left(1 - B^{*n}(x)\right) dx \end{aligned}$$

on integrating by parts. Thus

$$P_0(t) = 1 - \varrho + \sum_{n=0}^{\infty} \left(e^{-\lambda t}(\lambda t)^n/n!\right) t^{-1} \int_t^{\infty} \left(1 - B^{*n}(x)\right) dx.$$

From this expression the result at (7.1.8) follows immediately.

Proposition 7.1.3. *For $k = 1, 2$, suppose the* M/GI/1 *queueing systems Σ_k have the same arrival rate λ and that their service time d.f.s B_k have the same mean m_B. Then with $\varrho = m_B$, if $B_1 \leqq_{cv} B_2$ (equivalently, $B_2 \leqq_c B_1$) the probabilities of emptiness $P_{0,k}(t)$ satisfy*

$$P_{0,2}(t) - (1 - \varrho) \leqq P_{0,1}(t) - (1 - \varrho) \quad (\textit{all } t \geqq 0). \quad (7.1.8)$$

Thus, in Σ_1 the convergence of $P_{0,1}(t)$ to the limit $1 - \varrho$ if $\varrho < 1$ (or to 0 if $\varrho \geqq 1$), is faster than in Σ_2 when B_2 has a heavier tail than B_1.

7.2 Some applications in reliability and renewal theory

7.2.1 A two-unit system with repairable elements

Consider a system Σ consisting of two elements $\mathscr{E}_1$ and $\mathscr{E}_2$. At time $t = 0$, $\mathscr{E}_1$ commences work until failure at $t = X_0$, at which instant repair work on $\mathscr{E}_1$ is started, taking a time Y_1, while $\mathscr{E}_2$ commences work until its failure at time X_1 later. If $X_1 < Y_1$, the system terminates at $X_0 + X_1$, while otherwise $\mathscr{E}_1$ begins work again at $X_0 + X_1$ when repair on $\mathscr{E}_2$ is started.

Successive work times $X_0, X_1, \ldots$ are assumed i.i.d., and independent of the successive repair times $Y_1, Y_2, \ldots$ assumed i.i.d. Defining $N = \inf\{n: X_n < Y_n\}$, the life time T of Σ is the r.v. $T = X_0 + X_1 + \cdots + X_N$.

In the case of exponential work times with parameter λ and arbitrary repair time d.f. R, it is known (see e.g. section 6.2 of GNEDENKO, BELYAYEV and SOLOVYEV (1965)) that

$$\mathsf{E}T = \lambda^{-1}\left(1 + \left(1 - \int_0^\infty e^{-\lambda t}\, dR(t)\right)^{-1}\right). \tag{7.2.1}$$

The functional method can be used to study the influence of the nature of the d.f. R on $\mathsf{E}T$. It is trivially obvious, from Definition 1.8.1, that

Proposition 7.2.1. *For two such 2-unit standby systems as just described with work time parameters λ_1, λ_2, and repair time d.f.s R_1, R_2,*

$$\lambda_1 \leqq \lambda_2 \text{ and } R_1 \leqq_L R_2 \text{ imply } \mathsf{E}T_1 \geqq \mathsf{E}T_2. \tag{7.2.2}$$

Use of (7.2.2) facilitates obtaining bounds on $\mathsf{E}T$, in particular when R is known to belong to a class of d.f.s. KOZLOV and VASIL'EV (1969) considered the case where the first n ($n = 1, 2, 3, 4$) moments of R are specified. If in particular the mean m_R and variance σ_R^2 of repair times are specified, it follows from Section 1.9(c) that

$$1 + (1 - e^{-\lambda m_R})^{-1} \leqq \lambda \mathsf{E}T \leqq 1 + (1 + c_R^2)/\big(1 - \exp(-\lambda m_R(1 + c_R^2))\big) \tag{7.2.3}$$

where $c_R^2 = \sigma_R^2/m_R^2$. Similarly, if R is NBUE with mean m_R, it follows from Section 1.9(d) that

$$1 + (1 - e^{-\lambda m_R})^{-1} \leqq \lambda \mathsf{E}T \leqq 1 + (1 + \lambda m_R)^{-1}. \tag{7.2.4}$$

7.2.2 *Monotonicity properties and bounds for the renewal function*

Consider a system in which an element begins working at time $t = 0$ and fails at $t = X_1$, being replaced immediately by a new one which fails at $t = X_1 + X_2$, being in turn replaced by a new one, and so on. Assume $X, X_1, X_2, \ldots$ are i.i.d. non-negative r.v.s with the d.f. F having $F(0) < 1$ and finite mean $m = \mathsf{E}X = \lambda^{-1}$. Let ν_t denote the number of renewals in $(0, t]$, i.e., $\nu_t = \sup\{n: X_1 + \cdots + X_n \leqq t\}$, so that

$$\mathsf{P}(\nu_t \geqq i) = F^{*i}(t) \qquad (i = 1, 2, \ldots). \tag{7.2.5}$$

The mean of ν_t we denote by $H(t)$, and it is the *renewal function* (sometimes $1 + H(t)$ is called by this name), being the unique solution of the *renewal equation*

$$H(t) = F(t) + \int_0^t H(t - x)\, dF(x), \tag{7.2.6}$$

namely,

$$H(t) = \sum_{n=1}^\infty F^{*n}(t) \tag{7.2.7}$$

(see e.g. Chapter XI of FELLER (1966)). Further,

$$t/m = \lambda t = \sum_{n=1}^{\infty} (F_R * F^{*(n-1)})\,(t) \tag{7.2.8}$$

where

$$F_R(t) = \lambda \int_0^t \bar{F}(x)\,\mathrm{d}x, \qquad F^{*0}(t) = \Theta_0(t).$$

For the renewal function H and ν_t we mention two external monotonicity properties. First, since $\leqq_d$ has the convolution property (C), it follows from (7.2.5) that, with the numbers of renewals $\nu_{t,k}$ corresponding to the d.f.s F_k $(k = 1, 2)$,

$$F_1 \leqq_d F_2 \quad \text{implies} \quad \nu_{t,2} \leqq_d \nu_{t,1} \tag{7.2.9}$$

and hence

$$\mathsf{E}\nu_{t,1} = H_1(t) \leqq H_2(t) = \mathsf{E}\nu_{t,2} \quad (\text{all } 0 < t < \infty). \tag{7.2.10}$$

Second, it is also a consequence of property (C) that we have

Proposition 7.2.2. *If F is* NBUE (NWUE), *then*

$$H(t) \leqq (\geqq)\, \lambda t \quad (0 < t < \infty). \tag{7.2.11}$$

Proof. The NBUE property is equivalent (see (1.6.10′)) to $F_R \leqq_d F$, which by (7.2.8) and property (C) implies that

$$\lambda t = \sum_{n=1}^{\infty} (F_R * F^{*(n-1)})\,(t) \geqq \sum_{n=1}^{\infty} F^{*n}(t) = H(t).$$

The inequalities are reversed for NWUE, yielding (7.2.11).

The renewal function also has the following internal monotonicity property, that if $\{h_n\}$ $(n = 1, 2, \ldots)$ is a sequence of functions defined by specifying h_1, and thereafter by the recurrence relation

$$h_{n+1}(t) = F(t) + \int_0^t h_n(t - x)\,\mathrm{d}F(x) \qquad (\text{all } t > 0), \tag{7.2.12}$$

$h_1(t)$ being bounded on any finite interval and such that the integral at (7.2.12) with $n = 1$ is well-defined, then

$$h_1(t) \leqq (\geqq)\, h_2(t) \qquad (\text{all } t) \tag{7.2.13}$$

implies that for all $t > 0$,

$$h_n(t) \leqq (\geqq)\, h_{n+1}(t) \uparrow (\downarrow)\, H(t) \quad (n \to \infty). \tag{7.2.14}$$

The monotonicity of $\{h_n\}$ is easily shown from (7.2.12) and (7.2.13), while the limit property (and hence, boundedness) is a consequence of the convergence to 0 of $\int_0^t h_1(t - x)\,\mathrm{d}F^{*n}(x)$.

MARSHALL (1973) used the idea behind (7.2.14) to establish

Proposition 7.2.3. *Let A be the set of all $t \geqq 0$ with $\bar{F}(t) > 0$, and define*

$$b_L = \inf_{t \in A} \big(F(t) - F_R(t)\big)/\bar{F}(t), \quad b_U = \sup_{t \in A} \big(F(t) - F_R(t)\big)/\bar{F}(t). \tag{7.2.15}$$

Then for all $0 < t < \infty$

$$\lambda t + b_L \leqq H(t) \leqq \lambda t + b_U. \tag{7.2.16}$$

Proof. It follows from (7.2.15) that

$$b_L \bar{F}(t) \leqq F(t) - F_R(t) \leqq b_U \bar{F}(t).$$

Convolution with F^{*n} and summation over $n = 0, 1, \ldots$ yields (7.2.16).

Since $\lambda t + b_L$ has the property that (7.2.13) holds, it is possible in principle to improve the bound (7.2.16) by iteration as at (7.2.12), and similarly for the upper bound $\lambda t + b_U$. However, this upper bound need not necessarily be finite nor as good as Lorden's bound at (7.2.18) below.

Since $\big(F(t) - F_R(t)\big)/\bar{F}(t) = -1 + \bar{F}_R(t)/\bar{F}(t)$, Proposition 7.2.3 implies the well known bound

$$H(t) \geqq \lambda t - 1 \qquad (\text{all } t \geqq 0) \tag{7.2.17}$$

which can also be established by the sub-additivity of the function $H^{(0)}(t) = H(t) + 1$. Further, it follows from the proposition and the definition that when F is NBUE (NWUE), $b_L \leqq 0$ ($b_U \geqq 0$).

Brown (1980) showed that when F is DFR, the excess r.v. (or, forward recurrence time r.v.)

$$R(t) = S_{1+\nu_t} - t$$

where $S_0 = 0$, $S_n = X_1 + \cdots + X_n$, and ν_t as above (7.2.5), is stochastically increasing, i.e., $R(t) \leqq_d R(u)$ for all $t < u$. If F has only the IMRL property, then only $\mathsf{E}R(t) \leqq \mathsf{E}R(u)$ $(t \leqq u)$ necessarily holds. In either event, since $\lim_{t\to\infty} \mathsf{E}R(t) = \lambda \mathsf{E}X^2/2$, the bound at (7.2.18) can then be tightened. Brown's proof rests on a construction involving two coupled renewal processes. Brown (1981) subsequently showed that $R(t) \leqq_c R(u)$ for $t \leqq u$ when F has IMRL.

For a general upper bound, Lorden (1970) showed that

$$H(t) \leqq \lambda t + \lambda^2 \sigma_X{}^2 \qquad (\text{all } t \geqq 0), \tag{7.2.18}$$

the crucial step involving integration of the excess r.v. $R(t)$. This step yields

$$2 \int_0^t R(u)\,du = \sum_{i=1}^{1+\nu_t} X_i{}^2 - R^2(t),$$

and on taking expectations, using the sub-additivity of $1 + H(t)$, Wald's lemma, and the Cauchy-Schwarz inequality, (7.2.18) results.

This argument contrasts strongly with a Fourier analytic approach of STONE (1972) tightened in DALEY (1978a), and a comparison method for studying the solution of the general renewal equation referred to in DALEY (1976a) and refined by O'BRIEN (unpublished note). These latter papers established results weaker than Lorden's that

$$H(t) \leqq \lambda t + \lambda^2\sigma_X^2 + (C - 1)\,\lambda^2\sigma_X^2 \qquad \text{(all } t)$$

for some constant $C \geqq 1$ and $\leqq 2.86$ (STONE), $\leqq 1.32$ (DALEY), or $\leqq 1.299$ (O'BRIEN).

For any d.f. F of a not necessarily non-negative r.v. having a positive mean, the function

$$H(t) = \sum_{n=1}^{\infty} F^{*n}(t)$$

is still well-defined. DALEY (1980) used ladder variable techniques to extend the validity of (7.2.18) to all such r.v.s, and indeed found the slightly stronger bound at (7.2.18′) below. It should be remarked in connection with (7.2.18′) that the r.v. M there has the same distribution as the stationary waiting time r.v. W in a GI/GI/1 queueing system in which the d.f. of the difference between service and inter-arrival times is 1-$F(.)$. Consequently, the problem of finding upper bounds on $\mathsf{E}M$ is the same as finding lower bounds on m_W as in Chapter 5.

Theorem 7.2.4. *For i.i.d. r.v.s $X, X_1, X_2, \ldots$ with positive mean λ^{-1}, finite second moment, and partial sums $S_0 = 0$, $S_n = X_1 + \cdots + X_n = S_{n-1} + X_n$, the expectation $H(t) = \mathsf{E}\,\#\,\{n = 1, 2, \ldots : S_n \leqq t\}$ satisfies*

$$H(t) \leqq \lambda t + \lambda^2\sigma_X^2 - \lambda \mathsf{E}M \tag{7.2.18′}$$

where $M = -\inf_{n \geqq 0} S_n$.

Lorden's results on higher moments of $R(t)$ were also used by DALEY (1978b) to show for example that the variance function of a stationary renewal process var (ν_t'), where to define ν_t' we replace X_1 by a r.v. X_R independent of the i.i.d. r.v.s $X_2, X_3, \ldots$ and with d.f. F_R, satisfies

$$-4\mathsf{E}(\lambda X)^3/3 \leqq \operatorname{var}(\nu_t') - \operatorname{var}(\lambda X)\, t \leqq \left(\mathsf{E}(\lambda X)^2/2\right)^2. \tag{7.2.19}$$

Monotonicity properties of var (ν_t') with respect to $\leqq_c$ can be proved via

$$\operatorname{var}(\nu_t') = \lambda t - (\lambda t)^2 + 2\lambda \int_0^t (t - u)\, \mathrm{d}H(u).$$

7.2.3 *Reliability systems with coherent structure*

Let Σ be a reliability system which consists of n components, each of which is in one of the two states 0 (down) and 1 (up), and let the states $X_1, \ldots, X_n$ of the components determine the state σ of the system uniquely, so that there exists a $\{0, 1\}$-valued system function Φ say, such that $\sigma = \Phi(X_1, \ldots, X_n)$. While such systems frequently occur in reliability, not all systems can be described in this way. The system in Figure 7.1 can be so described.

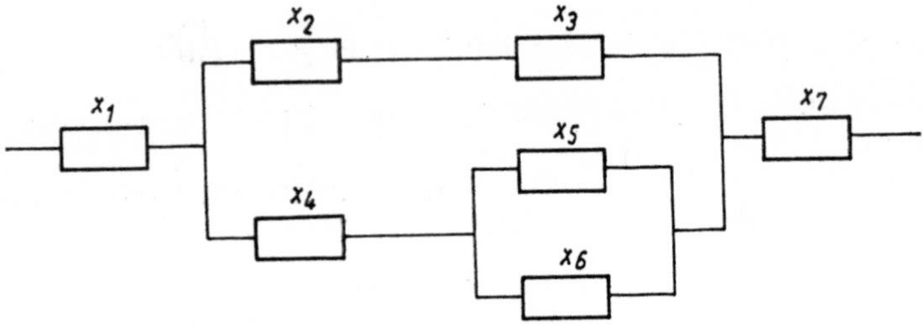

Fig. 7.1. Example of a reliability system of coherent structure

Its system function is

$$\Phi(X_1, \ldots, X_n) = X_1X_7\{X_2X_3 + X_4(1 - X_2X_3)\,[X_5 + (1 - X_5)\,X_6]\}.$$

Systems with system functions Φ having the following properties are said to be *systems with coherent structure* (see BARLOW and PROSCHAN (1975), PROSCHAN (1982)):

(a) $\Phi(0, \ldots, 0) = 0$;
(b) $\Phi(1, \ldots, 1) = 1$;
(c) if $x_i \leqq y_i$ (all $i = 1, \ldots, n$) then $\Phi(x_1, \ldots, x_n) \leqq \Phi(y_1, \ldots, y_n)$.

The system in Figure 7.1 is of coherent structure.

Now suppose that the component states are random and time-dependent, and can be described by the independent stochastic processes $\{X_{1,t}\}, \ldots, \{X_{n,t}\}$, so that the system state σ_t at time t is given by

$$\sigma_t = \Phi(X_{1,t}, \ldots, X_{n,t}).$$

Theorem 2.2.2b can be applied to systems with coherent structure by virtue of property (c) above, leading immediately to

Proposition 7.2.5. *For $k = 1, 2$, let Σ_k be reliability systems having the same coherent structure with component state processes $\{X_{i,t}^{(k)}\}$ $(i = 1, \ldots, n)$ and system state process $\{Z_t^{(k)}\}$. Then if*

$$\{X_{i,t}^{(1)}\} \leqq_d \{X_{i,t}^{(2)}\} \quad (\textit{all } i = 1, \ldots, n), \tag{7.2.20}$$

$$\{Z_t^{(1)}\} \leqq_d \{Z_t^{(2)}\}. \tag{7.2.21}$$

Certain corollaries follow for coherent systems in which all the components are new at $t = 0$ and none of which is replaced when it fails.

Corollary 7.2.5 a. *In Σ_k let the components have lifetime d.f.s $F_i^{(k)}$ $(i = 1, \ldots, n)$ and let G_k be the lifetime d.f. of Σ_k. Then if*

$$F_i^{(1)} \leqq_d F_i^{(2)} \quad (i = 1, \ldots, n), \tag{7.2.20'}$$

$$G_1 \leqq_d G_2 . \tag{7.2.21'}$$

Proof. The relations (7.2.20′) and (7.2.21′) restate the proposition in terms of d.f.s, since we can write

$$\bar{F}_i^{(k)}(t) = \mathsf{P}(X_{i,t}^{(k)} = 1), \quad \bar{G}_k(t) = \mathsf{P}(Z_t^{(k)} = 1).$$

Corollary 7.2.5 b. *If in a system Σ with coherent structure the component lifetime d.f.s have IFR with means $m_1, \ldots, m_n$, then the lifetime d.f. G of the system satisfies*

$$G(t) \leqq G_{\exp}(t) \ \text{for} \ t < \min(m_1, \ldots, m_n) \tag{7.2.22}$$

where $G_{\exp}$ here denotes the lifetime d.f. of a system with the same system function as Σ and exponential lifetime d.f.s with the means $m_1, \ldots, m_n$.

Proof. An IFR d.f. with mean m satisfies $F(t) \leqq 1 - e^{-t/m}$ for $0 \leqq t < m$ (see (1.6.13′)). The truth of (7.2.22) now follows from the proposition as in the previous corollary.

7.2.4 Lifetimes of systems with components in parallel or series and some in reserve

Let Σ be a reliability system whose component lifetimes are the independent r.v.s $X_1, \ldots, X_n$, and assume that no components are repaired.

Suppose first that some components are in series and the remainder are in "cold reserve". Then the lifetime T of Σ can be expressed as a function of the X_i,

$$T = \Psi(X_1, \ldots, X_n) \tag{7.2.23}$$

say, Ψ being constructed by a combination of additions (for components in reserve) and minimizations (for components in series). For example, Figure 7.2 shows a reliability system for which Ψ has the form

$$T = \min\{X_1 + \min(X_2, X_3, X_4), X_5 + X_6, X_7\}.$$

The mapping Ψ in (7.2.23) is nondecreasing and concave in the X_i.

Similarly, for systems with some components in parallel and others in cold reserve, Ψ is constructed by a combination of additions and maximizations (for components in parallel), so Ψ is nondecreasing and convex in the X_i.

In both cases, the mapping method justifies monotonicity statements such as

Proposition 7.2.6. *For $k = 1, 2$, let Σ_k be reliability systems of the same structure (hence, with the same Ψ) with components in series and reserve, the component lifetimes being $X_{i,k}$ ($i = 1, \ldots, n$). Then for the corresponding system lifetimes T_k,*

$$X_{i,1} \leqq_d (\leqq_{cv}) X_{i,2} \quad (i = 1, \ldots, n) \quad \textit{imply } T_1 \leqq_d (\leqq_{cv}) T_2. \tag{7.2.24'}$$

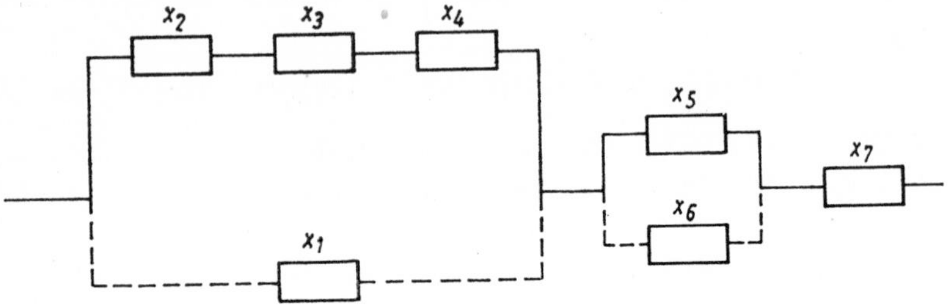

Fig. 7.2. Example of a reliability system with components in series (—) and cold reserve (– – –); lifetime $T = \min\{X_1 + \min(X_2, X_3, X_4), X_5 + X_6, X_7\}$

For systems with components in parallel and reserve,

$$X_{i,1} \leqq_d (\leqq_c) X_{i,2} \quad (i = 1, \ldots, n) \quad \textit{imply } T_1 \leqq_d (\leqq_c) T_2.. \tag{7.2.24''}$$

This result enables us to obtain bounds on the lifetime of systems with components in either parallel or series (but not both) and with reserves, the relations $\leqq_c$ and $\leqq_{cv}$ being of more interest than $\leqq_d$ because in using them it is possible to make comparisons via r.v.s with equal means.

7.3 Pert problem

Let T denote the duration of a project which can be described by a network $\mathcal{N}$ with the random independent activity durations $X_1, \ldots, X_n$,

$$Z = \Psi_{\mathcal{N}}(X_1, \ldots, X_n) \tag{7.3.1}$$

say. The mapping $\Psi_{\mathcal{N}}$ essentially depends on the structure of $\mathcal{N}$: it is generated by a combination of additions (activities in sequence) and maximizations (parallel activities). For example, for the network depicted in Figure 7.3,

$$T = \Psi_{\mathcal{N}}(X_1, \ldots, X_7) = \max\{\max(X_1 + X_5, X_2) + X_6, X_3, X_4 + X_7\}.$$

For arbitrary networks generated in this way, the mappings $\Psi_{\mathcal{N}}$ are non-decreasing and convex, so by appealing to the mapping method we have

Proposition 7.3.1. *For $k = 1, 2$, let $\mathcal{N}_k$ be networks of the same structure (so $\Psi_{\mathcal{N}_1} = \Psi_{\mathcal{N}_2}$) having activity durations $X_{i,k}$ ($i = 1, \ldots, n$). Then for the project durations T_k,*

$$X_{i,1} \leqq_d (\leqq_c) X_{i,2} \; (i = 1, \ldots, n) \; \textit{imply } T_1 \leqq_d (\leqq_c) T_2. \tag{7.3.2}$$

A particular case of this result is the known fact (see ROBILLARD and TRAHEN (1976)) that, when the mean activity durations are given, the least possible value for $\mathsf{E}T$ occurs when the X_i are constant (recall (1.3.5)). Proposition 7.3.1 also enables us, for example, to study the influence on T of the choice of different beta distributions, and to justify the use of normal d.f.s in obtaining approximate solutions of Pert problems, by interpreting the normal d.f.s as bounds with respect to $\leqq_c$ on the true activity d.f.s.

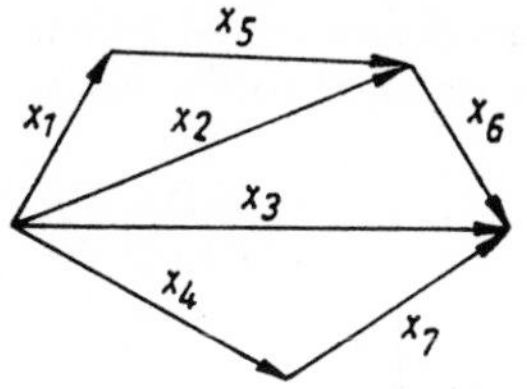

Fig. 7.3. Example of network with activities in sequence or in parallel

In HANISCH and STOYAN (1978) the use of uniformly distributed r.v.s to obtain upper bounds on $\mathsf{E}T$ via $\leqq_c$ is discussed; it can be described briefly as follows, using $U(a, b)$ to denote the d.f. of a r.v. uniformly distributed on the interval $[a, b]$. Suppose the d.f. F_i of each X_i in (7.3.1) is replaced by a uniform d.f. $U(a_i, b_i)$ such that $\mathsf{E}X_i = (a_i + b_i)/2$ and $F_i \leqq_c U(a_i, b_i)$. Then replace the d.f.s of r.v.s generated by maximization and addition of activity durations by $\leqq_c$-larger uniform d.f.s, ultimately approximating (bounding) the d.f. of T by a uniform d.f. In the process of doing this the following inequalities are useful:

(a) $U(a_1, b_1) * U(a_2, b_2) \leqq_c U(a_1 + a_2, b_1 + b_2)$;

(b) with V denoting the d.f. of max (X_1, X_2) where X_k are independent r.v.s with d.f.s $U(a_k, b_k)$, and $a_1 \leqq a_2$ without loss of generality

$$V \leqq_c U(c, d)$$

where

$$c = a_2 \quad \text{if} \quad b_1 \leqq b_2,$$

$$c < a_2 \text{ and } d = b_1 \text{ if } b_1 > b_2 \text{ and } (b_1 - b_2)(a_2 - a_1) > (b_2 - a_2)^2,$$

$$c = a_1 \text{ otherwise},$$

and $c + d = 2\mathsf{E}V$ then sufficing to determine the pair (c, d); note that

$$\mathsf{E}V - (a_2 + b_2)/2$$
$$= \begin{cases} 0 & \text{if } b_1 < a_2, \\ (b_1 - a_2)^3/6(b_1 - a_1)(b_2 - a_2) & \text{if } a_2 \leqq b_1 \leqq b_2, \\ [(b_2 - a_2)^2 + 3(b_1 - b_2)(b_1 - a_2)]/6(b_1 - a_1) & \text{if } b_2 < b_1. \end{cases}$$

It can be verified (e.g., by arguing graphically from plots of the tails of the d.f.s of V and $U(c, d)$) that $c = a_2$ when $b_2 \geqq b_1$, and also for some values of $b_2 < b_1$: $d = b_1$ and $c < a_2$ only if both $b_2 < b_1$ and $(b_1 - b_2)(a_2 - a_1) > (b_2 - a_2)^2$.

7.4 Monotonicity properties of inventory models

Most inventory models have an inventory policy that is demand-adapted in the sense that when the inventory is "small", a "large" order for re-stocking is made. Consequently we cannot in general expect that the inventory level Z_n at the beginning of the nth time period will depend monotonically on the demands $X_0, \ldots, X_{n-1}$ made in the preceding periods. For example, in an (s, S) inventory model with immediate delivery, meaning that the inventory level is restored to S so soon as it falls below s, we have the recurrence relation

$$Z_{n+1} = \Phi(Z_n, X_n) \tag{7.4.1}$$

where

$$\Phi(z, x) = \begin{cases} z - x & \text{if } z - x \geqq s, \\ S & \text{if } z - x < s. \end{cases}$$

This function Φ does not satisfy the conditions of Theorem 2.2.6, and it is easy to give examples showing that we can have

$$X_{i,1} \leqq_d X_{i,2} \quad (i = 0, \ldots, n-1) \tag{7.4.2}$$

without also having

$$Z_{n,1} \leqq_d Z_{n,2}. \tag{7.4.3}$$

On the other hand, we do have external monotonicity properties for the inventory level in certain "uncontrolled" inventory models, as for example for some dam models. We mention Moran's dam model (see e.g. section III. 5.1 of Cohen (1969)) in which a dam of capacity K with content Z_n at t_n has input X_n then, followed by a demand in (t_n, t_{n+1}) of Y_n, so that $\{Z_n\}$ satisfies the recursion formula

$$Z_{n+1} = \max\{\min(K, Z_n + X_n) - Y_n, 0\} \quad (n = 0, 1, \ldots). \tag{7.4.4}$$

With respect to the ordering $\leqq_d$, the content Z_n is nondecreasing in X_n and nonincreasing in Y_n, while the cumulative overflow $O_k = \sum_{n=1}^{k} (Z_n + X_n - K)_+$ is $\leqq_c$-monotone in X_n (see Stoyan (1972d)).

Monotonicity properties can also be proved for stochastic clearing systems involving Markov chains of the form

$$Z_{n+1} = \max(Z_n - Y_n, 0) + X_n \tag{7.4.5}$$

where Z_n is the content at t_n, Y_n the demand at t_n, and X_n the input between t_n and t_{n+1}. Of course, such a model is a disguised variant of a G/G/1 system, since the "residual stock" $V_n = \max(Z_n - Y_n, 0)$ after the nth clearance, of size $\min(Y_n, Z_n)$, satisfies $V_{n+1} = \max(V_n + X_n - Y_{n+1}, 0)$.

Using $\leqq_{cv}$, Whitt (1981b) studied the monotonicity properties of the stationary distributions of continuous time clearing processes.

Daley and Haslett (1982) studied a Markov chain $\{Z_n\}$ satisfying

$$Z_{n+1} = \max(Y_{n+1} + abZ_n, bZ_n) \tag{7.4.6}$$

for non-negative i.i.d. r.v.s $\{Y_n\}$ and $0 \leqq a \leqq 1$, $0 < b < 1$, and used a variety of the methods used to bound m_W in GI/GI/1 to bound the mean $\mathsf{E}Z$ of a stationary process satisfying (7.4.6). This relation arises as a simple thermal energy storage model with controlled input.

Notwithstanding our remarks above on the lack of monotonicity of certain inventory processes $\{Z_n\}$, the first stochastic models for which monotonicity properties were proved were inventory models. KARLIN (1960) and VEINOTT (1965) studied the effect on inventory processes of varying the demand d.f.s while keeping the other model parameters fixed, and showed that the total costs over n periods and the critical numbers (meaning the inventory levels below which orders are made) are nonincreasing and nondecreasing respectively in these d.f.s with respect to $\leqq_d$ (see also p. 63 in HOCHSTÄDTER (1969)).

7.5 Monotonicity properties and bounds for service systems with zero delays

7.5.1 Pure loss systems

For pure loss service systems, such monotonicity properties as have been found have mostly been given only where explicit formulae have been available, and consequently it has been possible to use the functional method. We use p_v to denote the stationary probability that an arriving customer finds all servers busy and is therefore lost.

For systems of the type GI/GI/1/0 it is not difficult to check that

$$p_v = 1 - \left[\int_0^\infty \left(\sum_{n=0}^\infty A^{*n}(t)\right) \mathrm{d}B(t)\right]^{-1}. \tag{7.5.1}$$

The integrand here is recognizable as $1 + H_A(.)$ where H_A is the renewal function as in Section 7.2.2, so the bounds given there are applicable to (7.5.1). For example, the lower bound at (7.2.17) and upper bound at (7.2.18) show that

$$1 - 1/\varrho \leqq p_v \leqq 1 - 1/(1 + \varrho + \lambda^2\sigma_B{}^2) \tag{7.5.2}$$

where $\lambda^{-1} = m_A$ and $\varrho = \lambda m_B$, while if A is NBUE (NWUE), (7.2.11) shows that the upper (lower) bound at (7.5.2) can be tightened to

$$p_v \leqq 1 - 1/(1 + \varrho) = \varrho/(1 + \varrho) \quad \bigl(\varrho/(1 + \varrho) \leqq p_v\bigr).$$

For systems of the type M/GI/1/0, (7.5.1) shows that

$$p_v = \varrho/(1 + \varrho).$$

For systems Σ_k ($k = 1, 2$) of the type GI/GI/1/0 and interarrival (service) time d.f.s A_k (B_k),

$$A_2 \leqq_d A_2 \quad \text{and} \quad B_1 \leqq_d B_2 \quad \text{imply} \quad p_{v,1} \leqq p_{v,2}, \tag{7.5.3}$$

while if H_A is convex or if B is concave then monotonicity properties with respect to $\leqq_c$ or $\leqq_{cv}$ respectively can be established.

For systems of the type GI/M/s/0, we have

$$p_v = \left(\sum_{i=0}^{s} \binom{s}{i} \gamma_i \right)^{-1}$$

where with $\alpha(.)$ denoting the Laplace-Stieltjes transform of $A(.)$ and $\mu^{-1} = m_B$, $\gamma_0 = 1$ and $\gamma_i = \gamma_{i-1}\alpha(i\mu)/(1 - \alpha(i\mu))$ $(i = 1, 2, \ldots)$. Since the function $f(x) = x/(1 - x)$ is nondecreasing in x, the statement at (7.5.3) also holds for GI/M/s/0 systems as well as the weaker assertion (recall the properties of $\leqq_L$ in Sections 1.8 and 1.9) that

$$A_2 \leqq_L A_1 \quad \text{and} \quad B_1 \leqq_d B_2 \quad \text{imply} \quad p_{v,1} \leqq p_{v,2}. \tag{7.5.4}$$

Since for any d.f. A with mean m, $A \leqq_L \Theta_m$, it follows from (7.5.4) that, in the class of GI/M/s/0 systems with given mean interarrival time, the loss probability p_v is least in D/M/s/0, a result due originally to Beneš (1959). The $\leqq_L$-extremal d.f.s in the class $\mathcal{M}_m^\sigma$ of d.f.s with mean m and variance σ^2 are known (see Section 1.9) so bounds on p_v for A in $\mathcal{M}_m^\sigma$ and specified μ can be computed (cf. Holtzmann (1973)).

Kirstein (1978) proved that the embedded Markov chain of GI/M/s/0, which in any case is stochastically monotone, is (for suitable initial distributions) nondecreasing and $\leqq_d$-comparable for $\leqq_d$-comparable interarrival time d.f.s and service distributions. The conclusion at (7.5.3) is then just a particular case of this comparability.

Kokotushkin (1974) considered GI/M/s/0 systems in which the arrival process is the so-called interrupted Poisson process input in which potential arrivals come from a Poisson process of intensity λ, there being no arrivals when an independent $\{0, 1\}$-valued alternating renewal process in the 0-state, whose durations have an arbitrary d.f. G with mean m, while arrivals during the 1-state, which has a negative exponential distribution with mean $1/\nu$, are not censored. The interarrival times are independent with mean $(1 + \nu m)/\lambda = 1/\lambda_I$ say, and the Laplace-Stieltjes transform of their common d.f. A_I is

$$\int_0^\infty \mathrm{e}^{-\theta t} \,\mathrm{d}A_I(t) = \lambda/\big(\lambda + \theta + \nu(1 - \gamma(\theta))\big) \quad (\mathrm{Re}\,(\theta) \geqq 0)$$

where $\gamma(.)$ is the L-S.T. of G. Since $1 - \gamma(\theta) \leqq \theta m$ for $\theta \geqq 0$, it follows that

$$A_I \leqq_L \mathrm{Exp}\,(\lambda_I), \tag{7.5.5}$$

and therefore by (7.5.4) the loss probability of such a system is greater than in M/M/s/0 with the same mean interarrival and service times.

Turning to systems with non-renewal inputs, monotonicity properties are known only when the service time d.f. is exponential, with mean $1/\mu$ say. In G/M/1/0, suppose the d.f. of the interarrival time $t_n - t_{n-1}$ is A_n, so by the loss of memory property of the service d.f., the probability that the arrival at t_n is lost equals

$$p_n = \int_0^\infty e^{-\mu t}\, dA_n(t) \qquad (n = 1, 2, \ldots),$$

it being assumed that the service times are i.i.d. and independent of the input. Then it is clear that

$$A_{n,1} \leqq_{cv} A_{n,2} \quad \text{or} \quad A_{n,1} \leqq_L A_{n,2} \quad \text{implies} \quad p_{n,1} \geqq p_{n,2}. \tag{7.5.6}$$

(By (1.8.4), we could just as easily omit $\leqq_{cv}$ here.) STOYAN (1973c) and FLEISCHMANN (1976) extended these $\leqq_{cv}$-monotonicity properties to G/M/1/n and G/M/2/0 respectively. It is an old conjecture that Beneš' result (below (7.5.4)) is also true for G/M/s/0 systems in which the input is a stationary point process.

Added in proof. Further bounds for loss systems are presented in the papers HALFIN (1981), HEYMAN (1980), NIU (1980b) and SOBEL (1980).

7.5.2 Systems with infinitely many servers

Consider a GI/GI/∞ system in which there is an arrival at $t = 0$ into an otherwise empty system. Let ν_n be the number of customers in the system immediately before the arrival at t_n. With $I(A)$ denoting the indicator r.v. for the event A, we have for $n = 0, 1, \ldots$ that

$$\begin{aligned} \nu_{n+1} &= I(\beta_n > \alpha_n) + I(\beta_{n-1} > \alpha_n + \alpha_{n-1}) + \cdots + I(\beta_0 > \alpha_n + \cdots + \alpha_0) \\ &\equiv \Phi_n(\alpha_0, \ldots, \alpha_n; \beta_0, \ldots, \beta_n) \end{aligned} \tag{7.5.7}$$

say. This mapping Φ_n is non-increasing in the α_i and nondecreasing in the β_i. Then the mapping method shows that for GI/GI/∞ systems Σ_k $(k = 1, 2)$,

$$A_2 \leqq_d A_1 \quad \text{and} \quad B_1 \leqq_d B_2 \quad \text{imply} \quad \nu_{n,1} \leqq_d \nu_{n,2} \quad (n = 1, 2, \ldots). \tag{7.5.8}$$

By reasoning similar to that used in Section 6.3.1.2, the numbers ν_t and ν_t^∞ of customers at time t in the systems GI/GI/s/0 and GI/GI/∞ with the same interarrival and service times and the same initial state satisfy

$$\nu_t \leqq \nu_t^\infty, \tag{7.5.9}$$

and this inequality can be used to obtain bounds for loss systems. For example, (6.3.23) yields a bound for the time dependent state probabilities in M/GI/s/0.

7.6 A heuristic approximation for open queueing networks

By an *open queueing network* we shall understand a service system such as is sketched in Figure 7.4 (for a general description of such networks, see e.g. KELLY (1980)). The nodes

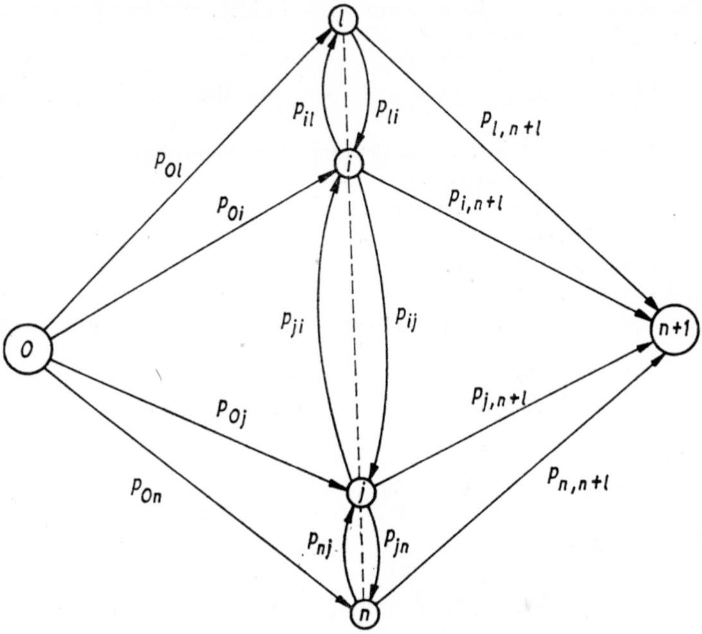

Fig. 7.4. Example of an open queueing network

N_i ($i = 1, \ldots, n$) of the graph represent FCFS queueing systems consisting of s_i ($1 \leqq s_i \leqq \infty$) servers with a common queue with infinite waiting room capacity, the service times at N_i being i.i.d. with common d.f. B_i for all s_i servers, mean $m_{B,i}$, variance $\sigma^2_{B,i}$, and these service times independent of all other nodes. Thus, N_i is a $G/GI/s_i$ ($i = 1, \ldots, n$) queueing system. On completing its service at N_i, a customer proceeds immediately to node N_j with probability p_{ij}. There is a (possibly fictitious) source node N_0 from which customers come at the epochs of a Poisson process at rate λ, proceeding immediately to node N_i with probability p_{0i} ($i = 1, \ldots, n$). Similarly, there is a (possibly fictitious) sink node N_{n+1} by which customers leave the network, the probability of a transition from N_i to N_{n+1} being $p_{i,n+1}$ ($i = 1, \ldots, n$). We assume that the matrix (p_{ij}) ($i, j = 1, \ldots, n + 1$) is the transition matrix of a homogeneous Markov chain having $n + 1$ as an absorbing state and no other closed subclass.

In view of this last assumption, the system of equations

$$y_i = p_{0i} + \sum_{j=1}^{n} y_j p_{ji} \qquad (i = 1, \ldots, n) \tag{7.6.1}$$

will have a unique non-negative solution y_i $(i = 1, \ldots, n)$, and provided that

$$\lambda y_i m_{B,i} < s_i \qquad (i = 1, \ldots, n),$$

the system then has a stationary state for which

$$\lambda_i \equiv \lambda y_i \qquad (i = 1, \ldots, n) \tag{7.6.2}$$

is the rate at which customers enter node N_i. Any further information concerning the network is known exactly only for all B_i exponential d.f.s or else $s_i = \infty$ (see GELENBE and MUNTZ (1976) and KELLY's (1980) monograph). Approximations can be found via diffusion techniques (see e.g. KOBAYASHI (1974), GELENBE and PUJOLLE (1976)), and via decompositions (for example, by assuming that each node has a Poisson input at rate λ_i, as in KLEINROCK (1964)).

Another method of approximation is to use the approach of Section 6.3.2.2. STOYAN (1978a) showed that if at each node the FCFS discipline is replaced by the pre-emptive resume modified LCFS discipline introduced in Section 6.3.2.2 for M/GI/s queues, the queueing networks as above are insensitive to the B_i (apart from the mean $m_{B,i}$). Arguing as for (6.3.37), this then suggests that

$$p_{W,i} \approx p_W(\mathrm{M/M}/s_i) \tag{7.6.3}$$

$$m_{W,i} \approx {}^1/_2(1 + \sigma^2_{B,i}/m^2_{B,i})\, m_W(\mathrm{M/M}/s_i), \tag{7.6.4}$$

where $p_{W,i}$ and $m_{W,i}$ denote the stationary probability of waiting and the mean waiting time of arrivals in N_i, and the other quantities are as around (6.3.37). The accuracy of these approximations is similar to that of diffusion approximations. Similar arguments can be applied to networks with totally or partially deterministic routes for customers, the insensitivity results like (7.6.2) to (7.6.4) again holding (see SCHASSBERGER (1978), JANSEN and KÖNIG (1980) and KELLY (1979)).

7.7 Monotonicity properties and bounds for Galton-Watson processes

For Galton-Watson processes several bounds and monotonicity properties are known, and the applications below have been selected primarily as being related to the methods of this monograph. For our purposes, a Galton-Watson process is a Markov chain $\{Z_n\}$ defined by

$$Z_0 = 1,\ Z_{n+1} = \sum_{i=1}^{Z_n} X_i^{(n)} \qquad (n = 0, 1, \ldots) \tag{7.7.1}$$

where Z_n denotes the number of individuals in the nth generation, the ith such individual (when $Z_n > 0$) having $X_i^{(n)}$ offspring, all such offspring comprising the $(n+1)^{\text{th}}$ generation. The $X_i^{(n)}$ are assumed to be i.i.d. with common distribution $\{p_j\}$, mean m, and probability generating function (p.g.f.) g so that, when X is a generic offspring r.v.,

$$g(s) = \mathsf{E}(s^X) = \sum_{j=0}^{\infty} p_j s^j \qquad (|s| \leqq 1),$$

and $m = g'(1)$. The p.g.f. g_n of Z_n is then given by $g_0(s) = s$ and for $n = 1, 2, \ldots$,

$$g_n(s) = \mathsf{E}(s^{Z_n}) = g_{n-1}(g(s)) = g(g_{n-1}(s)) \; (|s| \leqq 1, n = 1, 2, \ldots) \qquad (7.7.2)$$

(see e.g. Harris (1963) and Jagers (1975)). The *extinction time*

$$T \equiv \inf \{n: Z_n = 0\}$$

(with $T = \infty$ if $Z_n > 0$ for all n) then has distribution

$$\mathsf{P}(T \leqq n) = \mathsf{P}(Z_n = 0) = g_n(0) \equiv q_n \qquad (7.7.3)$$

and the *extinction probability* $q = \mathsf{P}(T < \infty)$, is a root of the equation

$$s = g(s), \qquad (7.7.4)$$

being equal to 1 for $m \leqq 1$ (except when $g(s) \equiv s$, i.e., $p_1 = 1$), and for $m > 1$ is less than one and is in fact the smaller non-negative root of (7.7.4).

P.g.f.s have been a valuable tool for studying Galton-Watson processes (cf. Seneta (1969)), and for non-negative integer-valued r.v.s X and Y with d.f.s F and G there is the partial ordering $\leqq_g$ defined in section 1.8 by

$$X \leqq_g Y \text{(equivalently, } F \leqq_g G\text{) when } \mathsf{E}(s^X) \geqq \mathsf{E}(s^Y) \text{ (all } 0 \leqq s \leqq 1). \qquad (7.7.5)$$

Of the properties cited there recall in particular that

$$X \leqq_g Y \text{ implies } \mathsf{P}(X = 0) \geqq \mathsf{P}(Y = 0) \text{ and } \mathsf{E}X \leqq \mathsf{E}Y \leqq \infty. \qquad (7.7.6)$$

It is easily checked that the Markov chain $\{Z_n\}$ is stochastically monotone, and it follows from (7.7.2) and the remarks around (7.7.1) that when $m \leqq 1$,

$$Z_{n+1} \leqq_g Z_n, \qquad (7.7.7)$$

i.e., Z_n is then $\leqq_g$-decreasing. If $m \geqq 1$, it follows from Proposition 2.2.5 that then

$$Z_n \leqq_c Z_{n+1}. \qquad (7.7.8)$$

External monotonicity properties are as follows.

Proposition 7.7.1. *For two Galton-Watson processes* $\{Z_n^{(k)}\}$ $(k = 1, 2)$ *with* $Z_0^{(k)} = 1$, *and with* $\prec$ *denoting one of* $\leqq_d, \leqq_c, \leqq_g$,

$$Z_1^{(1)} \prec Z_1^{(2)} \quad \text{implies} \quad Z_n^{(1)} \prec Z_n^{(2)}, \qquad (7.7.9)$$

and if $m \leqq 1$, *the extinction times* $T^{(k)}$ *satisfy*

$$Z_1^{(1)} \leqq_g Z_1^{(2)} \quad \text{implies} \quad T^{(1)} \leqq_d T^{(2)}. \qquad (7.7.10)$$

Proof. When $\prec$ denotes $\leqq_d$ or $\leqq_c$, (7.7.9) follows from Theorem 2.2.5, and for $\leqq_g$ it follows from (7.7.2) and the monotone nature of a p.g.f. on $0 \leqq s \leqq 1$. (7.7.10) follows from (7.7.3) and (7.7.9) in the case that $\prec$ denotes $\leqq_g$.

The implication at (7.7.10) is the starting point for approximations to the extinction time d.f. in work of SENETA (1967) and AGRESTI (1974). Fractional linear p.g.f.s (i.e., f.l. p.g.f.s),

$$f(b, c; s) = 1 - b/(1 - c) + bs/(1 - cs) \quad (|s| \leqq 1), \tag{7.7.11}$$

are used because their n^{th} functional iterates, such as occur in (7.7.2), can be explicitly stated. The f.l. p.g.f. at (7.7.11) arises from the zero-modified geometric offspring distribution $\{p_j\}$ for which

$$p_j = bc^{j-1}\ (j = 1, 2, \ldots),\ p_0 = (1 - b - c)/(1 - c),$$

where $0 \leqq b \leqq 1$, $0 \leqq c < 1$, $b + c \leqq 1$. The first two factorial moments are

$$m = f'(b, c; 1) = b/(1 - c)^2, \quad f''(b, c; 1) = 2bc/(1 - c)^3.$$

It is not difficult to verify that for $m \neq 1$,

$$f_n(b, c; s) = 1 - \frac{m^n(s_0 - 1)}{s_0 - m^n} + \frac{m^n[(s_0 - 1)/(s_0 - m^n)]^2\, s}{1 - s(1 - m^n)/(s_0 - m^n)}, \tag{7.7.12}$$

$$s_0 = (1 - b - c)/c(1 - c), \tag{7.7.13}$$

and the much simpler expression for $m = 1$, since then $s_0 = 1$,

$$f_n(b, c; s) = \frac{nc - [(n + 1)\, c - 1]\, s}{1 + (n - 1)\, c - ncs}. \tag{7.7.14}$$

AGRESTI (1974) showed by bounding a p.g.f. g for which $m < 1$ and $g''(1) < \infty$, that for $n = 1, 2, \ldots$,

$$\frac{[(1 - m)/g''(1)]\, m^{n+1}}{1 + m(1 - m)/g''(1) - m^n} \leqq \mathsf{P}(T > n) \leqq \frac{D(1 - m)\, m^n}{1 + D(1 - m) - m^n} \tag{7.7.15}$$

where $D = \max\big(2, m/(1 - m + p_0)\big)$, and hence deduced a more complex bound for $\mathsf{E}T$. AGRESTI gave formulae for the supercritical case by means of a duality argument. Also, for the Poisson offspring distribution with mean $m \leqq 1$, he showed that the best upper and lower bounding f.l. p.g.f.s have the parameters

$$b = (1 - e^{-m})^2/m, \qquad c = (e^{-m} - 1 + m)/m \quad \text{(upper)}, \tag{7.7.16}$$

$$b = 4m/(m + 2)^2, \qquad c = m(m + 2) \quad \text{(lower)}. \tag{7.7.17}$$

HWANG and WANG (1979) established the nature of the best upper and lower bounding f.l. p.g.f.s for a wider class of p.g.f.s.

Let g_k ($k = 1, 2$) be the offspring p.g.f.s of Galton-Watson processes with $m_k = g_k'(1) > 1$, and let q_k be their extinction probabilities. It is then an easy consequence of (7.7.4) that

$$g_1(s) \leqq g_2(s) \quad (\text{all } 0 \leqq s \leqq 1) \quad \text{implies} \quad q_1 \leqq q_2. \tag{7.7.18}$$

This statement is the basis of various inequalities that have been given for q for offspring p.g.f.s in classes defined by the first two or three moments. For example (see QUINE (1976)), since a p.g.f. g with first two factorial moments $m > 1$ and $\beta = g''(1)$ satisfies, for $0 \leqq s \leqq 1$,

$$\begin{aligned} g(s) &= 1 + m(s - 1) + \beta(s)\,(s - 1)^2/2 \\ &\leqq 1 + m(s - 1) + \beta(s - 1)^2/2 \end{aligned} \tag{7.7.19}$$

where $0 \leqq \beta(s) \leqq \beta$, it follows that

$$q \leqq 1 + m(q - 1) + \beta(q - 1)^2/2$$

and hence that

$$1 - q \geqq 2(m - 1)/\beta. \tag{7.7.20}$$

This inequality is in fact a weaker one than a result of TURNBULL (1973) who showed by finding the extremal p.g.f. (cf. Section 1.9) with first two factorial moments m and β, that

$$q \leqq q_{\max}(m, \beta) \tag{7.7.21}$$

where $q_{\max}$ is the root in (0, 1) of $q_{\max} = f_{m,\beta}(q_{\max})$,

$$f_{m,\beta}(s) \equiv 1 - p_j - p_{j+1} + p_j s^j + p_{j+1} s^{j+1} \tag{7.7.22}$$

where j = largest integer $\leqq 1 + \beta/m$, $p_j = m - \beta/j$, and $p_{j+1} = \big(\beta - m(j - 1)\big)/(j + 1)$, so $p_{j+1} = 0$ if β/m is an integer.

NARAYAN (1981) has given the tightest lower bound $q_{\min}$ that depends on the first three moments. DALEY and NARAYAN (1980) also give some further bounds, as well as clarifying an incomplete argument in QUINE (1976).

One technique not considered here bounds p.g.f.s by polynomials that need not themselves be p.g.f.s (see BRAUN (1975) and BROOK (1966)).

7.8 A bibliographic table of further applications

In the Table 7.2 there are given some further references to papers in which monotonicity properties are established or bounds derived by methods similar to those of this monograph.

SIEGEL (1978), STOYAN and STOYAN (1980) and STOYAN (1980) have studied problems of stochastic geometry by means of partial orderings. In the second paper partial orderings of random sets are considered and such partial orderings as enable the comparison of real-valued r.v.s with given

n^{th} moments, these being needed for the comparison of random bodies of equal n-dimensional volume.

Table 7.2. Further monotonicity results

Model	System Quantity	Partial Orderings	References
G/M/∞	Binomial moments of number of customers	$\leqq_d$, $\leqq_{cv}$	KERSTAN and FRANKEN (1968), STOYAN and STOYAN (1972), KIRSTEIN (1978)
GI/GI/1 with warm-up time after idle periods	Waiting time d.f. and its moments	$\leqq_d$, $\leqq_c$	ROSSBERG and SIEGEL (1974a, b), SIEGEL (1974), KREININ (1979a)
Cyclic single-server system	Waiting time d.f.	$\leqq_d$, $\leqq_c$	STOYAN (1972c)
Queues with finite waiting room	Number of customers	$\leqq_d$	SONDERMAN (1979a, b)
Repairman model with one server	Utilisation of machines	$\leqq_d$, $\leqq_{cv}$	ROLSKI and STOYAN (1974), KIRSTEIN (1978)
Systems of 2 components, one in warm reserve	Mean life time, availability	$\leqq_L$	FRANKEN and HEUSER (1977)
System of $n + 1$ components with one repair unit and n components in hot reserve	Availability	$\leqq_d$	GAEDE (1974a, b)

The partial orderings $\leqq_d$, $\leqq_c$, and $\leqq_{cv}$ (and variations of them) have been employed as tools in the theory of economic decision making: see BRUMELLE and VICKSON (1975), KOFLER and MENGES (1976), LEVHARI, PARONSH and PELEG (1975), ROTHCHILD and STIGLITZ (1970, 1971), VICKSON (1977), and references therein. KALL and STOYAN (1982) used $\leqq_c$ for constructing two-sided error bounds for stochastic programming problems.

Notes and bibliographic comments to Chapter 7

7.1

All bounds given for the rate of convergence are based on the functional method. It may be asked whether it is possible to prove the qualitative part of the statements (for example, that the convergence is faster for d.f.s with smaller tails) by a method similar to the mapping method.

KINGMAN (1962) proved bounds of the form

$$m_W - m_{W,n} \leqq ca^{n+1}/(n+1) \qquad (n = 0, 1, \ldots)$$

for certain constants $c > 0$ and $0 < a < 1$. For bounds on the rate of convergence see also Section 5.2 and BOROVKOV (1972, 1976a, 1980).

An inequality similar to (7.1.4) is true for the variances σ_W^2 (see BERGMANN, DALEY, ROLSKI, and STOYAN (1979)).

7.2

MARSHALL and PROSCHAN (1972) gives a good coverage of implications in renewal and replacement theory of the notions NBUE (NWUE) etc. In MARSHALL and PROSCHAN (1970) they gave Proposition 7.2.6, and this was probably the first application of $\leqq_c$ to a reliability problem.

EL-NEWEIHI, PROSCHAN and SETHURAMAN (1978) and EL-NEWEIHI and PROSCHAN (1978) studied qualitative properties of coherent systems at length; see also BUTLER (1982).

WALK (1975) showed that for every $\varepsilon > 0$ there exists a constant $C(\varepsilon)$ such that

$$H(t) \leqq (\lambda + \varepsilon)\, t + C(\varepsilon) \; (t \geqq 0),$$

and used this with (7.2.18) to give an elementary proof of the "elementary renewal theorem" (that $H(t)/\lambda t \to 1$ $(t \to \infty)$). Further bounds and approximations for reliability systems with renewal can be found in the papers of SOLOV'EV and coworkers (see KOZLOV and SOLOV'EV (1978) for further references). SIEGEL and WÜNSCHE (1979) have surveyed the literature on bounds for $H(t)$.

WALDMANN (1980) used methods similar to MARSHALL (1973) to obtain bounds on $H(t)$.

Open problem. Do relations similar to those of Proposition 7.2.1 also hold for the life time d.f.s?

7.3

Comparison with Section 7.2.4 shows that every network plan is equivalent to a reliability system with components in parallel and cold reserve, so the methods used in investigating such reliability systems can be applied to the study of Pert networks.

MEILIJSON and NADAS (1979) have also discussed applications of $\leqq_c$ to Pert problems.

7.5

ROGOZIN (1966) gave bounds for p_v in GI/GI/1/0, and he considered its extremal values in the class of service time d.f.s with given first and second moments. A statement similar to (7.5.4) with respect to $\leqq_{cv}$ is true for the case of a concave service time d.f. (see also KERSTAN and FRANKEN (1969)). Bounds for p_v in GI/M/s/0 are given in BOROVKOV (1972), while KIRSTEIN (1978) showed that when the inter-arrival times are $\leqq_L$-comparable, so also are the interdeparture times. BENEŠ (1965) proved inequalities for networks of loss systems. ROSS (1978) showed that (7.5.5) may be sharpened to Exp $(\lambda_I) \leqq_c A_I$ (equivalently, $A_I \leqq_{cv}$ Exp (λ_I)).

Open problems 1. Is (7.5.2) true for multi-server loss systems of the type GI/GI/s/0?

2. Is it true that for GI/GI/s/0 loss systems with NBUE interarrival times,

$$p_v \leqq \varrho^s \bigg/ \left(s! \sum_{i=1}^{s} \varrho^i/i! \right) ?$$

The truth of this inequality was asserted on p. 134 of the German edition, but the proof relied on an incorrect formula of Le Gall.

3. Is it possible to give useful bounds for the difference $\nu_t - \nu_t^\infty$?

7.6

A further example of a good system approximation is the use of processor-sharing systems as approximating systems for time-sharing systems (see p. 156 of KLEINROCK (1976)). Whereas formulae for time-sharing systems are very complicated, there are very simple formulae for processor-sharing systems with Poisson input and arbitrary service time d.f.s.

7.7

The $\leqq_g$-inequality implications of Proposition 7.7.1 are in SENETA (1967); for the rest see STOYAN (1974b). The lower bound (7.7.17) was derived by SENETA (1967). AGRESTI (1975) considered the case of variable offspring d.f.s.

In addition to the references in the text, some other papers which are concerned with inequalities for branching processes include BRAUN (1975), HEYDE and SCHUH (1978), MORAN (1960) and WHITTLE (1955). A more complete bibliography on inequalities for branching processes can be found from the references of all these papers.

Added in proof. POLLETT (1983) showed by simulation experiments that (7.6.3) and (7.6.4) are fairly accurate, and also applied an approximation method to obtain distributional approximations. MASSEY (1983) proved comparison theorems for open networks of queues with exponential service times and approximating M/M/1 queues.

MASSEY, W. A. (1983) Open networks of queues: Their algebraic structure and estimating their transient behaviour. *Adv. Appl. Prob.* **15**.

POLLETT, P. K. (1983) Residual life approximations. *Elektron. Informationsverarb. Kybernet.* **19**.

8. Robustness properties of stochastic models

8.0 Fundamentals

8.0.1 The problem

In the investigation of stochastic models it is a common technique to replace complicated distributions by simpler ones which are in some sense "close" to the real distributions. For example, approximation by Erlang d.f.s has often been used in queueing theory, and approximation of d.f.s by discrete d.f.s is common in simulation methods. A similar situation occurs if statistical data are used for the determination of component distributions. Clearly, in this case the only way to proceed is via approximations of (unknown) distributions. In both cases it is proper to ask whether such approximations are justified or not. This chapter will help to find an answer to this question by studying *robustness* properties. We shall consider only simpler models and methods, and in particular, Markov chains; for more general models, see the references in Section 8.3.

Formally, the problem which we study can be formulated as follows. Let Σ_k $(k = 1, 2, \ldots)$ and Σ be stochastic models of the same structure with the component distributions F_m and $F_{m,k}$ $(m = 1, 2, \ldots)$. Let c_Σ be a system quantity in Σ. We denote classes of component distributions by $\mathscr{D}_m$ so that $F_{m,k}$ and $F_m \in \mathscr{D}_m$. Types of convergence are denoted by $\to_m$ when on $\mathscr{D}_m$ and by $\to_c$ when on the image set $c_\Sigma(\mathscr{D}_1, \mathscr{D}_2, \ldots)$. Below we consider certain models and prove some theorems which ensure that as $k \to \infty$,

$$F_{m,k} \to_m F_m \quad (m = 1, 2, \ldots) \text{ imply } c_\Sigma(F_{1,k}, F_{2,k}, \ldots) \to_c c_\Sigma(F_1, F_2, \ldots), \tag{8.0.1}$$

where in general (8.0.1) is true for certain classes of convergent sequences, and is proved without reference to the special form of the limit d.f.s F_m.

When (8.0.1) holds we speak of *robustness* (or *continuity*) of c_Σ in F_m. If a system quantity c_Σ is robust in F_m, then "small" perturbations of component distribution F_m result in only "small" changes in c_Σ, and it is then admissible to use approximations for F_m.

Approximations have been and are often used in queueing theory and elsewhere in applied probability without demanding robustness. In many cases this causes no difficulties because the models are robust or the approximation methods are chosen with sufficient care that the resultant errors are small. Two examples of approximation methods for d.f.s are given in Section 8.0.3. Some general guidelines for successful approximation include the following:

(a) the approximations should be so constructed that the distributions to be approximated and the approximating distributions differ little in the sense of weak convergence;

(b) the corresponding (low order) moments should be equal, or nearly so;
(c) particular care is needed in approximating a distribution which is not continuous or when the system quantity of interest is discrete in character (for example a loss probability, or the distribution of an integer-valued r.v.).

We start the subject matter of this chapter by giving some fundamental results of the theory of weak convergence and of metrics for distributions. We then consider robustness problems for nonstationary queueing systems in order to demonstrate the methods. We study at length the stability of stationary distributions of Markov chains, where the queues GI/GI/1 and GI/GI/s are of special interest. Finally, some examples of robustness theorems for other stochastic models are given.

8.0.2 Weak convergence in metric spaces

We take our basic space E to be Polish, i.e., a complete separable metric space, remarking that some of the statements below hold without the assumptions of separability and completeness. Let $\mathcal{M}$ be the set of all Borel subsets of E, and $P_{\mathcal{M}}$ the class of probability measures on $(E, \mathcal{M})$. Let P, P_n be elements of $P_{\mathcal{M}}$.

Definition 8.0.1. *A sequence* $\{P_n\}$ *is* weakly convergent *to* P *if for every continuous bounded functional* f *on* E

$$\lim_{n\to\infty} \int_E f(x)\, P_n(\mathrm{d}x) = \int_E f(x)\, P\,(\mathrm{d}x). \tag{8.0.2}$$

A sequence $\{X_n\}$ *of r.v.s with values in* $(E, \mathcal{M})$ *is* convergent in distribution *if the induced distributions on* $\mathcal{M}$ *are weakly convergent.*

We use the usual symbol $\Rightarrow$ for weak convergence, writing

$$P_n \Rightarrow P \quad \text{and} \quad X_n \Rightarrow X.$$

Important references for weak convergence theory are Billingsley (1968), Prokhorov (1956) and Borovkov (1976b, 1980).

The following spaces E are of special importance for us:

1. $E = \mathbb{R}^n$

$\mathbb{R}^n$ is the n-dimensional Euclidean space of vectors $\boldsymbol{x} = (x_1, \ldots, x_n)$ with the metric ϱ_n given by

$$\varrho_n(\boldsymbol{x}, \boldsymbol{y}) = \left[\sum_{i=1}^{n} (x_i - y_i)^2\right]^{1/2}.$$

Here a distribution P is given by its d.f. F on $\mathbb{R}^n$,

$$F(\boldsymbol{x}) = P\big(\{\boldsymbol{y} \in \mathbb{R}^n : \boldsymbol{y} = (y_1, \ldots, y_n), y_i \leq x_i;\ i = 1, \ldots, n\}\big) \text{ (all } \boldsymbol{x} \text{ in } \mathbb{R}^n),$$

$F_n \Rightarrow F$ if for all continuity points $\boldsymbol{x}$ of F

$$\lim_{n\to\infty} F_n(\boldsymbol{x}) = F(\boldsymbol{x})$$

and $P_n \Rightarrow P$ iff $F_n \Rightarrow F$. (See Section 8.0.4 for the Levy metric connected with weak convergence.) Further, the convergence of Laplace-Stieltjes transforms of probability measures on $\mathbb{R}_+{}^n = \{\boldsymbol{x} \in \mathbb{R}^n : x_i \geqq 0 \ (i = 1, \ldots, n)\}$ is equivalent to the weak convergence of distributions on $\mathbb{R}^n{}_+$, i.e.

$$F_n \Rightarrow F \quad \text{iff} \lim_{n\to\infty} F_n{}^*(s) = F^*(s) \qquad \text{(all } s \text{ in } \mathbb{R}_+{}^n). \tag{8.0.3}$$

2. $E = \mathbb{R}^\infty$

$\mathbb{R}^\infty$ is the space of all infinite sequences $\{x_n\}$ of real numbers, while the set of all such sequences with $x_n \geqq 0$ (all n) is denoted by $\mathbb{R}_+{}^\infty$. The usual topology on $\mathbb{R}^\infty$ is defined so that for a point $\boldsymbol{x} \in \mathbb{R}^\infty$ the sets

$$\mathcal{N}_{k,\varepsilon}(\boldsymbol{x}) = \{\boldsymbol{y} \colon |x_i - y_i| < \varepsilon; \quad i = 1, 2, \ldots, k\} \qquad (k = 1, 2, \ldots)$$

form a neighbourhood base. It corresponds to the metric ϱ_∞ defined by

$$\varrho_\infty(\boldsymbol{x}, \boldsymbol{y}) = \sum_{i=1}^{\infty} 2^{-i} \, |x_i - y_i|/(1 + |x_i - y_i|),$$

for if $\varrho_\infty(\boldsymbol{x}, \boldsymbol{y}) < 2^{-k}/(1 + \varepsilon)$, then $\boldsymbol{y} \in \mathcal{N}_{k,\varepsilon}(\boldsymbol{x})$.

With this topology $\mathbb{R}^\infty$ (and also $\mathbb{R}_+{}^\infty$) is a Polish space. A sequence of distributions on the Borel σ-algebra of $\mathbb{R}^\infty$ is weakly convergent towards P iff the corresponding projections on $\mathbb{R}^m$ are weakly convergent. Thus, when

$$Y_n = \{X_{1,n}, X_{2,n}, \ldots\} \quad \text{and} \quad Y = \{X_1, X_2, \ldots\}$$

are r.v.s with values in $\mathbb{R}^\infty$, we have $Y_n \Rightarrow Y$ iff

$$\{X_{1,n}, \ldots, X_{m,n}\} \Rightarrow \{X_1, \ldots, X_m\} \qquad \text{for all } m = 1, 2, \ldots$$

If the components of Y_n and Y are sequences of independent r.v.s then $Y_n \Rightarrow Y$ iff $X_{m,n} \Rightarrow X_m$ for all m.

3. $E = D[0, a]$

$D[0, a]$ is the set of all right-continuous real functions on $[0, a]$ having left limits at all points (see Billingsley (1968)). On $D[0, a]$ the Skorokhod metric d is used, being defined as follows. Let Λ be the set of all continuous increasing real functions λ on $[0, a]$ with

$$\lambda(0) = 0, \qquad \lambda(a) = a.$$

Then

$$d(x, y) = \inf_{\lambda\in\Lambda} \Big\{ \sup_{0\leq t\leq a} \big|x(t) - y\big(\lambda(t)\big)\big| + \sup_{0\leq t\leq a} |t - \lambda(t)| \Big\}.$$

The space $D[0, a]$ with the Skorokhod metric d is a Polish space.

For $D[0, a]$ the criteria for weak convergence are not as simple as $\mathbb{R}^n$ or $\mathbb{R}^\infty$; but this is no great problem for us here because the only distributions on $D[0, a]$ we consider are those induced by mappings from other spaces into $D[0, a]$.

The following conditions are equivalent to the weak convergence $P_n \Rightarrow P$ in Polish spaces:

(1) $\limsup_{n\to\infty} P_n(A) \leqq P(A)$ for all closed $A \in \mathcal{M}$;

(2) $\liminf_{n\to\infty} P_n(A) \geqq P(A)$ for all open $A \in \mathcal{M}$;

(3) $\lim_{n\to\infty} P_n(A) = P(A)$ for all P-continuity sets $A \in \mathcal{M}$, i.e. for all sets A whose boundary ∂A satisfies $P(\partial A) = 0$;

(4) every subsequence $\{P_{n'}\}$ of $\{P_n\}$ contains a subsequence $\{P_{n''}\}$ with $P_{n''} \Rightarrow P$.

By $E = E' \times E''$ we denote the product of the Polish spaces E' and E'' and by $P' \times P''$ the product measure on $\mathcal{M}' \times \mathcal{M}''$ of two distributions on $\mathcal{M}'$ and $\mathcal{M}''$. Then $P_n' \Rightarrow P'$ and $P_n'' \Rightarrow P''$ imply $P_n' \times P_n'' \Rightarrow P' \times P''$, and, conversely, $P_n' \times P_n'' \Rightarrow P' \times P''$ implies $P_n' \Rightarrow P'$ and $P_n'' \Rightarrow P''$. (The same is also true for infinite products of Polish spaces.) Theorems on the weak convergence of distributions induced by mappings are important for applications. Let E and E' be Polish spaces with the Borel σ-algebras $\mathcal{M}$ and $\mathcal{M}'$. Let h be a measurable mapping from E into E', let X be a r.v. with values in $(E, \mathcal{M})$ and let P be the distribution induced by X on $\mathcal{M}$. Then $h(X)$ is a r.v. with values in $(E', \mathcal{M}')$, the distribution of which we denote by Ph^{-1}, being defined for all $A \in \mathcal{M}'$ by

$$Ph^{-1}(A) = P(h^{-1}(A)).$$

We then have the following result known as the *continuous mapping theorem* (Theorem 5.1 of BILLINGSLEY (1968)):

If $P_n \Rightarrow P$ and $P(D_h) = 0$ where D_h denotes the set of discontinuities of h, then $P_n h^{-1} \Rightarrow Ph^{-1}$.

There is also the following more general statement (Theorem 5.5 of BILLINGSLEY (1968)) involving a sequence $\{h_n\}$ of mappings; we write $\tilde{D}_h$ for the set of all $x \in E$ for which there exists a sequence $\{x_i\}$ such that

$$\lim_{i\to\infty} x_i = x \quad \text{but} \quad \lim_{i\to\infty} h_i(x_i) \neq h(x).$$

Then if $P_n \Rightarrow P$ and $P(\tilde{D}_h) = 0$, $P_n h_n^{-1} \Rightarrow Ph^{-1}$.

For example, the continuity of convolution of d.f.s is a consequence of the continuous mapping theorem, i.e.,

$$F_n \Rightarrow F \text{ and } F_n' \Rightarrow F' \text{ imply } F_n * F_n' \Rightarrow F * F' \tag{8.0.4}$$

Finally, we need the notion of weak compactness of sets of distributions (called relative compactness in BILLINGSLEY (1968)).

Definition 8.0.2. *A family $\Pi \subset P_{\mathcal{M}}$ is said to be* weakly compact *if every sequence of elements of Π contains a weakly convergent subsequence.*

An important characterization of weak compactness in a Polish space E was proved by PROKHOROV, who showed (Theorems 6.1 and 6.2 of BILLINGS-

LEY (1968)) *that a set Π is weakly compact if and only if it is tight*, i.e. for every $\varepsilon > 0$ there exists a compact $K_\varepsilon \subset E$ such that $P(K_\varepsilon) > 1 - \varepsilon$ for all $P \in \Pi$.

8.0.3 Two approximation methods for d.f.s

To compute approximations to characteristics of stochastic models it is usual to approximate their component distributions by "simpler" distributions. Here we consider the two methods of approximation, by discrete distributions, and by mixtures of Erlang distributions. The first method is used in connection with simulation methods, the second with the Erlang phase method.

One method of discretizing a given d.f. F is to take a step length $\Delta > 0$, and approximate F by the discrete d.f. F_Δ defined by

$$F_\Delta(t) = F(k\Delta) \qquad \big(k\Delta \leqq t < (k+1)\,\Delta,\ k = 0, \pm 1, \pm 2, \ldots\big). \tag{8.0.5}$$

By sketching graphs of the d.f.s concerned, it is clear that for Δ_2 an integer multiple of Δ_1,

$$F \leqq_d F_{\Delta_1} \leqq_d F_{\Delta_2}, \tag{8.0.6}$$

and $F_\Delta \Rightarrow F$ for $\Delta \to 0$, in which case also all such moments as exist converge to those of F. MAL'CEVA (1970) and SCHASSBERGER (1970, 1973) suggested for any d.f. F on $\mathbb{R}_+$ the following approximations F_n based on mixtures of Erlang d.f.s Erl (. , .) defined in Example 1.2(b) above:

$$F_n(t) = F(0) + \sum_{k=1}^{\infty} \big(F(k/n) - F((k-1)/n)\big) \operatorname{Erl}(n, k)(t). \tag{8.0.7}$$

As shown on p. 32 of SCHASSBERGER (1973), we now prove that

$$F_n \Rightarrow F. \tag{8.0.8}$$

(The essence of the argument is outlined in Section 11 of KINGMAN (1966).)

Start by observing that (8.0.7) is equivalent to

$$F_n(t) = \sum_{k=0}^{\infty} F(k/n)\, \mathrm{e}^{-nt}(nt)^k/k! = \int_0^\infty F(x)\, \mathrm{d}G_{n,t}(x) \tag{8.0.9}$$

where $G_{n,t}$ is a scaled Poisson d.f. with probability mass $\mathrm{e}^{-nt}(nt)^k/k!$ at the point k/n ($k = 0, 1, \ldots$). $G_{n,t}$ has mean t and variance t/n, and thus $G_{n,t} \Rightarrow \Theta_t$. In order to establish (8.0.8) it is then enough to show that

$$\lim_{n\to\infty} \int_0^\infty F(x)\, \mathrm{d}G_{n,t}(x) = F(t) = \int_0^\infty F(x)\, \mathrm{d}\Theta_t(x) \tag{8.0.10}$$

for all continuity points t of F. This last is an elementary fact.

It is not difficult to check that the moments of F_n converge to the corresponding moments of F when finite.

Analogous to (8.0.6), it can be shown that for positive integers m and n,

$$F \leqq_c F_{mn} \leqq_c F_n . \tag{8.0.11}$$

The proof relies on the two properties of Erlang distributions (see Equation (1.5.7)) that $\text{Erl}\,(mn, mk) \leqq_c \text{Erl}\,(n, k)$ and for non-negative $j \leqq k$, $\text{Erl}\,(n, k - j) \leqq_c \text{Erl}\,(n, k)$. To see that a more general result (i.e., relaxing the integer-valued nature of m) need not hold, suppose that $F = \Theta_{0.6}$, so that $F_2 = \text{Erl}\,(2, 2)$ with mean 1, $F_3 = \text{Erl}\,(3, 2)$ with mean 2/3, and $F_4 = \text{Erl}\,(4, 3)$ with mean 3/4; $F_4 \leqq_c F_3$ is then impossible.

Observe that F_Δ at (8.0.5) can be expressed as

$$F_\Delta = \sum_{k=-\infty}^{\infty} \big(F(k\Delta) - F((k-1)\,\Delta)\big)\,\Theta_{k\Delta}, \tag{8.0.5'}$$

so since $\Theta_{k/n} \leqq \text{Erl}\,(mn, mk)$, (8.0.11) can be amplified to the form

$$F \leqq_d F_\Delta \leqq_c F_{mn} \leqq_c F_n \qquad (\Delta \equiv 1/mn). \tag{8.0.11'}$$

In conclusion we note that Bux and Herzog (1977) studied aspects of the optimal approximation of a d.f. F by mixtures of Erlang d.f.s such as at (8.0.7).

8.0.4 Some metrics for distributions

The distance between any two d.f.s F and G in $\mathbb{R}$ is measured by the Levy metric λ defined by

$$\lambda(F, G) = \inf \{\varepsilon : F(t - \varepsilon) - \varepsilon < G(t) < F(t + \varepsilon) + \varepsilon \quad (\text{all } t \in \mathbb{R})\} \tag{8.0.12}$$

(cf. p. 21 of Billingsley (1968)). It is not difficult to check that $F_n \Rightarrow F$ $(n \to \infty)$ if and only if $\lambda(F, F_n) \to 0$ $(n \to \infty)$, and that the space $\mathscr{D}$ of all d.f.s on $\mathbb{R}$ is a Polish space with λ as its metric.

For distributions on a more general metric space E with metric ϱ and associated Borel σ-algebra $\mathscr{M}$, the natural generalization of λ is the Levy-Prokhorov metric π defined as follows.

First, let $C^\varepsilon = \{x \in E : \varrho(x, C) < \varepsilon\}$ denote the ε-neighbourhood of any $C \in \mathscr{M}$, and for distributions P, Q on $(E, \mathscr{M})$, set

$$\pi_1(P, Q) = \inf \{\varepsilon : P(C) \leqq Q(C^\varepsilon) + \varepsilon \quad (\text{all } C \in \mathscr{M})\}. \tag{8.0.13a}$$

Then define

$$\pi(P, Q) = \min \big(\pi_1(P, Q), \quad \pi_1(Q, P)\big). \tag{8.0.13b}$$

Observe that π generalizes λ in the sense that, apart from the particular space $\mathbb{R}$ being involved, λ makes reference only to the half-lines $(-\infty, t]$ while π refers to all Borel sets.

For d.f.s F, G on $\mathbb{R}$, the uniform or supremum metric u is defined by

$$u(F, G) = \sup_{t \in \mathbb{R}} |F(t) - G(t)|, \tag{8.0.14}$$

and if the positive tails of the d.f.s have finite expectations, then we may also define, by analogy with $\leqq_c$, the metric

$$\varrho_c(F, G) = \sup_{t \in \mathbb{R}} \left| \int_t^\infty \big(F(x) - G(x)\big)\, dx \right|. \tag{8.0.15}$$

Other metrics that have been used include the variation metric (for distributions P, Q on $(E, \mathcal{M})$ with metric ϱ)

$$\sigma(P, Q) = \sup_{A \in \mathcal{M}} |P(A) - Q(A)|, \tag{8.0.16}$$

and Dudley's metric

$$D(P, Q) = \sup_{f \in \mathcal{F}} \left| \int_E f(x)\, P(dx) - \int_E f(x)\, Q(dx) \right| \tag{8.0.17}$$

where $\mathcal{F}$ is the set of all measurable functionals f on E for which

$$\sup_{x \in E} |f(x)| \leqq 1, \qquad \sup_{x, y \in E, x \neq y} |f(x) - f(y)|/\varrho(x, y) \leqq 1.$$

The metrics λ, u and ϱ_c are closely connected with the partial orderings $\leqq_d$ and $\leqq_c$, in that

$$F \leqq_d G \leqq_d H \quad \text{and} \quad \varrho'(F, H) < \varepsilon$$

imply

$$\varrho'(F, G) < \varepsilon \quad \text{and} \quad \varrho'(G, H) < \varepsilon \tag{8.0.18}$$

where ϱ' is λ or u. On replacing $\leqq_d$ by $\leqq_c$, the same statement is true with ϱ_c as the metric ϱ'.

Between the metrics we also have, amongst others, the relations

$$\pi^2(P, Q) \leqq D(P, Q) \leqq 3\pi(P, Q), \tag{8.0.19}$$

$$\lambda(F, G) \leqq u(F, G) \leqq \sigma(F, G). \tag{8.0.20}$$

For d.f.s F and G with means m_F, m_G, it is readily observed that

$$\varrho_c(F, G) \geqq |m_F - m_G|, \tag{8.0.21}$$

equality holding if either $F \leqq_d G$ or $G \leqq_d F$.

For extensive discussions concerning these metrics, see for example Zolotarev (1976b, 1977a, 1977c, 1979) and Kalashnikov (1978), and for ϱ_c, Kotzurek and Stoyan (1976).

8.1 Robustness properties of nonstationary stochastic models

8.1.1 Discrete time processes

Many stochastic models, for example recursive models, can be described by Markov chains $\{X_n\}$, by considering their state at discrete (often random) instants. In the following we consider some theorems which ensure the weak convergence of one-dimensional, finite-dimensional and infinite-dimensional sample path distributions of such processes. We will mostly study the distributions of the full sequence $\{X_1, X_2, \ldots\}$, for if we have convergence for them, then of course the finite-dimensional distributions also converge.

Let $\{X_n\}$, $\{X_{n,k}\}$ $(k = 1, 2, \ldots)$ be homogeneous Markov chains with the state space $(E, \mathcal{M})$, where E is a Polish space with metric ϱ and $\mathcal{M}$ the corresponding Borel σ-algebra. Let P_1, $P_{1,k}$ and p, p_k be the initial distributions and the transition functions respectively. Let E^∞ be the space of all infinite sequences in E with the product topology, so that sets of the form $\mathcal{N}_{k,\varepsilon}(x) = \{y \in E^\infty \colon \varrho(x_i, y_i) < \varepsilon_i \quad (i = 1, \ldots, k)\}$ are neighbourhoods of $x \in E^\infty$, and let $\mathcal{M}^\infty$ be the corresponding Borel σ-algebra. By $\mathcal{P}_\infty$ and $\mathcal{P}_{\infty,k}$ we denote the distributions of $\{X_n\}$ and $\{X_{n,k}\}$ on $\mathcal{M}^\infty$ $(k = 1, 2, \ldots)$. Karr (1975) proved

Theorem 8.1.1. *If*

$$P_{1,k} \Rightarrow P_1 \qquad (k \to \infty) \tag{8.1.1}$$

and if for all $x \in E$ and for all sequences $\{x_k\}$ converging towards x

$$p_k(x_k, .) \Rightarrow p(x, .) \tag{8.1.2}$$

then

$$\mathcal{P}_{\infty,k} \Rightarrow \mathcal{P}_\infty. \tag{8.1.3}$$

Remark. (8.1.2) *is equivalent to the following conditions* (a) *and* (b):

(a) *for every bounded, uniformly continuous functional g on E the functional pg defined by*

$$pg(x) = \int_E g(y)\, p(x, \mathrm{d}y) \qquad (x \in E),$$

is continuous;

(b) *for every bounded uniformly continuous functional g on E the sequence $\{p_k g\}$ converges to pg uniformly on every compact subset of E.*

Proof. The weak convergence (8.1.3) is equivalent to

$$\mathcal{P}_{\infty,k}\pi_m^{-1} \Rightarrow \mathcal{P}_\infty\pi_m^{-1} \tag{8.1.4}$$

for all $n = 1, 2, \ldots$, where π_m is thr projection of E^∞ on the first m factors of the infinite product E^∞. We prove (8.1.4) for all m by induction.

(8.1.4) holds for $m = 1$ by (8.1.1). Let (8.1.4) be true for some given positive integer m. Let $M_1 \subset E^m$ be closed and let g be any bounded, uniformly continuous functional on E with $\sup_{x \in E} g(x) = \|g\| > 0$. If we denote by χ_A the indicator function of the set A we obtain by the Markov property for $k = 1, 2, \ldots$

$$\left| \int_{E^{m+1}} \chi_{M_1}(y_1, \ldots, y_m)\, g(y_{m+1})\, \mathscr{P}_{\infty,k}\pi_{m+1}^{-1}(\mathrm{d}\boldsymbol{y}) - \int_{E^{m+1}} \chi_{M_1}(y_1, \ldots, y_m)\, g(y_{m+1})\, \mathscr{P}_{\infty}\pi_{m+1}^{-1}(\mathrm{d}\boldsymbol{y}) \right|$$

$$= \left| \int_{M_1} \int_E g(x)\, p_k(z_m, \mathrm{d}x)\, \mathscr{P}_{\infty,k}\pi_m^{-1}(\mathrm{d}\boldsymbol{z}) - \int_{M_1} \int_E g(x)\, p(z_m, \mathrm{d}x)\, \mathscr{P}_{\infty}\pi_m^{-1}(\mathrm{d}\boldsymbol{z}) \right|$$

$$\leq \int_{E^m} \left| \int_E g(x)\, p_k(z_m, \mathrm{d}x) - \int_E g(x)\, p(z_m, \mathrm{d}x) \right| \mathscr{P}_{\infty,k}\pi_m^{-1}(\mathrm{d}\boldsymbol{z})$$

$$+ \left| \int_{M_1} \int_E g(x)\, p(z_m, \mathrm{d}x)\, \mathscr{P}_{\infty,k}\pi_m^{-1}(\mathrm{d}\boldsymbol{z}) - \int_{M_1} \int_E g(x)\, p(z_m, \mathrm{d}x)\, \mathscr{P}_{\infty}\pi_m^{-1}(\mathrm{d}\boldsymbol{z}) \right|.$$

Because of (a) and the induction assumptions the second term $\to 0$ as $k \to \infty$. By (8.1.4) and Prokhorov's theorem there exists for every $\eta > 0$ a compact set $K_\eta \subset E^m$ with

$$\mathscr{P}_{\infty,k}\pi_m^{-1}(K_\eta) > 1 - \eta/(2\, \|g\|) \qquad (k = 1, 2, \ldots). \tag{8.1.5}$$

If K_η' denotes the projection of K_η on E (interpreted as the m-th factor of E^m) we obtain for the first term

$$\int_{E^m} \left| \int_E g(x)\, p_k(z_m, \mathrm{d}x) - \int_E g(x)\, p(z_m, \mathrm{d}x) \right| \mathscr{P}_{\infty,k}\pi_m^{-1}(\mathrm{d}\boldsymbol{z})$$

$$= \int_{K_\eta} |\ldots|\, \mathscr{P}_{\infty,k}\pi_m^{-1}(\mathrm{d}\boldsymbol{z}) + \int_{E^m \setminus K_\eta} |\ldots|\, \mathscr{P}_{\infty,k}\pi_m^{-1}(\mathrm{d}\boldsymbol{z})$$

$$\leq \sup_{z_m \in K_\eta'} \left| \int_E g(x)\, p_k(z_m, \mathrm{d}x) - \int_E g(x)\, p(z_m, \mathrm{d}x) \right| + 2\, \|g\| \sup_k \mathscr{P}_{\infty,k}\pi_m^{-1}(E^m \setminus K_\eta).$$

Because of (8.1.5) the last term here is smaller than η, and by (b) the previous term $\to 0$ as $k \to \infty$. (K_η' is compact because it is the continuous image of a compact set.) It thus follows that for closed $M_1 \subset E^m$,

$$\lim_{k \to \infty} \int_{E^{m+1}} \chi_{M_1}(y_1, \ldots, y_m)\, g(y_{m+1})\, \mathscr{P}_{\infty,k}\pi_{m+1}^{-1}(\mathrm{d}\boldsymbol{y})$$

$$= \int_{E^{m+1}} \chi_{M_1}(y_1, \ldots, y_m)\, g(y_{m+1})\, \mathscr{P}_{\infty}\pi_{m+1}^{-1}(\mathrm{d}\boldsymbol{y}). \tag{8.1.6}$$

Now let M_2 be an arbitrary closed subset of E. By p. 8 of BILLINGSLEY (1968) there exists a sequence $\{g_l\}$ of bounded, uniformly continuous functionals on E converging pointwise from above to χ_{M_2}. Thus for all k and l,

$$\mathscr{P}_{\infty,k}\pi_{m+1}^{-1}(M_1 \times M_2) \leq \int_{E^{m+1}} \chi_{M_1}(y_1, \ldots, y_m)\, g_l(y_{m+1})\, \mathscr{P}_{\infty,k}\pi_{m+1}^{-1}(\mathrm{d}\boldsymbol{y})$$

and for fixed l using (8.1.6), the limit supremum as $k \to \infty$ of the left-hand side is bounded by the right-hand side with $\mathscr{P}_{\infty,k}$ replaced by $\mathscr{P}_\infty$. Finally, the Lebesgue monotone convergence theorem is applicable to yield

$$\limsup_{k\to\infty} \mathscr{P}_{\infty,k}\pi_{m+1}^{-1}(M_1 \times M_2) \leqq \mathscr{P}_\infty \pi_{m+1}^{-1}(M_1 \times M_2),$$

which ensures (8.1.4) for $m+1$ by appealing to (1) on p. 153.

The assumptions of Theorem 8.1.1 are satisfied in the following special case. Let $\{Y_n\}$ and $\{Y_{n,k}\}$ be sequences of i.i.d. r.v.s with values in a Polish space F with the Borel σ-algebra $\mathscr{N}$, and let X_1 and $X_{1,k}$ be r.v.s independent of all Y_n and $Y_{n,k}$ having values in the Polish space E with the Borel σ-algebra $\mathscr{M}$ $(k = 1, 2, \ldots)$. Let ϕ be a continuous mapping from $E \times F$ into E, and let $\{X_n\}$ and $\{X_{n,k}\}$ be homogeneous Markov chains given by

$$X_{n+1} = \phi(X_n, Y_n), \tag{8.1.7}$$

$$X_{n+1,k} = \phi(X_{n,k}, Y_{n,k}) \tag{8.1.8}$$

and let

$$X_{1,k} \Rightarrow X_1, \tag{8.1.1'}$$

$$Y_{1,k} \Rightarrow Y_1. \tag{8.1.2'}$$

If we denote the distributions of Y_1 and $Y_{1,k}$ by P and p_k, we have

$$p(x, .) = P\phi^{-1}(x, .), \quad p_k(x_k, .) = P_k\phi^{-1}(x_k, .) \qquad (k = 1, 2, \ldots).$$

On account of the continuity of ϕ, the continuous mapping theorem yields (8.1.2).

Of course, for processes of the form (8.1.7) and (8.1.8) with continuous ϕ, the weak convergence of all finite-dimensional distributions (and, consequently, those of sample paths) may be proved directly by the continuous mapping theorem. Without the assumption that the Y_n and $Y_{n,k}$ are i.i.d. we can still show the following

Theorem 8.1.2. $X_{1,k} \Rightarrow X_1$ *and* $\{Y_{1,k}, \ldots, Y_{n,k}\} \Rightarrow \{Y_1, \ldots, Y_n\}$ *imply*

$$\{X_{1,k}, \ldots, X_{n+1,k}\} \Rightarrow \{X_1, \ldots, X_{n+1}\} \tag{8.1.9}$$

for all n.

Proof. Consider the mappings Ψ_n from $E \times F^n$ into E defined by

$$\Psi_1(x_1, y_1) = \phi(x_1, y_1),$$

$$\Psi_n(x_1, y_1, \ldots, y_n) = \phi\big(\Psi_{n-1}(x_1, y_1, \ldots, y_{n-1}), y_n\big) \quad (x \in E,\ y_i \in F).$$

These are continuous by the continuity of ϕ. Thus, by the continuous mapping theorem,

$$\begin{aligned}\{X_{1,k}, \ldots, X_{n+1,k}\} &= \Psi_n(X_{1,k}, Y_{1,k}, \ldots, Y_{n,k}) \Rightarrow \Psi_n(X_1, Y_1, \ldots, Y_n)\\ &= \{X_1, \ldots, X_{n+1}\}.\end{aligned}$$

In particular, for the queueing model GI/GI/s we obtain the

Theorem 8.1.3. *Let Σ and Σ_k be* GI/GI/s *queues with interarrival time d.f.s of the first customer W_1 and $W_{1,k}$ $(k = 1, 2, \ldots)$. Then*

$$A_k \Rightarrow A_1, \quad B_k \Rightarrow B \quad \text{and} \quad W_{1,k} \Rightarrow W_1 \tag{8.1.10}$$

imply

$$W_{n,k} \Rightarrow W_n \tag{8.1.11}$$

for the waiting time d.f.s W_n and $W_{n,k}$ of nth customer in Σ and Σ_k.

8.1.2 Continuous time processes

In what follows we consider processes whose sample paths $\{X_t\}$ and $\{X_{t,k}\}$ $(0 \leqq t \leqq a;\ k = 1, 2, \ldots)$ are assumed to belong to the space $D[0, a]$. This assumption is usually fulfilled in queueing theory because many processes like the virtual waiting time and queue length have piecewise continuous sample paths. These processes can be interpreted as r.v.s with values in $D[0, a]$. Our aim is to prove that under suitable conditions,

$$\{X_{t,k}\} \Rightarrow \{X_t\} \qquad (k \to \infty). \tag{8.1.12}$$

Suppose we describe the processes $\{X_t\}$ and $\{X_{t,k}\}$ as images of stochastic processes whose weak convergence properties are known. If the corresponding mappings are continuous (except possibly for sets of measure zero) then (8.1.12) follows by means of the continuous mapping theorem.

The theoretical basis of this method was worked out by Kennedy (1972) and Whitt (1974a), see also Whitt (1980c). We sketch its use in proving the robustness of the number of customers ν_t in G/G/1.

Let Σ and Σ_k be G/G/1 queues with interarrival times α and $\alpha_{n,k}$ and service times β_n and $\beta_{n,k}$ $(k = 1, 2, \ldots;\ n = 1, 2, \ldots)$. Denote the arrival epochs of the nth customer by t_n and $t_{n,k}$, so

$$t_{n+1} = \alpha_1 + \cdots + \alpha_n, \quad t_{n+1,k} = \alpha_{1,k} + \cdots + \alpha_{n,k}; \quad t_1 = t_{1,k} = 0, \tag{8.1.13}$$

and let the corresponding departure epochs be d_n and $d_{n,k}$. Obviously, d_n is a function of the sequence $\{\alpha_n, \beta_n\}$,

$$d_n = d_n(\{\alpha_n, \beta_n\}).$$

Let the number of customers in Σ and Σ_k at time t be ν_t and $\nu_{t,k}$, where we assume $\nu_0 = \nu_{0,k} = 1$. Define subsets A, J, and K of $\mathbb{R}_+^\infty \times \mathbb{R}_+^\infty$ and $\mathbb{R}_+^\infty$ by

$$A = \left\{(\boldsymbol{x}, \boldsymbol{y}) : \left|\sum_{i=1}^{r} x_i - d_s(\boldsymbol{x}, \boldsymbol{y})\right| > 0 \text{ for all } r \geqq 1 \text{ and } s \geqq 1\right\},$$

$$J = \left\{\boldsymbol{x} : x_i > 0 \text{ for } i = 1, 2, \ldots \text{ and } \sum_{i=1}^{\infty} x_i > a\right\},$$

$$K = \left\{\boldsymbol{x} : \boldsymbol{x} \in J, \sum_{i=1}^{r} x_i \neq a \text{ for all } r = 1, 2, \ldots\right\}.$$

Theorem 8.1.4. *If*

$$\mathsf{P}(\{\alpha_n, \beta_n\} \in K \times K) = \mathsf{P}(\{\alpha_{n,k}, \beta_{n,k}\} \in J \times J) = \mathsf{P}(\{\alpha_n, \beta_n\} \in A) = 1 \quad (8.1.14)$$

and

$$\{\alpha_{n,k}, \beta_{n,k}\} \Rightarrow \{\alpha_n, \beta_n\} \qquad (k \to \infty), \quad (8.1.15)$$

then

$$\{\nu_{t,k}\} \Rightarrow \{\nu_t\}. \quad (8.1.16)$$

Remark. In the case of GI/GI/1 queues, (8.1.15) is equivalent to the weak convergence of interarrival and service time d.f.s $A_k \Rightarrow A$ and $B_k \Rightarrow B$, and (8.1.14) is true for example, if

$$A(0+) = A_k(0+) = B(0+) = B_k(0+) = 0$$

and if one of the d.f.s A or B is continuous. It is possible to give examples which show that if (8.1.14) is not true then (8.1.16) need not be true (see Whitt (1974a)). However, no continuity assumption for A or B is needed to deduce the robustness of waiting times (see Theorem 8.1.3).

Proof. We use the following notation:

- $n_A(t)$ and $n_{A,k}(t)$ = number of arrivals in $[0, t]$ in Σ and Σ_k;
- $n_D(t)$ and $n_{D,k}(t)$ = number of departures in $[0,1]$ in Σ and Σ_k;
- $n_B(t)$ and $n_{B,k}(t)$ = number of customers (in Σ and Σ_k) who could be served in $[0, t]$ if the server worked without interruptions;
- $l(t)$ and $l_k(t)$ = sum of all service times of customers arriving in $[0, t]$ in Σ and Σ_k;
- $b(t)$ and $b_k(t)$ = total busy time in $[0, t]$ of the server in Σ and Σ_k.

Obviously,

$$\nu_t = n_A(t) - n_D(t), \qquad v_{t,k} = n_{A,k}(t) - n_{D,k}(t).$$

We now construct a mapping $\psi\colon \mathbb{R}_+^\infty \times \mathbb{R}_+^\infty \to D[0, a]$ with the following properties

(1) $\psi(\{\alpha_n, \beta_n\}) = \{\nu_t\}$, $\psi(\{\alpha_{n,k}, \beta_{n,k}\}) = \{\nu_{t,k}\}$ $(k = 1, 2, \ldots)$; (8.1.17)
(2) ψ is measurable;
(3) ψ is continuous on $K \times K \cap A$.

We write ψ in the form

$$\psi((\boldsymbol{x}, \boldsymbol{y})) = \psi_1(\boldsymbol{x}) - \psi_2\Big(\psi_3\big(\psi_4(\psi_5(\boldsymbol{x}, \boldsymbol{y}))\big)\Big), \quad (8.1.17')$$

where $\psi_1, \ldots, \psi_5$ are mappings

$$\psi_1\colon \mathbb{R}_+^\infty \to D[0, a];$$
$$\psi_2\colon D[0, a] \times D_0[0, a] \to D[0, a];$$
$$\psi_3\colon D[0, a] \times D[0, a] \to D[0, a] \times D[0, a];$$
$$\psi_4\colon D[0, a] \times D[0, a] \times \mathbb{R}_+^\infty \to D[0, a] \times D[0, a];$$
$$\psi_5\colon \mathbb{R}_+^\infty \times \mathbb{R}_+^\infty \to D[0, a] \times D[0, a] \times \mathbb{R}_+^\infty;$$

here, $D_0[0, a]$ is the set of all nondecreasing elements $x \in D[0, a]$ with $x(t) \in [0, a]$.

For $\boldsymbol{x} \in K$ and $x_1 \leqq t$ define

$$\psi_1(\boldsymbol{x})\,(t) = \sup \left\{k : \sum_{i=1}^{k} x_i \leqq t\right\};$$

otherwise (i.e., $\boldsymbol{x} \in K$ and $x_1 > t$, or $\boldsymbol{x} \notin K$) define $\psi_1(\boldsymbol{x})\,(t) = 0$. Then

$$\psi_1(\{\alpha_n\}) = \{n_A(t)\}, \qquad \psi_1(\{\beta_n\}) = \{n_B(t)\}.$$

Define ψ_5 by

$$\psi_5(\boldsymbol{x}, \boldsymbol{y}) = \big(\psi_1(\boldsymbol{x}), \psi_1(\boldsymbol{y}), \boldsymbol{y}\big).$$

Define ψ_4 by $\psi_4(x, y, z) = \big(y, h(x, z)\big)$ where

$$h(x, z) = \sum_{i=1}^{[g(x)(t)+1/2]} z_i,$$

the empty sum being equal to 0, $g(x)\,(t) = \sup_{0 \leqq s \leqq t} |x(s)|$, and $[y]$ = greatest integer less y. In particular, we have

$$h\big(n_A(t), \quad \{\beta_n\}\big) = l(t).$$

ψ_3 is defined by $\psi_3(y, z) = \big(y, m(z)\big)$ where

$$m(x)\,(t) = t + \inf_{0 \leqq s \leqq t} \big(x(s) - s\big),$$

and in particular, $m(l)\,(t) = b(t)$. Finally $\psi_2(y, z) \equiv y(z)$, and we have

$$n_B\big(b(t)\big) = n_D(t).$$

These mappings can be summarized in the diagram

$$\begin{array}{ccccc}
\{\alpha_n, \beta_n\} & \xrightarrow{\psi_5} & \big\{n_A(t), n_B(t), \{\beta_n\}\big\} & \xrightarrow{\psi_4} & \{n_B(t), l(t)\} \\
\downarrow{\scriptstyle \psi_1} & & & & \downarrow{\scriptstyle \psi_3} \\
\{n_A(t)\} & & \{n_D(t)\} & \xleftarrow{\psi_2} & \{n_B(t), l(t)\}
\end{array}$$

The mapping ψ has the property desired at (8.1.17′).

We shall shortly show that ψ_1 is measurable and continuous on K, referring to WHITT (1974a) for proof of the continuity of mappings from $D[0, a]$ into $D[0, a]$ (or from corresponding product spaces into product spaces) as needed to show similar properties of $\psi_2, \ldots, \psi_5$. We also note that WHITT (1971) showed that the mapping

$$g(x)\,(t) = \sup_{0 \leqq s \leqq t} |x(s)| \qquad (x \in D[0, a], \quad 0 \leqq t \leqq a)$$

is continuous. While addition in $D[0, a]$ need not necessarily be continuous, WHITT (1974a) showed that the condition $\mathsf{P}(\{\alpha_n, \beta_n\} \in A) = 1$ does ensure that the difference

$$\psi_1(.) - \psi_2\big(\psi_3(\psi_4(\psi_5(.)))\big)$$

is continuous.

We now follow KENNEDY (1972) in showing that ψ_1 is continuous. Take $\boldsymbol{x} \in K$, so that $\boldsymbol{x} = (x_1, x_2, \ldots)$, and let n be the integer for which

$$\sum_{i=1}^{n} x_i < a < \sum_{i=1}^{n+1} x_i,$$

with $n = 0$ if $x_1 > a$. Consequently $\psi_1(x)(a) = n_A(a) = n$. Take $\varepsilon > 0$ and such that

$$0 < \varepsilon < \min\left\{\sum_{i=1}^{n+1} x_i - a,\ a - \sum_{i=1}^{n} x_i\right\},$$

and let $\boldsymbol{y}$ be an arbitrary element of K for which

$$\varrho_\infty(\boldsymbol{x}, \boldsymbol{y}) < \delta(\varepsilon) \equiv \varepsilon/2^{n+1}(n + 1 + \varepsilon). \tag{8.1.18}$$

This condition ensures that for $i = 1, \ldots, n + 1$, $|x_i - y_i| < \varepsilon/(n + 1)$, and thus that

$$\left|\sum_{i=1}^{k} (x_i - y_i)\right| < \varepsilon \quad \text{for} \quad k = n,\ n + 1,$$

from which, by the definition of ψ_1 and the integer n above, we have $\psi_1(y)(a) = n$ also.

Now define a function λ: $[0, a] \to [0, a]$ by $\lambda(0) = 0$, $\lambda(a) = a$,

$$\lambda\left(\sum_{j=1}^{i} x_j\right) = \sum_{j=1}^{i} y_j \qquad (i = 1, \ldots, n),$$

and interpolate linearly between the points $0, x_1, x_1 + x_2, \ldots, a$. It is clear that $\lambda \in \Lambda$, that

$$\sup_{0 \leq t \leq a} |\lambda(t) - t| < \varepsilon,$$

and that for $0 \leq t \leq a$, $\psi_1(x)(t) = \psi_1(y)\big(\lambda(t)\big)$. Consequently

$$d\big(\psi_1(\boldsymbol{x}), \psi_1(\boldsymbol{y})\big) < \varepsilon \quad \text{whenever} \quad \varrho_\infty(\boldsymbol{x}, \boldsymbol{y}) < \delta(\varepsilon),$$

i.e., ψ_1 is continuous on K.

To establish the measurability of ψ_1, recall that for metric spaces X_1, X_2 and Z an open subset of X_1, if: ϕ: $X_1 \to X_2$ is continuous on Z and constant on $X_1 \setminus Z$, then ϕ is measurable. Since K is an open subset of $\mathbb{R}_+^\infty$ and ψ_1 is continuous on K, it follows that ψ_1 is measurable.

We remark (see KENNEDY (1972) for details) that under the conditions of Theorem 8.1.4, we also have for $k \to \infty$,

$$\{n_{A,k}(t)\} \Rightarrow \{n_A(t)\}, \qquad \{n_{D,k}(t)\} \Rightarrow \{n_D(t)\},$$
$$\{l_k(t)\} \Rightarrow \{l(t)\}, \qquad \{b_k(t)\} \Rightarrow \{b(t)\}.$$

8.2 Robustness properties of stationary distributions of homogeneous Markov chains

8.2.0 Preliminary remarks

In this section we describe methods which can be used to prove the robustness of stationary distributions of homogeneous Markov chains, formulating the robustness problem as follows. Let $\boldsymbol{T}$ and $\boldsymbol{T}_k$ $(k = 1, 2, \ldots)$ be the transition operators of homogeneous Markov chains $\{X_n\}$ and $\{X_{n,k}\}$ on the state space $(E, \mathscr{M})$, and assume that stationary distributions Q and Q_k exist and are uniquely determined. By $\to$ we denote a suitable type of convergence on the set $P_{\mathscr{M}}$. We seek conditions on the operators $\boldsymbol{T}$ and $\boldsymbol{T}_k$ which will ensure the convergence

$$Q_k \to Q. \tag{8.2.1}$$

Note that for stochastic models Σ and Σ_k the operators are determined by the component distributions F_m and $F_{m,k}$ $(m = 1, 2, \ldots)$, so for these we are interested in condition (8.2.1) in terms of conditions on F_m and $F_{m,k}$.

The question of robustness for stationary system quantities is in general more complicated than that of robustness for nonstationary quantities because two limit processes $(n \to \infty, k \to \infty)$ are involved. In general conditions ensuring the robustness in the nonstationary case are not sufficient for the stationary case, as is reflected in the difference in the proofs where in addition to showing the continuity of mappings, the compactness of sets of distributions must be shown, and this is often a much more complicated problem.

We consider three methods for the proof of robustness theorems. Results in the theory of weak convergence, in particular the continuous mapping theorem and Prokhorov's theorem, are the basis for the method of *weak convergence* in Section 8.2.2. The method of *test functions* in Section 8.2.3 is of a quite different nature, being similar to the direct method of Lyapunov in classical stability theory. Finally, in Section 8.2.4 we use *d.f. metrics* to prove "quantitative" robustness theorems, i.e., we find bounds for the distance between stationary distributions in Σ and Σ_k in terms of the distances of the component distributions.

Before describing these methods, known robustness theorems will be given for GI/GI/I, this model being of special interest here as it is used in examples illustrating the application of the various methods.

8.2.1 Robustness properties of the queue GI/GI/1

We consider first robustness properties of the waiting times, and then use these properties with (5.0.15) and (5.0.16) to establish similar results for the number of customers in the system. The first result is in Section 21 of Borovkov (1976a).

Theorem 8.2.1. *Let Σ and Σ_k be* GI/GI/1 *queues with difference d.f.s K and K_k and stationary waiting time d.f.s W and W_k $(k = 1, 2, \ldots)$. If*

$$K_k \Rightarrow K \tag{8.2.2}$$

and if the infinite series

$$R_k = \sum_{n=1}^{\infty} n^{-1}[1 - K_k^{*n}(0)]$$

are uniformly convergent in k then

$$W_k \Rightarrow W. \tag{8.2.3}$$

Proof. By (5.0.4) we can write for the Laplace-Stieltjes transforms $W^*(\theta)$ and $W_k^*(\theta)$ of W and W_k,

$$\log\big(W^*(\theta)\big) = \sum_{n=1}^{\infty} \int_0^{\infty} \big(1 - \exp(-\theta x)\big)\,\mathrm{d}K^{*n}(x) \equiv \sum_{n=1}^{\infty} s_n(\theta) = -S(\theta),$$

$$\log\big(W_k^*(\theta)\big) = \sum_{n=1}^{\infty} \int_0^{\infty} \big(1 - \exp(-\theta x)\big)\,\mathrm{d}K_k^{*n}(x) \equiv \sum_{n=1}^{\infty} s_{n,k}(\theta) = -S_k(\theta), \text{ say.}$$

By (8.0.3), (8.2.3) holds if

$$W_k^*(\theta) \to W^*(\theta) \quad \text{for all} \quad \theta \geqq 0,$$

equivalently, $S_k(\theta) \to S(\theta)$. Now by elementary analysis, this holds if for each n and all $\theta \geqq 0$,

$$s_{n,k}(\theta) \to s_n(\theta) \qquad (k \to \infty) \tag{8.2.4}$$

and if the series $S_k(\theta)$ converge uniformly in k. (8.2.2) ensures (8.2.4), and since the terms of the series for R_k dominate those of S_k, the assumed uniform convergence of R_k implies that of $S_k(\theta)$.

Borovkov also shows that when (8.2.2) holds, the convergence of the mean service times,

$$m_{B,k} \to m_B \tag{8.2.5}$$

is sufficient for the uniform convergence of R_k and hence sufficient for (8.2.3) also.

By contrast with the nonstationary case (see Theorem 8.1.3) (8.2.2) is not sufficient on its own for the robustness of stationary waiting times, as the following example shows.

Example 8.2 (a). Let Σ and Σ_k be systems of the type M/GI/1 with the same arrival rate $\lambda < .5$ and the service time d.f.s

$$B = \Theta_1, \quad B_k = (1 - k^{-1})\,\Theta_1 + k^{-1}\Theta_k \qquad (k = 1, 2, \ldots).$$

Then while (8.2.2) holds, (8.2.5) does not because $m_B = 1$ and $m_{B,k} = 2 - k^{-1} \to 2$ as $k \to \infty$. Recall (see e.g. Section 4.1.6 of Gnedenko and Kovalenko (1966)) that in an M/GI/1 system, the Laplace-Stieltjes transforms B^* and W^* of the service time and stationary waiting time d.f.s are related, with $\varrho = \lambda m_B$, by

$$W^*(\theta) = (1 - \varrho)/\big(1 + \lambda(1 - B^*(\theta))/\theta\big) \qquad (\text{all Re}\,(\theta) \geqq 0).$$

For the systems Σ and Σ_k, since $B_k \Rightarrow B$ implies $B_k^*(\theta) \to B^*(\theta)$ (all Re $(\theta) \geqq 0$), $W_k \Rightarrow W$ is impossible unless $\varrho_k \to \varrho$, which does not hold here.

Borovkov (1976a) also showed for general GI/GI/1 systems, that even

$$\int_{-\infty}^{\infty} t\,\mathrm{d}K_k(t) = \int_{-\infty}^{\infty} t\,\mathrm{d}K(t) \qquad (\text{all } k) \tag{8.2.6}$$

together with (8.2.2) is not sufficient for (8.2.3) to hold.

Now let ν_t and $\nu_{t,k}$ denote the number of customers at time t in Σ and Σ_k, and let t_n and $t_{n,k}$ be the respective arrival epochs of the nth customers. Let L, L_k, L_k^0, L^0 be the stationary d.f.s which, when the limits exist, are given by

$$L(x) = \lim_{t\to\infty} \mathsf{P}(\nu_t \leqq x), \quad L_k(x) = \lim_{t\to\infty} \mathsf{P}(\nu_{t,k} \leqq x),$$

$$L^0(x) = \lim_{n\to\infty} \mathsf{P}(\nu_{t_n} \leqq x), \quad L_k^0(x) = \lim_{n\to\infty} \mathsf{P}(\nu_{t_{n,k}} \leqq x).$$

Theorem 8.2.2. *For* GI/GI/1 *systems Σ and Σ_k let* (8.2.2) *hold. Then for $k \to \infty$, $m_{A,k} \to m_A$ implies*

$$L_k \Rightarrow L, \tag{8.2.7}$$

and if also A or W is continuous on $[0, \infty)$ or $(0, \infty)$ respectively, then

$$L_k^0 \Rightarrow L^0. \tag{8.2.8}$$

The proof of the theorem is based on (5.0.15) and (5.0.16) and the fact that convolution is a continuous operation with respect to weak convergence.

The following example shows that without $m_{A,k} \to m_A$ holding in addition to (8.2.2), (8.2.7) need not hold.

Example 8.2 (b). Consider GI/GI/1 queues Σ and Σ_k with $A = \text{Exp}\,(1)$, $A_k = (1 - k^{-1})\,\text{Exp}\,(1) + k^{-1}\,\text{Exp}\,(k^{-1})$, $B = B_k = \text{Exp}\,(2)$ $(k = 1, 2, \ldots)$. Obviously, we have $m_A = 1$, $m_{A,k} = 2 - k^{-1} \to 2$, and

$$A_k \Rightarrow A, \qquad B_k = B, \qquad m_{B,k} = m_B.$$

Consequently, (8.2.3) must hold. By (5.0.14), the probability $L(0)$ that Σ is empty is 0.5, while $L_k(0) = (3 - k^{-1})/(4 - 2k^{-1})$, so (8.2.7) does not hold.

8.2.2 The method of weak convergence

Perhaps the most natural way of proving robustness properties of stationary distributions of homogeneous Markov chains is by weak convergence methods. It was first used for queueing systems by Rossberg (1965) and

Borovkov (1976a). It uses standard analytical techniques, in particular those of weak convergence theory, and has a wide field of applications; because of this generality we formulate the method somewhat vaguely.

Using $\boldsymbol{T}_k$ and $\boldsymbol{T}$ to denote the transition operators of the Markov chains considered, assume that the stationary distributions Q and Q_k are the unique solutions of the equations

$$Q_k = \boldsymbol{T}_k Q_k, \tag{8.2.9'}$$

$$Q = \boldsymbol{T} Q. \tag{8.2.9''}$$

Proving the weak convergence relation

$$Q_k \Rightarrow Q \tag{8.2.10}$$

and determining conditions on $\boldsymbol{T}$ and $\boldsymbol{T}_k$ so that (8.2.10) holds, can be accomplished in two steps. First weak compactness of $\{Q_k\}$ is established by showing that every subsequence $\{Q_{k_n}\}$ of $\{Q_k\}$ has a further subsequence $\{Q_{k_n'}\}$ for which there is a measure Q_∞ (which may depend on the subsequence $\{Q_{k_n}\}$) satisfying the equation $Q_\infty = \boldsymbol{T} Q_\infty$ and to which $Q_{k_n'}$ converge weakly. Then by the uniqueness of Q as the solution of (8.2.9'') $Q_\infty = Q$, and by a result quoted in Section 8.0.2 (8.2.10) follows. In general the second step is more complicated than the first.

One possibility for proving the weak compactness of $\{Q_k\}$ is the use of monotonicity properties of the processes considered (cf. the proof of Proposition 8.2.5a). Another avenue is the result that a sequence $\{Q_k\}$ is weakly compact if the family of distributions $\{p_k(x, .)\}$ $(k = 1, 2, \ldots)$ indexed by $x \in E$, is weakly compact, p_k being the transition function of the process with the transition operator $\boldsymbol{T}_k$. That this is so follows by appealing to Prokhorov's theorem which for every $\eta > 0$ asserts the existence of a compact $K_\eta \subset E$ for which

$$p_k(x, K_\eta) > 1 - \eta \qquad (k = 1, 2, \ldots;\ x \in E).$$

For the stationary distribution Q_k corresponding to p_k we then have

$$Q_k(K_\eta) = \int_E p_k(x, K_\eta)\, Q_k(\mathrm{d}x) > (1 - \eta) \int_E Q_k(\mathrm{d}x) = 1 - \eta,$$

and repeated use of Prokhorov's theorem yields the weak compactness of $\{Q_k\}$.

The second step may be formalized for homogeneous Markov chains $\{X_n\}$ and $\{X_{n,k}\}$ with state space $(E, \mathscr{M})$, where $\mathscr{M}$ is the Borel σ-algebra on the Polish space E, by relating the transition functions p and p_k as in the following theorem (Karr (1975)).

Theorem 8.2.3. *Let the set* $\{Q_k\}$ *be weakly compact, and for any* $x \in E$ *let* $\{x_k\}$ *be any sequence converging to* x. *Then*

$$p_k(x_k, .) \Rightarrow p(x, .) \tag{8.1.2}$$

implies (8.2.10).

Proof. Let $\{Q_{k_n}\}$ be any weakly convergent subsequence of $\{Q_k\}$, and let Q_∞ be its limit. If we show that

$$\boldsymbol{T}Q_\infty = Q_\infty \tag{8.2.11}$$

then (8.2.10) is true. Now (8.2.11) is true if for all bounded, uniformly continuous functionals g on E the equation

$$\int_E g(x)\, Q_\infty(\mathrm{d}x) = \int_E \int_E g(y)\, p(x, \mathrm{d}y)\, Q_\infty(\mathrm{d}x)$$

holds (see Billingsley (1968), p. 9). For brevity, write

$$\int_E \int_E g(y)\, p(x, \mathrm{d}y)\, Q_\infty(\mathrm{d}x) = \int \int g\, \mathrm{d}p\, \mathrm{d}Q_\infty .$$

Then for all k_n,

$$\begin{aligned}
&\left|\int g\, \mathrm{d}Q_\infty - \int\int g\, \mathrm{d}p\, \mathrm{d}Q_\infty\right| \\
&\leq \left|\int g\, \mathrm{d}Q_\infty - \int g\, \mathrm{d}Q_{k_n}\right| + \left|\int g\, \mathrm{d}Q_{k_n} - \int\int g\, \mathrm{d}p_{k_n}\, \mathrm{d}Q_{k_n}\right| \\
&\quad + \left|\int\int g\, \mathrm{d}p_{k_n}\, \mathrm{d}Q_{k_n} - \int\int g\, \mathrm{d}p\, \mathrm{d}Q_{k_n}\right| + \left|\int\int g\, \mathrm{d}p\, \mathrm{d}Q_{k_n} - \int\int g\, \mathrm{d}p\, \mathrm{d}Q_\infty\right|.
\end{aligned}$$

The second term vanishes because Q_{k_n} is stationary, the first and fourth converge to zero as $k_n \to \infty$ (this is because (8.1.2) is equivalent to the conditions (a) and (b) above (8.1.4)). The third term can be shown to $\to 0$ as $k_n \to \infty$ in similar fashion to the proof of the boundedness of the first term in the proof of Theorem 8.1.1.

For a chain with complete connections the condition (8.1.2) may be replaced by a simpler one. Let $\{X_n\}$ and $\{X_{n,k}\}$ be Markov chains of the form, for $k, n = 1, 2, \ldots,$

$$X_{n+1} = \phi(X_n, Y_n), \tag{8.2.12}$$

$$X_{n+1,k} = \phi_k(X_{n,k}, Y_{n,k}). \tag{8.2.12$'$}$$

Here we assume that X_n and $X_{n,k}$ are r.v.s with $(E, \mathscr{M})$ as their common state space. The Y_n and $Y_{n,k}$ are i.i.d. $(F, \mathscr{N})$-valued r.v.s and independent of X_1 and $X_{1,k}$, F being a Polish space with Borel σ-algebra $\mathscr{N}$. Let F^∞ be the space of all sequences $\boldsymbol{x} = (x_1, x_2, \ldots,)$, $x_i \in F$, with the product space topology, and let $\mathscr{N}^\infty$ be the corresponding Borel σ-algebra. ϕ and the ϕ_k are measurable mappings from $E \times F$ into E. The robustness problem is then to state what conditions are sufficient, in addition to the weak convergence

$$P_k \Rightarrow P \tag{8.2.13}$$

of the distributions P and P_k of Y and Y_k, to ensure the weak convergence $Q_k \Rightarrow Q$ of the stationary distributions Q and Q_k of $\{X_n\}$ and $\{X_{n,k}\}$ whose existence and uniqueness are assumed. A partial solution is as follows, the proof being a simple application of the continuous mapping theorem using (8.2.13) and the continuity properties of ϕ and the ϕ_k to imply (8.1.2).

Theorem 8.2.4. *Let the set* $\{Q_k\}$ *be weakly compact and let* (8.2.13) *hold. Let* π_F *be the projection operator of* $E \times F$ *onto* F, *and let* $\tilde{D}_\phi$ *denote the set of all* $(x, y) \in E \times F$ *for which there exists a sequence* $\{(x_k, y_k)\}$ *with* $\lim_{k\to\infty} (x_k, y_k) = (x, y)$ *and* $\lim_{k\to\infty} \phi_k(x_k, y_k) \neq \phi(x, y)$. *Then* (8.2.10) *holds if*

$$P\big(\pi_F(\tilde{D}_\phi)\big) = 0. \tag{8.2.14}$$

Note that if $\phi_k = \phi$, (8.2.14) reduces to

$$P\big(\pi_F(D_\phi)\big) = 0 \tag{8.2.14'}$$

where D_ϕ is the set of all discontinuity points of ϕ.

By way of example, consider the waiting times in GI/GI/s for which the sequence of waiting time vectors $\boldsymbol{w}_n = \{w_{1,n}, \ldots, w_{s,n}\}$ satisfies the recursion, for $n = 1, 2, \ldots$,

$$\boldsymbol{w}_{n+1} = \big(\boldsymbol{R}(\boldsymbol{w}_n + \beta_n \boldsymbol{e}) - \alpha_n \boldsymbol{i}\big)_+, \tag{6.0.1}$$

this being the analogue of (8.2.12). Interpret $\boldsymbol{w}_n$ as X_n, (α_n, β_n) as Y_n, $\big(\boldsymbol{R}(\boldsymbol{w}_n + \beta_n \boldsymbol{e}) - \alpha_n \boldsymbol{i}\big)_+$ as $\phi\ (X_n, Y_n)$; then Theorem 8.2.4 applies to the process $\{\boldsymbol{w}_n\}$.

Proposition 8.2.5a. *Let* Σ *and* Σ_k *be* GI/GI/s *queues with the interarrival and service time d.f.s* A, A_k, B, *and* B_k $(k = 1, 2, \ldots)$ *for which*

$$A_k \Rightarrow A \tag{8.2.15}$$

and

$$B_k \Rightarrow B. \tag{8.2.16}$$

Let the corresponding stationary waiting time d.f.s be W *and* W_k. *If there exists a further* GI/GI/s *queue* Σ_0 *with d.f.s* A_0 *and* B_0 *for which*

$$A_0 \leqq_d A_k, \tag{8.2.17}$$

$$B_k \leqq_d B_0 \tag{8.2.18}$$

and for which the stationary waiting time d.f. W_0 *exists, then*

$$W_k \Rightarrow W. \tag{8.2.19}$$

Proof. Because of the independence of interarrival and service times, (8.2.15) and (8.2.16) are the same as the assumption (8.2.13) of Theorem 8.2.4. Let $\boldsymbol{W}_k$ and $\boldsymbol{W}$ be the stationary d.f.s of waiting time vectors in Σ_k and Σ. From the external monotonicity of GI/GI/s,

$$\boldsymbol{W}_k \leqq_d \boldsymbol{W}_0 \quad (k = 1, 2, \ldots). \tag{8.2.20}$$

Consequently, the set $\{W_k\}$ is weakly compact by Prokhorov's theorem.

(For example, we can choose for K_ε a set of the form

$$K_\varepsilon = \{\boldsymbol{x} \in \mathbb{R}_+{}^s : x_i \leqq y_i \ (i = 1, \ldots, s), \quad \boldsymbol{W}_0(y_1, \ldots, y_s) \geqq 1 - \varepsilon\}.)$$

Since ϕ is continuous, Theorem 8.2.4 can be applied to deduce (8.2.19).

The boundedness conditions (8.2.17) and (8.2.18) of this Proposition can be weakened.

Proposition 8.2.5b. (8.2.19) *holds if* (8.2.15) *and* (8.2.16) *hold and, additionally,*

$$m_{B,k} \to m_B. \tag{8.2.5}$$

The proof is similar to the preceding: the compactness of $\{\boldsymbol{W}_k\}$ can be proved via GI/GI/1 queues as bounding systems, using the cyclic queue Σ_c as in Wolff (1977a) or Section 6.3.2.1 above.

8.2.3 The method of test functions

Quite different from weak convergence methods is the suggestion of Kalashnikov and Tsitsiashvili (1972) (see also Kalashnikov (1978)) involving the use of test functions. Formally, there are some similarities with Lyapunov's direct method in classical stability theory, where stability is proved via a test function without knowledge of the solution. The question of proving continuity and compactness as in Section 8.2.2 is replaced by the search for a suitable test function V. First the theoretical basis of the method of test function is described, and then we give some practical guidelines concerning the choice of suitable V. While we assume, as in the preceding section, for ease of exposition, that the processes are of the form

$$X_{n+1} = \phi(X_n, Y_n), \tag{8.2.12}$$

this involves no loss generality; we will consider real X_n and Y_n only. All processes studied by Kalashnikov and Tsitsiashvili have the form (8.2.12), where in some cases the X_n and Y_n are random vectors. For more complicated state spaces, for example, $\mathbb{R}^n$, we refer to the papers of Tsitsiashvili (1972a, 1972b, 1973) and the book of Kalashnikov (1978).

Let $\{X_n\}$ and $\{X_{n,k}\}$ $(k = 1, 2, \ldots)$ be ergodic, homogeneous Markov chains of the form

$$X_{n+1} = \phi(X_n, Y_n), \quad X_{n+1,k} = \phi(X_{n,k}, Y_{n,k}). \tag{8.2.12'}$$

where $\{Y_n\}$ and $\{Y_{n,k}\}$ are sequences of i.i.d. real-valued r.v.s with d.f.s P and P_k, and independent of X_1 and $X_{1,k}$, and ϕ is a measurable mapping from $\mathbb{R}^2$ into $\mathbb{R}$. Let the stationary d.f.s of the processes be Q and Q_k.

A crucial assumption is that all r.v.s Y_n, $Y_{n,k}$, X_n, $X_{n,k}$ are defined on the same probability space $(\Omega, \mathscr{A}, \mathsf{P})$, this being no essential restriction because our interest is in the distributions and it is easy to construct such processes for given P and P_k. To see this, let $\{Z_n\}$ be a sequence of i.i.d.

r.v.s uniformly distributed on $[0, 1]$ and defined on $(\Omega, \mathscr{A}, \mathsf{P})$. Then, in the usual functional inverse notation, the r.v.s

$$Y_n^* \equiv P^{-1}(Z_n) \quad \text{and} \quad Y_{n,k}^* \equiv P_k^{-1}(Z_n) \tag{8.2.21}$$

form sequences of i.i.d. r.v.s on $(\Omega, \mathscr{A}, \mathsf{P})$ with d.f.s P and P_k. We can put $X_1 = X_{1,k} = 0$, because by assumption the processes are ergodic and only the stationary d.f.s are of interest, and by the construction

$$X_{n+1}^* = \phi(X_n^*, Y_n^*), \quad X_{n+1,k}^* = \phi(X_{n,k}^*, Y_{n,k}^*)$$

the processes $\{X_n^*\}$ and $\{X_{n,k}^*\}$ are homogeneous Markov chains on $(\Omega, \mathscr{A}, \mathsf{P})$, whose one-dimensional distributions converge to Q and Q_k. Denote the joint two-dimensional distribution of Y_n^* and $Y_{n,k}^*$ by $\boldsymbol{P}_k$, and assume that the homogeneous Markov chains $\{(X_n^*, X_{n,k}^*)\}$ are ergodic, denoting their stationary distributions by $\boldsymbol{Q}_k$.

Using these r.v.s but omitting the asterisks for convenience, we shall show that

$$Q_k \Rightarrow Q \tag{8.2.10}$$

via the properties of homogeneous Markov chains and not, as in Sections 8.2.2 and 8.2.3, by working with their distributions. This is done by showing that for every $\eta_1, \eta_2 > 0$, there exists $k(\eta_1, \eta_2)$ for which

$$\lim_{n\to\infty} \mathsf{P}\{|X_n - X_{n,k}| < \eta_1\} > 1 - \eta_2 \quad (\text{all } k \geq k(\eta_1, \eta_2)), \tag{8.2.22}$$

from which (8.2.10) follows. To establish conditions which ensure that (8.2.22) holds, KALASHNIKOV and TSITSIASHVILI used the quantities

$$A_k V(x, y) = \int_{-\infty}^{\infty} \int_{-\infty}^{\infty} \left[V\big(\phi(x, u), \phi(y, v)\big) - V(x, y)\right] \boldsymbol{P}_k(du, dv),$$

$$A_0 V(x, y) = \int_{-\infty}^{\infty} \left[V\big(\phi(x, u), \phi(y, u)\big) - V(x, y)\right] P(du),$$

where the real function V on $\mathbb{R}^2$ plays the role of the test function. If $V(x, y)$ is thought of as measuring the distance between the points x and y then $A_k V(x, y)$ may be interpreted as the difference between this distance and the distance between the "point sets" that are the result of translating x and y according to ϕ, P and P_k. $A_0 V$ may be interpreted analogously. (In passing, observe that

$$A_k V(x, y) = \mathsf{E}\big(V(X_{n+1}, X_{n+1,k}) - V(X_n, X_{n,k}) \mid (X_n, X_{n,k}) = (x, y)\big),$$
$$A_0 V(x, y) = \mathsf{E}\big(V(X'_{n+1}, X''_{n+1}) - V(X_n', X_n'') \mid (X_n', X_n'') = (x, y)\big)$$

where $\{X_n'\}$, $\{X_n''\}$ are realizations of $\{X_n\}$ with the same $\{Y_n\}$ but different initial distribution.) We expect that for robust processes, $A_0 V(x, y)$ and $A_k V(x, y)$ should be small: more precisely we have the following.

Theorem 8.2.6. (8.2.22) *holds if for arbitrary* $\eta > 0$, *there exists a non-negative function* V *on* $\mathbb{R}^2$ *such that*

$$\sup_{(x,y)} A_k V(x, y) \leqq \varepsilon_{1,k} \tag{8.2.23}$$

and

$$\sup_{|x-y|>\eta} A_k V(x, y) < -c(\eta) + \varepsilon_{2,k} \tag{8.2.24}$$

for some $c(\eta) > 0$ *and non-negative null sequences* $\{\varepsilon_{1,k}\}$ *and* $\{\varepsilon_{2,k}\}$.

A key step in the proof is a lemma concerning a homogeneous Markov chain $\{Z_n\}$ with state space $(E, \mathscr{M})$, transition function

$$p(x, B) = \mathsf{P}(Z_{n+1} \in B \mid Z_n = x),$$

and having a stationary distribution Q with

$$Q(B) = \lim_{n\to\infty} \mathsf{P}(Z_n \in B)\ (B \in \mathscr{M}).$$

Let f be a non-negative function on E and define

$$Af(x) = \int_E f(y)\, p(x, \mathrm{d}y) - f(x) = \mathsf{E}\big(f(Z_{n+1}) - f(Z_n) \mid Z_n = x\big),$$

$$\overline{Af} = \int_E Af(y)\, Q(\mathrm{d}y),$$

$$\mathsf{E}_z f(Z_n) = \int_E f(y)\, \mathsf{P}(Z_n \in \mathrm{d}y \mid Z_0 = z) = \mathsf{E}\big(f(Z_n) \mid Z_0 = z\big).$$

It is assumed that these integrals are well-defined.

Lemma 8.2.6 a. *If* $c \geqq 0$ *exists for which* $Af(x) \leqq c$ *for all* $x \in E$, *then*

$$\overline{Af} \geqq 0. \tag{8.2.25}$$

Proof. If the integral $\int_E f(y)\, Q(\mathrm{d}y)$ is finite then $\overline{Af} = 0$ and (8.2.25) holds trivially, so we assume the integral is infinite. First, since

$$\begin{aligned}\mathsf{E}_z Af(Z_i) &= \mathsf{E}_z\big(\mathsf{E}(f(Z_{i+1}) - f(Z_i) \mid Z_i)\big)\\ &= \mathsf{E}_z f(Z_{i+1}) - \mathsf{E}_z f(Z_i),\end{aligned}$$

$$\mathsf{E}_z f(Z_n) = f(z) + \sum_{i=0}^{n-1} \mathsf{E}_z Af(Z_i) \quad (n = 1, 2, \ldots). \tag{8.2.26}$$

From the non-negativity of f and Markovian nature of $\{Z_n\}$,

$$\begin{aligned}0 \leqq \int_E f(y)\, \mathsf{P}\{Z_n \in \mathrm{d}y\} &= \mathsf{E}f(Z_n) = \mathsf{E}\big(\mathsf{E}(f(Z_n) \mid Z_{n-1})\big)\\ &= \mathsf{E}\big(Af(Z_{n-1}) + f(Z_{n-1})\big)\\ &= \mathsf{E}f(Z_0) + \sum_{i=0}^{n-1} \mathsf{E}\big(Af(Z_i)\big)\end{aligned} \tag{8.2.27}$$

where $\mathsf{E}\big(Af(Z_i)\big) = \int_E Af(y)\, \mathsf{P}\{Z_i \in \mathrm{d}y\}$.

We now prove (8.2.25) by contradiction, so assume that $-\infty < \overline{Af} < 0$ (the case $\overline{Af} = -\infty$ can be treated similarly). For arbitrary $\eta > 0$ choose K such that

$$\left| \int_{-K \leqq Af(y) \leqq c} Af(y)\, Q(\mathrm{d}y) - \overline{Af} \right| < \eta/2.$$

Since $\mathsf{P}\{Z_n \in .\} \Rightarrow Q(.)$, there exists N such that for all $n > N$,

$$\left| \int_{-K \leqq Af(y) \leqq c} Af(y)\, Q(\mathrm{d}y) - \int_{-K \leqq Af(y) \leqq c} Af(y)\, \mathsf{P}\{Z_n \in \mathrm{d}y\} \right| < \eta/2.$$

Thus, for all $n > N$, since $Af(x) \leqq c$ (all $x \in E$),

$$\begin{aligned} \mathsf{E}\big(Af(Z_n)\big) &\leqq \mathsf{E}\big(Af(Z_n);\, -K \leqq Af(Z_n) \leqq c\big) \\ &\leqq \eta/2 + \int_{-K \leqq Af(y) \leqq c} Af(y)\, Q(\mathrm{d}y) \\ &\leqq \eta + \overline{Af}. \end{aligned}$$

Since we may choose η arbitrarily small but positive, it follows that when $\overline{Af} < 0$, we must have $\mathsf{E}\big(Af(Z_n)\big) < 0$ for $n > N$, contradicting (8.2.27).

Proof of Theorem 8.2.6. Apply the lemma to the chain $\{(X_n, X_{n,k})\}$ with $f = V$, so that for non-negative V and all k,

$$\overline{A_k V} = \int_{-\infty}^{\infty} \int_{-\infty}^{\infty} A_k V(x, y)\, \boldsymbol{Q}_k(\mathrm{d}x, \mathrm{d}y) \geqq 0.$$

From this, with (8.2.23) and (8.2.24), we obtain for all $\eta_1 > 0$ that

$$0 \leqq \overline{A_k V} \leqq \varepsilon_{1,k} \lim_{n\to\infty} \mathsf{P}\{|X_n - X_{n,k}| < \eta_1\} + \big(\varepsilon_{2,k} - c(\eta_1)\big) \lim_{n\to\infty} \mathsf{P}\{|X_n - X_{n,k}| > \eta_1\}$$

and hence that

$$\mathsf{P}\{|X_n - X_{n,k}| < \eta_1\} \geqq 1 - \varepsilon_{1,k}/\big(c(\eta_1) + \varepsilon_{1,k} - \varepsilon_{2,k}\big).$$

Since $\{\varepsilon_{1,k}\}$ and $\{\varepsilon_{2,k}\}$ are null sequences we can choose $k(\eta_1, \eta_2)$ for every η_1 and η_2 such that the right hand of the last inequality exceeds $1 - \eta_2$ for all $k \geqq k(\eta_1, \eta_2)$.

The conditions of Theorem 8.2.6 are relatively complex, and KALASHNIKOV and TSITSIASHVILI have given other sets of conditions, including the following.

Theorem 8.2.7. (8.2.22) *holds if there exists a non-negative function* V *on* $\mathbb{R}^2$ *such that for all* $x, y \in \mathbb{R}$,

$$A_0 V(x, y) \leqq 0 \tag{8.2.28}$$

and

$$A_k V(x, y) \leqq A_0 V(x, y) + \varepsilon_k \tag{8.2.29}$$

for some non-negative null sequence $\{\varepsilon_k\}$, *and*

$$\sup_{\substack{|x-y|>\eta,\\ x<M}} A_0 V(x, y) < -c(\eta, M) \tag{8.2.30}$$

for some function $c(\eta, M)$ *that is strictly positive for all* η, $M > 0$.

Proof. Much as in the proof of Theorem 8.2.6, we have for all η_1 and M that

$$0 \leqq \varepsilon_k Q((M, \infty)) + (\varepsilon_k - c(\eta_1, M)) (Q(-\infty, M]) - \lim_{n\to\infty} \mathsf{P}\{|X_n - X_{n,k}| < \eta_1\}$$
$$+ \varepsilon_k \lim_{n\to\infty} \mathsf{P}\{|X_n - X_{n,k}| < \eta_1\},$$

so that

$$\lim_{n\to\infty} \mathsf{P}\{|X_n - X_{n,k}| < \eta_1\} \geqq Q((-\infty, M]) - \varepsilon_k/c(\eta_1, M).$$

Obviously, by suitable choice of k and M we can ensure that the right hand side of the last inequality exceeds $1 - \eta_2$ for arbitrary $\eta_2 > 0$.

Relatively simple sufficient conditions for (8.2.28) and (8.2.29) are

$$V(\phi(x, u), \phi(y, u)) \leqq V(x, y) \tag{8.2.28'}$$

and

$$V(\phi(x, u), \phi(y, v)) \leqq V(\phi(x, u), \phi(y, u)) + \delta(u, v) \tag{8.2.29'}$$

for all x, y, u, and v; here $\delta(u, v)$ is a function with

$$\lim_{k\to\infty} \int_{-\infty}^{\infty} \int_{-\infty}^{\infty} \delta(u, v)\, \boldsymbol{P}_k(\mathrm{d}u, \mathrm{d}v) = 0. \tag{8.2.31}$$

(8.2.28′) expresses a contraction property of the mapping ϕ for fixed u, while (8.2.29′) states that ϕ varies in the second variable only "weakly".

Practical guidelines on the use of the method of test functions.

(1) Choose V to be a metric such that ϕ has the property (8.2.28) with respect to V. Tsitsiashvili (1979b) suggested a general method for constructing V for a wide class of Markov chains, namely the so-called piece-wise linear chains.

(2) Use the construction (8.2.21) and the following type of convergence for d.f.s:

$$P_k \to P \text{ if and only if } \lim_{k\to\infty} \int_{-\infty}^{\infty} |P_k(t) - P(t)|\, \mathrm{d}t = 0. \tag{8.2.32}$$

This is useful because, using (8.2.21),

$$\int_{-\infty}^{\infty} |P_k(t) - P(t)|\, dt = \int_0^1 |P_k^{-1}(x) - P^{-1}(x)|\, \mathrm{d}x = E\,|Y_{n,k}^* - Y_n^*|,$$

from which we obtain (8.2.31) for δ of the form

$$\delta(u, v) = f(|u - v|), \tag{8.2.33}$$

where f is a convex function on $\mathbb{R}_+$ with

$$\lim_{t \to 0} f(t) = 0; \tag{8.2.34}$$

this assertion holds because Jensen's inequality yields

$$\int_{-\infty}^{\infty} \int_{-\infty}^{\infty} \delta(u, v) \, \boldsymbol{P}_k(\mathrm{d}u, \mathrm{d}v) = \mathsf{E} f(|Y_{n,k}^* - Y_n^*|) \geqq f(\mathsf{E}\, |Y_{n,k}^* - Y_n^*|).$$

Example 8.2.3(a).

So as to compare the test function method with the other methods we prove the following theorem of Kalashnikov and Tsitsiashvili (1972) for waiting times in GI/GI/1.

Theorem 8.2.8. *Let K and K_k be difference distributions for stable* GI/GI/1 *systems and let*

$$\lim_{k \to \infty} \int_{-\infty}^{\infty} |K(t) - K_k(t)| \, \mathrm{d}t = 0 \tag{8.2.35}$$

with $K(t) > 0$ for all t. Then $W_k \Rightarrow W$ where W and W_k are the stationary waiting time d.f.s corresponding to K and K_k.

Proof. We use the recursion formulae for the waiting times in GI/GI/1, for $n, k = 1, 2, \ldots$

$$w_{n+1} = \max(0, w_n + U_n), \quad w_{n+1,k} = \max(0, w_{n,k} + U_{n,k}).$$

Here, w_n and $w_{n,k}$ play the role of X_n and $X_{n,k}$ and U_n and $U_{n,k}$ that of Y_n and $Y_{n,k}$, and we assume that U_n and $U_{n,k}$ are constructed according to (8.2.21). As test function we choose

$$V(x, y) = |x - y|,$$

and show that it has the properties (8.2.28), (8.2.29) and (8.2.30). Assuming without loss of generality that $x > y$, we have

$$A_0 V(x, y) = \int_{-\infty}^{\infty} |(x + t)_+ - (y + t)_+| \, \mathrm{d}K(t) - |x - y|$$

$$= \int_{-y}^{-x} t \, \mathrm{d}K(t) - xK(-x) + yK(-y) = -\int_{-y}^{-x} K(t) \, \mathrm{d}t.$$

Because K is non-negative, the condition (8.2.28) of Theorem 8.2.7 is satisfied. For η, $M > 0$ and general x, y, we obtain

$$\sup_{\substack{|x-y|>\eta,\\ x<M}} A_0 V(x, y) = \sup_{\substack{|x-y|>\eta,\\ x<M}} \left(-\int_{-\max(x,y)}^{-\min(x,y)} K(t) \, \mathrm{d}t \right) \leqq -\eta K(-M - \eta).$$

Because K is positive, we have $\eta K(-M - \eta) > 0$ for all η and $M > 0$, i.e. we have shown (8.2.30). Since also

$$|(x + u)_+ - (y + v)_+| \leqq |(x + u)_+ - (y + u)_+| + |u - v|,$$

we may take $\delta(u, v) = |u - v|$. Thus, δ has the form (8.2.31), from which (8.2.29) follows.

8.2.4 *Quantitative robustness theorems*

The robustness theorems discussed so far have been of a qualitative nature: they have been devoid of any information on the rate of convergence. Of course, more interest attaches to theorems which give estimates for the distance, via a suitable metric, of the distributions of system characteristics in terms of the distances of the component distributions. Such theorems have been given by BOROVKOV (1972, 1977, 1980), DELBROUCK (1973), KALASHNIKOV (1978, 1979), KALASHNIKOV and ANICHKIN (1981), KOTZUREK and STOYAN (1976), STOYAN (1972f) and ZOLOTAREV (1975a, b, 1976a, 1977b). One possible way of proving quantitative robustness theorems is the use of bounds for the rate of convergence to a stationary distribution, as we now show for the waiting times in GI/GI/1.

Let ϱ be a metric for d.f.s and for $i = 1, 2$ let Σ_i be GI/GI/1 queues with the characteristic d.f.s K_i, $W_{n,i}$, and W_i. Then since

$$\varrho(W_1, W_2) \leqq \varrho(W_1, W_{n,1}) + \varrho(W_{n,1}, W_{n,2}) + \varrho(W_{n,2}, W_2),$$

we need bounds for $\varrho(W_{n,1}, W_{n,2})$ and for the rate of convergence of $W_{n,i}$ to W_i.

Theorem 8.2.9. *Let* $W_{0,1} = W_{0,2} = \Theta_0$. *Then for the metrics* $\varrho = \lambda$, u *and* ϱ_c, $\varrho(K_1, K_2) < \varepsilon$ *implies*

$$\varrho(W_{n,1}, W_{n,2}) < n\varepsilon. \qquad (8.2.36)$$

Proof. We give Borovkov's proof for λ; the proofs for the other metrics are similar (see BOROVKOV (1977), ZOLOTAREV (1975a) and KOTZUREK and STOYAN (1976)). For $i = 1, 2$, let X_i and Y_i be independent r.v.s with d.f.s F_i and G_i with $\lambda(F_1, F_2) < \varepsilon_1$, $\lambda(G_1, G_2) < \varepsilon_2$, and let M_i be the d.f. of $Z_i = \max(X_i, X_i + Y_i)$. Then

$$\begin{aligned} M_1(x) &= \int_{-\infty}^{x} G_1(x - t)\, \mathrm{d}F_1(t) \leqq \int_{-\infty}^{x} G_2(x + \varepsilon_2 - t)\, \mathrm{d}F_1(t) + \varepsilon_2 \\ &= \mathsf{P}\big(X_1 + (Y_2)_+ \leqq x + \varepsilon_2\big) + \varepsilon_2 \\ &\leqq \mathsf{P}\big(X_2 + (Y_2)_+ \leqq x + \varepsilon_2 + \varepsilon_1\big) + \varepsilon_2 + \varepsilon_1 \\ &= M_2(x + \varepsilon_1 + \varepsilon_2) + \varepsilon_1 + \varepsilon_2. \end{aligned}$$

The inequality $M_2(x) \leqq M_1(x + \varepsilon_1 + \varepsilon_2) + \varepsilon_1 + \varepsilon_2$ follows similarly, showing that $\lambda(M_1, M_2) \leqq \varepsilon_1 + \varepsilon_2$. (8.2.36) follows easily by induction.

We omit here any discussion of bounds for $\varrho(W_{n,i}, W_i)$; the lengthy details are in the papers cited above. By way of example, the bound (5.3.15) yields for $t \geqq 0$ and $n = 1, 2, \ldots,$

$$W_{n,i}(t) - W_i(t) \leqq [e_i(C)]^n \tag{8.2.37}$$

or

$$\varrho(W_{n,k}, W_k) \leqq [e_k(C)]^n \tag{8.2.37'}$$

for $\varrho = u$ and λ, where

$$e_i(C) = \int_0^\infty \exp(Ct) \, \mathrm{d}K_i(t) < 1.$$

(Of course, we have to assume that such a $C > 0$ exists.) With $e(C) = \max\big(e_1(C), e_2(C)\big)$ we obtain for $\varrho = u$ and λ

$$\varrho(W_1, W_2) \leqq (n-1)\, \varrho(K_1, K_2) + 2[e(C)]^n. \tag{8.2.38}$$

Because the second term on the right hand side of (8.2.38) $\to 0$ as $n \to \infty$, there exists, for every $\varepsilon > 0$, some $\delta > 0$ for which

$$\varrho(K_1, K_2) < \delta \quad \text{implies} \quad \delta(W_1, W_2) < \varepsilon. \tag{8.2.39}$$

Using the metric ϱ_c and bounds for $m_W - m_{W,n}$, (8.2.39) can be proved for $\varrho = \varrho_c$ also (see KOTZUREK and STOYAN (1976)). Because of (8.0.18) this yields quantitative robustness for the mean stationary waiting times.

BOROVKOV (1977, 1980) has given further quantitative robustness theorems for the waiting times in GI/GI/1 with respect to the metrics u and σ. Using similar methods and equation (7.5.7), ZOLOTAREV (1977b) studied the number of customers in GI/GI/∞ systems with respect to σ and other metrics. Finally, KALASHNIKOV (1978, 1979) proved "uniform" robustness theorems for some queueing systems, such as the following.

Theorem 8.2.10. *Let Σ and Σ' be initially empty* GI/GI/s *queues with interarrival time d.f.s A and A', service time d.f.s B and B', and waiting time vector d.f.s $\mathbf{W}_n$ and $\mathbf{W}_n'$ ($n = 1, 2, \ldots$). If there exists $\gamma > 1$ and $g < \infty$ for which*

$$\int_0^\infty t^\gamma \, \mathrm{d}B(t) \leqq g, \quad \int_0^\infty t^\gamma \, \mathrm{d}B'(t) \leqq g, \tag{8.2.40}$$

then

$$\sup_n \varrho(\mathbf{W}_n, \mathbf{W}_n') \leqq c[\varrho(A, A') + \varrho(B, B')]^{(\gamma-1)/\gamma} \tag{8.2.41}$$

for $\varrho = \sigma$ and D and suitable c.

8.3 Further robustness theorems

8.3.1 Robustness theorems for queueing systems with non-renewal input

In recent years some robustness theorems have been proved for stationary G/G/. systems. Here we consider two typical examples of such theorems, namely the results of FRANKEN and KALÄHNE (1978) for G/G/∞ and those of BOROVKOV (1978, 1980) for the G/G/s queue. (Note also the paper of BRANDT and LISEK (1983).) The proofs use "construction points" or "innovation points", namely, that under some additional assumptions on the stationary sequence $\{(\alpha_n, \beta_n)\}$ of interarrival and service times it is possible to neglect the influence of the "distant" part of system behaviour on the system state at $t = 0$ which is the instant at which we look at the system. Weak convergence theorems then lead to robustness theorems.

We note that LISEK (1981) avoided the use of such "construction points".

Let Σ and Σ_k ($k = 1, 2, \ldots$) be G/G/∞ systems whose pairs $\{(\alpha_n, \beta_n)\}$ give stationary marked point processes with the arrival instants as points and service times as marks, and having distributions P and P_k on the set of all marked point processes on $\mathbb{R}$ with real-valued marks. Let $P(j)$ and $P_k(j)$ denote the probability that at $t = 0$ there are j customers in Σ and Σ_k respectively.

Theorem 8.3.1. *If*

$$\text{(i)} \quad P_k \Rightarrow P, \tag{8.3.1}$$

$$\text{(ii)} \quad \lim_{k\to\infty} m_{A,k} = m_A, \tag{8.3.2}$$

and

$$\text{(iii)} \quad \lim_{k\to\infty} m_{B,k} = m_B, \tag{8.2.5}$$

then

$$\lim_{k\to\infty} P_k(j) = P(j) \qquad (j = 0, 1, \ldots). \tag{8.3.3}$$

Additional assumptions, such as continuity properties which exclude simultaneous arrivals and departures, ensure the convergence of state probabilities in arrival and departure instants. Similar conditions and the ergodicity of P and P_k are needed for the proof of robustness theorems for general loss systems (see FRANKEN, KÖNIG, ARNDT, and SCHMIDT (1981)). Within the framework of stationary $\{X_n\}$ of the form

$$X_{n+1} = \phi(X_n, Y_n) \tag{8.2.12}$$

with independent Y_n, BOROVKOV (1978, 1980) has given conditions under which the sequences $\{X_{n+l}\}_{l=1,2,\ldots}$ converge as $n \to \infty$ to stationary sequences $\{X^l\}$ such that the latter are robust. For G/G/s ($s \geqq 2$) queues Σ and Σ_k ($k = 1, 2, \ldots$) where the sequences $\{(\alpha_n, \beta_n)\}$ and $\{(\alpha_{n,k}, \beta_{n,k})\}$ are stationary and metrically transitive, let

$$s\mathsf{E}\alpha_n = sm_A > m_B = \mathsf{E}\beta_n,$$

$$s\mathsf{E}\alpha_{n,k} = sm_{A,k} > m_{B,k} = \mathsf{E}\beta_{n,k} \quad (n = 0, \pm 1, \pm 2, \ldots; k = 1, 2, \ldots),$$

let $\{\boldsymbol{w}_n\}$ and $\{\boldsymbol{w}_{n,k}\}$ be the corresponding sequences of stationary waiting time vectors, and let A_0 be the event

$$A_0 = \{\alpha_{-1} > \gamma_{-1}, \alpha_0 + \cdots + \alpha_{s-2} > \beta_0\},$$

where $\gamma_n = \sup\limits_{k \leq n} \left\{(s-1)^2 \beta_{k-1} + (s-1) v_{k-1} + \sum\limits_{j=k}^{n-1} X_j\right\}$,

$X_k = \beta_k - s\alpha_k$, $v_k = \sup\limits_{l \leq k} \left\{\max\left(0, \sum\limits_{j=l}^{k} Z_j\right)\right\}$, $Z_j = -\beta_j + (s-1)(\beta_{j-1} - \beta_j)$.

With this notation BOROVKOV proved the following.

Theorem 8.3.2. *If*

(i) $\{(\alpha_{n,k}, \beta_{n,k})\} \Rightarrow \{(\alpha_n, \beta_n)\}$, (8.3.1′)

(ii) $\lim\limits_{k\to\infty} m_{B,k} = m_B$, (8.2.5)

and

(iii) $\mathsf{P}(A_0) > 0$, (8.3.4)

then

$$\{\boldsymbol{w}_{n,k}\} \Rightarrow \{\boldsymbol{w}_n\}. \quad (8.3.5)$$

The condition (8.3.4) is what facilitates the application of the construction point method, while the sequence $\{\gamma_n\}$ is a majorant for the sequence $\{\tau_n\}$ of sums of components of the waiting time vector $\boldsymbol{w}_n$ in Σ, i.e., for

$$\tau_n = w_{n,1} + \cdots + w_{n,s}.$$

For loss systems see FRANKEN (1970), BOROVKOV (1976a) and LISEK (1981).

8.3.2 *Robustness theorems for reliability and inventory problems*

(i) Life times of systems without repair

We consider a reliability system Σ which consists of M components with life times $X_1, \ldots, X_n$, the components being either in series or in parallel or in cold reserve. Let T be the system life time, so that, as in Section 7.2.4, T can be written in the form

$$T = \Psi(X_1, \ldots, X_M), \quad (8.3.6)$$

where Ψ is generated by a combination of additions (cold reserve), minimizations (series) and maximizations (parallel), and hence Ψ is a continuous mapping from $\mathbb{R}_+{}^M$ into $\mathbb{R}_+$. The continuous mapping theorem then yields the following.

Theorem 8.3.3. *Let Σ_k be reliability systems consisting of M components with life times $X_{1,k}, \ldots, X_{M,k}$ $(k = 1, 2, \ldots)$. Let Σ_k and Σ have the same structure, their system life times T_k and T being determined by the mapping Ψ. Then*

$$X_{i,k} \Rightarrow X_i \qquad (i = 1, \ldots, M) \tag{8.3.7}$$

implies

$$T_k \Rightarrow T. \tag{8.3.8}$$

(ii) **Renewal processes**

We consider the number ν_t of renewals in $[0, t)$ and the renewal function $H(t)$ and their robustness properties. Let $\{F_k\}$ $(k = 1, 2, \ldots)$ be a sequence of d.f.s of life time and let $\nu_{t,k}$ and $H_k(t)$ be the corresponding numbers of renewals and renewal functions.

Theorem 8.3.4. *Suppose*

$$F_k \Rightarrow F \qquad (k \to \infty). \tag{8.3.9}$$

Then as $k \to \infty$, for all continuity points t of $H(t)$,

$$H_k(t) \to H(t) \tag{8.3.10}$$

and if F is continuous,

$$\nu_{t,k} \Rightarrow \nu_t \qquad (\textit{all } t \geqq 0). \tag{8.3.11}$$

Proof. By (8.0.4) we obtain from (8.3.9) the weak convergence $F_k^{*m} \Rightarrow F^{*m}$ for all m, and when F is continuous, so also are the F^{*m}. Then (8.3.11) follows from (7.2.5).

To prove (8.3.10) we use the Laplace-Stieltjes transforms F^* and H^* of F and H, in terms of which it is well known that the renewal function has the form

$$H^*(s) = F^*(s)/\bigl(1 - F^*(s)\bigr) \qquad (\text{all } \operatorname{Re}(s) > 0).$$

Because weak convergence is equivalent to convergence of Laplace-Stieltjes transforms, (8.3.10) follows.

Boylan (1969) proved a robustness theorem for inventory problems by showing that the solution of the optimal inventory equation

$$f(x) = \inf_{y \geqq x} \left\{ g(y) + h(y - x) + a \int_0^\infty f(y - z)\, dF(z) \right\}$$

depends continuously on the quantities g, h, a, and F. Here $f(x)$ denotes the mean stationary optimal inventory costs in a period with inventory x, g represents the cost of storage or of loss in case of emptiness, h denotes the ordering cost, a is a discount factor, and F is the d.f. of demand.

Notes and bibliographic comments to Chapter 8

8.1

The robustness of homogeneous Markov chains and of queueing systems in the nonstationary case has been studied extensively by KALASHNIKOV (1974, 1978) using methods similar to those of Section 8.2.3. He also considered relations between this type of robustness and that of stationary distributions (see KALASHNIKOV (1975, 1978, 1979)).

It would be interesting to prove robustness theorems for the sample paths of loss systems as in Theorem 8.1.4; probably, the same conditions as for the single server queue are sufficient.

8.2

The proofs in Section 8.2.1 of robustness theorems for stationary waiting times are mainly based on the relatively deep formula (8.0.4). It would be desirable to have methods of proof applicable to general homogeneous Markov chains, because we may then be able to establish robustness properties without knowing the limit distributions. — ROSSBERG (1965) proved the first robustness theorem for GI/GI/1 using weak convergence methods, by assuming the weak convergence of K_n and the convergence of mean interarrival and service times. FRANKEN and STOYAN (1974) proved Theorem 8.2.4 for the case $\phi_k = \phi$ and $\mathsf{P}(\pi_F(D_\phi)) = 0$; KOTZUREK and STOYAN (1977) studied the general case. Robustness properties of processes of the form (8.2.12) are thoroughly studied in BOROVKOV (1978, 1980). TSITSIASHVILI (1972a, b, 1973, 1975, 1979b) and KALASHNIKOV (1978) used the method of test functions to study GI/GI/s and multi-stage GI/GI/.../1 queueing systems, reliability systems with renewal, and piecewise linear Markov chains. KRUPIN (1974, 1975) considered the robustness of multi-server systems with bounded delay times (see also KOTZUREK and STOYAN (1977)) or bounded queue lengths. Similar single-server queues are studied in BOROVKOV (1980). KARR (1975) also proved robustness theorems for semi-Markovian and Markovian renewal processes, WHITT (1980d) that of so-called generalized semi-Markovian processes. STOYAN (1972b) and STOYAN (1973) used internal and external monotonicity properties to prove robustness properties of stationary distributions in GI/GI/1 and GI/GI/s; this method is similar to that of Section 8.2.4, and the relations (8.0.13) and (8.0.14) are key steps in the proofs. NECHEPURENKO (1971) obtained quantitative robustness theorems for GI/GI/1 using a method in which the theory of contraction operators in quasi metric spaces plays a role.

The quantitative robustness theorems so far established are very rough and of less value for obtaining bounds. A combination of robustness and monotonic properties may yield better results.

TSITSIASHVILI (1979a) contains an excellent survey of recent results in the robustness theory of queueing models.

Open problem. Give conditions which ensure the robustness of moments of the stationary waiting time d.f.s in GI/GI/1 and GI/GI/s. For GI/GI/1 this should be possible by means of (5.0.5), and it may be the case that the means are robust if, in addition to the conditions (8.2.2) and (8.2.5), the second moments of service times converge.

References

The titles of the references below are given in the original language if English, French, or German, and otherwise translated into English and the original language identified.

Where bibliographic details are known of a published English translation of a reference, these are given in the form: details of original publication (= details of translation). Occasionally the details of an abstract in *Mathematical Reviews* are listed, but this information has not been sought out systematically.

AGRESTI, A. (1974) Bounds on the extinction time distribution of a branching process. *Adv. Appl. Prob.* **6**, 322–335.

— (1975) On the extinction time of varying and random environment branching processes. *J. Appl. Prob.* **12**, 39–46.

AHMED, A.-H. N., LEON, R., & PROSCHAN, F. (1981) Generalized association, with applications in multivariate statistics. *Ann. Statist.* **9**, 168–176.

AL-KHAYYAL, F. A., & GROSS, D. (1977) On approximating and bounding queues. In: *Algorithmic Methods in Probability* (Ed. M. F. NEUTS), North Holland, Amsterdam, 233–246.

ANDERSON, W. J. (1972) Local behaviour of solutions of stochastic differential equations. *Trans. Amer. Math. Soc.* **164**, 309–321.

ARJAS, E., & LEHTONEN, T. (1978) Approximating many server queues by means of single server queues. *Math. Operat. Res.* **3**, 205–223.

AVIS, D. M. (1977) Computing waiting times in $GI/E_k/c$ queueing systems. In: *Algorithmic Methods in Probability* (Ed. M. F. NEUTS), North Holland, Amsterdam, 215–232.

BANDEMER, H., BELLMANN, A., JUNG, W., & RICHTER, K. (1976) *Theorie und Anwendung der optimalen Versuchsplanung, I.* Akademie-Verlag, Berlin.

BANDEMER, H., & STOYAN, D. (1971) Optimaler Stichprobenumfang bei a-priori-Information über gewisse Parameter. *Biom. Zeit.* **13**, 376–383.

BARLOW, R. E., BARTHOLOMEW, D. J., BREMNER, J. M., & BRUNK, H. D. (1972) *Statistical Inference under Order Restrictions.* Wiley, New York.

BARLOW, R. E., & PROSCHAN, F. (1965) *Mathematical Theory of Reliability.* Wiley, New York.

— (1975) *Statistical Theory of Reliability and Life Testing: Probability Models.* Holt, Rinehart & Winston, New York.

— (1976) Theory of Maintained Systems: Distribution of time to first system failure. *Math. Operat. Res.* **1**, 32–42.

— (1980) Data analysis and inference in reliability and biometry. Chapter in a book of the same title 1983 being in preparation.

BARTFAI, P. (1973) Große Abweichungen der Wartezeit. In: *Proc.* 4th *Conf. Probab. Theory* (Brasov (1971), Ed. Acad. R.S. Romania, Bucharest, 97–105. [*Math. Rev.* 53 # 6779.]

BASHARIN, G. P., KOKOTUSHKIN, V. A., & NAUMOV, V. A. (1979) On a method of substituting equivalent fragments for calculations in a communications network for a CPU, I (in Russian). *Izv. Akad. Nauk SSSR Tehn. Kibernet.* **1979,** 6: 92–99 (= *Engrg. Cybernetics* **17,** 6: 66–73).

— (1980) Ibid., II. *Izv. Akad. Nauk SSSR Tehn Kibernet.* **1980,** 1: 55–61 (= *Engrg. Cybernetics* **18,** 1: 48–53).

BASHARIN, G. P., & LYSENKOVA, V. T. (1972) The servicing of several inhomogeneous flows by a fully accessible complex with a limited queue (in Russian). In: *Sistemy Raspredelenija Informacii* (Ed. A. D. HARKEVICH & V. E. GARMAS), Nauka, Moskva, 3–16. [*Math. Rev.* 51 # 15012.]

BENEŠ, V. E. (1959) On trunks with negative exponential holding times serving a renewal process. *Bell System Tech. J.* **38,** 211–258.

— (1965) Some inequalities in the theory of telephone traffic. *Bell. System Tech. J.* **44,** 1941–1975.

BERGMANN, R. (1974) Qualitätsunterschiede bei Wartemodellen vom Typ G/G/1. *Math. Operat.-forschung Statist.* **5,** 709–724.

— (1978) Some classes of semi-ordering relations for random vectors and their use for comparing covariances. *Math. Nachr.* **82,** 103–114.

— (1979a) Qualitative properties and bounds for the serial covariances of waiting times in single-server queues. *Operat. Res.* **27,** 1168–1179.

— (1979b) Some classes of distributions and their application in queueing. *Math. Operat.-forschung Statist., Ser. Statist.* **10,** 583–600.

BERGMANN, R., DALEY, D. J., ROLSKI, T., & STOYAN, D. (1979) Bounds for cumulants of waiting times in GI/GI/1 queues. *Math. Operat.-forschung Statist., Ser. Optim.* **10,** 257–263.

BERGMANN, R., & STOYAN, D. (1976) On exponential bounds for the waiting time distribution function in GI/G/1. *J. Appl. Prob.* **13,** 411–417.

— (1978) Monotonicity properties of second order characteristics of stochastically monotone Markov chains. *Math. Nachr.* **82,** 99–102.

BESSLER, S., & VEINOTT, T. (1966) Optimal policy for a dynamic multi-echelon inventory model. *Naval Res. Logist. Quart.* **13,** 355–389.

BHAT, U. N., SHALABY, M., & FISCHER, M. J. (1979) Approximation techniques in the solution of queueing problems. *Naval Res. Log. Quart.* **26,** 311–326.

BILLINGSLEY, P. (1968) *Convergence of Probability Measures.* Wiley, New York.

BLACKWELL, D. (1953) Equivalent comparisons of experiments. *Ann. Math. Statist.* **24,** 265–272.

BLOCK, H. W., & SAVITS, T. H. (1980) Multivariate increasing failure rate average distributions. *Ann. Prob.* **8,** 793–801.

BLOMQVIST, N. (1970) On the transient behaviour of the GI/G/1 waiting-times. *Skand. Akt.* **1970,** 118–129.

BLOOMFIELD, P., & COX, D. R. (1972) A low traffic approximation for queues. *J. Appl. Prob.* **9,** 832–840.

BOHL, E. (1974) *Monotonie: Lösbarkeit und Numerik bei Operatorgleichungen.* Springer-Verlag, Berlin.

BOROVKOV, A. A. (1970) Factorization identities and properties of the distribution of the supremum of sequential sums (in Russian). *Teor. Verojat. Primen.* **15,** 377–418 (= *Theory Prob. Appl.* **15,** 359–402).

— (1972) Continuity theorems for multichannel systems with refusals (in Russian). *Teor. Verojat. Primen.* **17,** 458–467 (= *Theory Prob. Appl.* **17,** 434–444).

— (1976a) *Stochastic Processes in Queueing Theory.* Springer-Verlag, New York (Translation by K. WICKWIRE of 1972 Russian edition with additional material).

— (1976b) Convergence of measures and random processes (in Russian). *Uspehi Mat. Nauk* **31,** 2: 3–68 (= *Russian Math. Surveys* **31,** 2: 1–69).

— (1977) Some estimates of the convergence rate in stability theorems (in Russian). *Teor. Verojat. Primen.* **22,** 689–699 (= *Theory Prob. Appl.* **22,** 668–678).

— (1978) Ergodic and stability theorems for a class of stochastic equations and their application (in Russian). *Teor. Verojat. Primen.* **23**, 241—262 (= *Theory Prob. Appl.* **23**, 227—247).

— (1980) *Asymptotic Methods in Queueing Theory* (in Russian), Nauka, Moskva.

Boxma, O. J., Cohen, J. W., & Huffels, N. (1979) Approximations in the mean waiting time in a M/G/s queueing system. *Operat. Res.* **27**, 1115—1127.

Boylan, E. S. (1969) Stability theorems for solutions to the optimal inventory equation. *J. Appl. Prob.* **6**, 211—217.

Brandt, A., & Lisek, B. (1983) On the continuity of G/GI/m queues. *Math. Operat.-forschung Statist., Ser. Statist.* **14**.

Braun, H. (1975) Polynomial bounds for probability generating functions. *J. Appl. Prob.* **12**, 507—514.

Brook, D. (1966) Bounds for moment generating functions and extinction probabilities. *J. Appl. Prob.* **3**, 171—178.

Brown, M. (1980) Bounds, inequalities, and monotonicity properties for some specialized renewal processes. *Ann. Probab.* **8**, 227—240.

— (1981) Further monotonicity properties for specialized renewal processes. *Ann. Prob.* **9**, 891—895.

Brumelle, S. L. (1971a) Some inequalities for parallel-server queues. *Operat. Res.* **19**, 402—413.

— (1971b) On the relation between customer and time averages in queues. *J. Appl. Prob.* **8**, 508—520.

— (1973) Bounds on the wait in a GI/M/k queue. *Management Sci.* **19**, 773—777.

Brumelle, S. L., & Vickson, R. G. (1975) A unified approach to stochastic dominance. In: *Stochastic Optimization: Models in Finance* (Ed. W. T. Ziemba & R. G. Vickson), Academic Press, New York, 101—113.

Butler, D. A. (1982) Bounding the reliability of multistate systems. *Operat. Res.* **10**, 530—544.

Bux, W., & Herzog, U. (1977) The phase concept: Approximation of measured data and performance analysis. In: *Computer Performance* (Ed. K. M. Chandy & M. Reiser), North Holland, Amsterdam, 23—38.

Calo, S. B., & Schwarz, S. C. (1977) Bounds and bounding functions for a stationary GI/G/1 queue. *Math. Operat. Res.* **2**, 240—243.

Cambanis, S., & Simons, G. (1982) Probability and expectation inequalities. *Zeit. Wahrscheinlichkeitsth.* **59**, 1—25.

Cambanis, S., Simons, G., & Stout, W. (1976) Inequalities for Ek (X, Y) when the marginals are fixed. *Zeit. Wahrscheinlichkeitsth.* **36**, 285—294.

Chan, M. M. W. (1971) System simulation and maximum entropy. *Operat. Res.* **19**, 1553—1556.

Chong, Kong-Ming (1974) Some extensions of a theorem of Hardy, Littlewood and Polya and their applications. *Can. J. Math.* **26**, 1321—1340.

— (1975) Spectral orders, conditional expectations and martingales. *Zeit. Wahrscheinlichkeitsth.* **31**, 329—332.

Chung, K. L. (1967) *Markov Chains with Stationary Transition Probabilities.* 2nd. ed. Springer-Verlag, Berlin.

Cohen, J. W. (1969) *The Single-Server Queue.* North Holland, Amsterdam.

— (1973) Asymptotic relations in queueing theory. *Stoch. Processes Appl.* **1**, 107 to 124.

— (1976) *On Regenerative Processes in Queueing Theory.* Lect. Notes Econ. Math. Syst. **121**, Springer-Verlag, Berlin.

Collatz, L. (1966) *Functional Analysis and Numerical Mathematics.* Academic Press, New York (= Translation by H. Oser of 1964 German edition).

Cosmetatos, G. L. (1974) Approximate equilibrium results for the multi-server queue GI/M/r. *Operat. Res. Quart.* **25**, 625–634.

— (1976) Some approximate equilibrium results for the multi-server queue (M/G/r). *Operat. Res. Quart.* **27**, 615–620.

Cosmetatos, G. P., & Godsave, S. A. (1980) Approximations in the multi-server queue with hyperexponential inter-arrival times and exponential service times. *Operat. Res. Quart.* **31**, 57–62.

Cox, D. R. (1962) *Renewal Theory.* Methuen, London.

Daley, D. J. (1968) Stochastically monotone Markov chains. *Zeit. Wahrscheinlichkeitsth.* **10**, 305–317.

— (1976a) Another upper bound for the renewal function. *Ann. Prob.* **4**, 109–114.

— (1976b) Queueing output processes. *Adv. Appl. Prob.* **8**, 395–415.

— (1976c) Bounds on mean waiting times in GI/G/1 queues. *Research Report* No. 9, School of Mathematical Sciences, Univ. Melbourne.

— (1977) Inequalities for moments of tails of random variables with a queueing application. *Zeit. Wahrscheinlichkeitsth.* **41**, 139–143.

— (1978a) Upper bounds for the renewal functions via Fourier methods. *Ann. Prob.* **6**, 876–884.

— (1978b) Bounds for the variance of certain stationary point processes. *Stoch. Processes Appl.* **7**, 255–264.

— (1980) Tight bounds for the renewal function of a random walk. *Ann. Prob.* **8**, 615–621.

Daley, D. J., & Haslett, J. (1982) A thermal energy storage processes with controlled input. *Adv. Appl. Prob.* **14**, 257–271.

Daley, D. J., & Moran, P. A. P. (1968) Two-sided inequalities for waiting time and queue size distribution in GI/G/1. *Teor. Verojat. Primen.* **13**, 356–358 (= *Theory Prob. Appl.* **13**, 338–341).

Daley, D. J., & Narayan, P. (1980) Series expansions of probability generating functions and bounds for the extinction probability of a branching process. *J. Appl. Prob.* **17**, 939–947.

Daley, D. J. & Rolski, T. (1983) A light traffic approximation for a single-server queue. To appear in *Math. Operat. Res.*

Daley, D. J., & Tong, Y. L. (1980) On a multivariate generalization of Tchebychev's covariance inequality and some important stochastic processes. (Preprint).

Daley, D. J., & Trengove, C. D. (1977) Bounds for mean waiting times in single-server queues: a survey. Statistics Dept. (IAS), Australian National University (Preprint).

Daley, D. J. & Whitt, W. (1983) Queueing systems with extremal mean waiting times: a review. (Unpublished).

David, F. N., Kendall, M. G., & Barton, D. E. (1966) *Symmetric Function and Allied Tables.* Cambridge University Press, Cambridge.

Delbrouck, L. E. N. (1973) A set of Wiener-Hopf integral equations with common solution in fluctuation theory. *J. Math. Anal. Appl.* **44**, 100–112.

— (1975) Convexity properties, moment inequalities and asymptotic exponential approximations for delay distributions in GI/G/1 systems. *Stoch. Processes Appl.* **3**, 193–207.

Derman, C., & Ignall, E. (1975) On the stochastic ordering of Markov chains. *Operat. Res.* **23**, 574–576. Priority acknowledgement **24**, 592.

Disney, R. L., & Cherry, W. P. (1974) Some topics in queueing network theory. In: *Mathematical Methods in Queueing Theory.* (Ed. A. B. Clarke.) Lect. Notes Econ. Math. Syst. **98**, Springer-Verlag, Berlin, 23–44.

Eckberg, A. E. (1977) Sharp bounds on Laplace-Stieltjes transforms, with applications to various queueing problems. *Math. Operat. Res.* **2**, 135–142.

EDWARDS, R. E. (1965) *Functional Analysis: Theory and Applications*, Holt, Rinehart and Winston, New York.

EL-NEWEIHI, E. (1981) Stochastic ordering and a class of multivariate new better then used distributions. *Commun. Statist.-Theor. Meth.* **10 (16)**, 1655–1672.

EL-NEWEIHI, E., & PROSCHAN, F. (1978) Multistate reliability models: a survey. In: *Multivariate Analysis* V., North Holland, Amsterdam–New York, 523 to 541.

EL-NEWEIHI, E., PROSCHAN, F., & SETHURAMAN, J. (1978) Multistate coherent systems. *J. Appl. Prob.* **15**, 675–688.

ESARY, J. D., & MARSHALL, A. W. (1979) Multivariate distributions with increasing hazard rate average. *Ann. Prob.* **7**, 359–370.

ESARY, J. D., MARSHALL, A. W., & PROSCHAN, F. (1970) Some reliability applications of the hazard transform. *SIAM J. Appl. Math.* **18**, 849–860.

ESARY, J. D., PROSCHAN, F., & WALKUP, D. W. (1967) Association of random variables, with applications. *Ann. Math. Statist.* **38**, 1466–1474.

FAHADY, K. S., QUINE, M. P., & VERE-JONES, D. (1971) Heavy traffic approximations for the Galton-Watson process. *Adv. Appl. Prob.* **3**, 282–300.

FAHMY, S., PEREIRA, C. A. B., PROSCHAN, F. & SKAKED. M. (1982) The influence of the sample on the posterior distribution. Commun. Statist.-Theor. Meth. **11 (16)**, 1757–1768.

FAINBERG, M. A. (1976) Estimates of the parameters of a queue in multiline queueing systems of the type E_k/G/s (in Russian). *Izv. Akad. Nauk SSSR Tehn. Kibernet.* **1976**, 5: 71–78 (= *Engrg. Cybernetics* **14**, 5: 66–73).

— (1979) Bounds on the waiting time in single-server queues (in Russian), *Izv. Akad. Nauk SSSR Tehn. Kibernet.* **1979**, 2: 217–219 (= *Engrg. Cybernetics* **17**, 2: 144–149).

FAINBERG, M. A., & STOYAN, D. (1978) A further remark on a paper of K. T. Marshall on bounds for GI/G/1. *Math. Operat.-forschung Statist., Ser. Optim.* **9**, 145–150.

FELLER, W. (1966) *An Introduction to Probability Theory and Its Applications*, Vol. **2**. Wiley, New York.

FERGUSON, T. F. (1967) *Mathematical Statistics: A Decision Theoretic Approach*. Academic Press, New York.

FINCH, P. D. (1959) On the distribution of queue size in queueing problems. *Acta Math. Acad. Sci. Hungar.* **10**, 327–336.

FLEISCHMANN, K. (1976) Optimal input for the loss system G/M/2. *Math. Operat.-forschung Statist.* **7**, 129–138.

FLIPO, D. (1981) Comparison des disciplines de service des files d'attente G/G/1. *Ann. Inst. H. Poincaré* **B17**, 191–212.

FRANKEN, P. (1970) Ein Stetigkeitssatz für Verlustsysteme. In: *Operationsforschung und Mathematische Statistik*, II (Ed. O. BUNKE), Akademie-Verlag, Berlin, 9–23. [Maths. Rev. **41** # 4690.]

— (1976) Einige Anwendungen der Theorie zufälliger Punktprozesse in der Bedienungstheorie I. *Math. Nachr.* **70**, 303–319.

FRANKEN, P., & HEUSER, K. (1977) Estimates of indexes of reliability for redundant systems with renewal (in Russian). *Izv. Akad. Nauk. SSSR Tehn. Kibernet.* **1977**, 4: 100–105 (= *Engrg. Cybernetics* **15**, 4: 64–69).

FRANKEN, P., & KALÄHNE, U. (1978) Existence, uniqueness and continuity of stationary distributions for queueing systems without delay. *Math. Nachr.* **86**, 97–115.

FRANKEN, P., & KIRSTEIN, B. M. (1977) Zur Vergleichbarkeit zufälliger Prozesse. *Math. Nachr.* **78**, 197–205.

FRANKEN, P., KÖNIG, D., ARNDT, U., & SCHMIDT, V. (1981) *Queues and Point Processes*. Akademie-Verlag, Berlin. J. WILEY & SONS, Chichester 1982.

FRANKEN, P., & STOYAN, D. (1974) Stabilitätssätze für eine Klasse homogener Markovscher Prozesse. *Math. Nachr.* **60**, 311–316.

— (1975) Einige Bemerkungen über monotone und vergleichbare Markovsche Prozesse. *Math. Nachr.* **66**, 201–209.

FREDERICKS, A. A. (1980) A class of approximations for the waiting time distribution in a G/G/1 queueing system. *Bell. Syst. Techn.* J. **61**, 295–326.

GAEDE, K.-W. (1965) Konfidenzgrenzen bei Warteschlangen- und Lagerhaltungs-Problemen. *Zeit. angew. Math. Mech.* **45**, T 91–92.

— (1973) Einige Abschätzungen in der Bedienungstheorie. In: *Proceedings in Operations Research 2* (Ed. H. JACOB et al.), Physica-Verlag, Würzburg, 241–255.

— (1974a) Monotonieeigenschaften bei Zuverlässigkeitsproblemen. In: *Proceedings in Operations Research 3* (Ed. P. GESSNER et al.), Physica-Verlag, Würzburg, 240–249.

— (1974b) Sensibilitätsanalyse für einen Semi-Markov-Prozeß. *Zeit. Operat. Res. Ser. A-B* **18**, A197–204. [Maths. Rev. **50** # 11515.]

GELENBE, E. (1975) On approximate computer models. *J. Ass. Comp. Mach.* **22**, 261–269.

GELENBE, E., & MUNTZ, R. R. (1976) Probabilistic models of computer systems — Part I (exact results). *Acta Informatica* **7**, 35–60.

GELENBE, E., & PUJOLLE, G. (1976) The behaviour of a single queue in a general queueing network. *Acta Informatica* **7**, 123–136.

GITTINS, J. C. (1971) Stochastic monotonicity and queues subject to tidal interruptions. *Proc. Camb. Phil. Soc.* **70**, 61–75.

— (1978) A comparison of service disciplines for GI/G/m queues. *Math. Operat.-forschung Statist., Ser. Optim.* **9**, 255–260.

GNEDENKO, B. V., BELYAYEV, J. K., & SOLOVYEV, A. D. (1965) *Mathematical Methods of Reliability Theory* (in Russian). Nauka, Moskva. (English translation (1969), Academic Press, New York; German translation in 2 Vols. (1968), Akademie-Verlag, Berlin.)

GNEDENKO, B. V., & KOVALENKO, I. N. (1966) *Introduction to Queueing Theory*, Nauka, Moskva (English translation (1968), Israel Program for Scientific Translations, Jerusalem; German translation (1971), with appendix by D. KÖNIG, K. MATTHES & K. NAWROTZKI, Akademie-Verlag, Berlin.)

GOLENKO, D. I. (1972) *Statistische Methoden in der Netzplantechnik.* Teubner-Verlag, Leipzig.

GOROFF, D., & WHITT, W. (1980) Approximating the admissible set in stochastic dominance. *J. Econ. Theory* **23**, 218–235.

GRANOT, D., GRANOT, F., & LEMOINE, A. J. (1975) Approximations for a service system with nonindependent interarrival times. *Operat. Res.* **23**, 162–166.

GREEN, L. (1982) A limit theorem on subintervals of interrenewal times. *Operat. Res.* **30**, 210–216.

GROENEVELT, H., VAN HOORN, M. H. & TIJMS, H. C. (1982) Tables for M/G/c queueing systems with phase type service. (Submitted).

HAEZENDONCK, J., & DE VYLDER, F. (1980) A comparison criterion for explosions in point processes. *J. Appl. Prob.* **17**, 1102–1107.

HALACHMI, B., & FRANTA, W. R. (1977) A diffusion approximation solution to the G/G/k queueing system. *Comp. Operat. Res.* **4**, 37–46.

— (1978) A diffusion approximation to the multi-server queue. *Management Sci.* **24** 522–529.

HALFIN, S. (1981) Distribution of the interoverflow time for the GI/G/1 loss system. *Math. Operat. Res.* **6**, 563–570.

HALFIN S., & WHITT, W. (1981) Heavy traffic limits for queues with exponential servers. *Operat. Res.* **29**, 567–588.

HALMOS, P. (1958) *Measure Theory*. Van Nostrand, Princeton.

HANISCH, K.-H., & STOYAN, D. (1978) Abschätzungen für die mittlere Gesamtvorgangsdauer bei PERT-Netzwerken. *Proceedings 8th IKM Weimar* 1978.

HARDY, G. H., LITTLEWOOD, J. E., & POLYA, G. (1929) Some simple inequalities satisfied by convex functions. *Messenger of Math.* **58**, 145–152.

HARRIS, B. (1962) Determining bounds on expected values of certain functions. *Ann. Math. Statist.* **33**, 1454–1457.

HARRIS, T. E. (1963) *The Theory of Branching Processes*. Springer-Verlag, Berlin.

HARRISON, J. M. (1977) Some stochastic bounds for dams and queues. *Math. Operat. Res.* **2**, 54–63.

HARTLEY, H. O., & DAVID, H. A. (1954) Universal bounds for mean range and extreme observation. *Ann. Math. Statist.* **25**, 85–99.

HEATHCOTE, C. R. (1967) Complete exponential convergence and some related topics. *J. Appl. Prob.* **4**, 217–256.

HEATHCOTE, C. R., & WINER, P. (1969) An approximation for the moments of waiting times. *Operat. Res.* **17**, 175–186.

HEIN, O. (1973) Wartezeiten für beliebige Bedienungszeiten. *Angew. Informatik* **15**, 281–289.

HEYDE, C. C., & SCHUH, H.-J. (1978) Uniform bounding of probability generating functions and the evolution of reproduction rates in birds. *J. Appl. Prob.* **15**, 243–250.

HEYMAN, D. P. (1980) Comments on a queueing inequality. *Management Sci.* **26**, 956–959.

— (1982) On Ross's conjectures about queues with non-stationary Poisson arrivals. *J. Appl. Prob.* **19**, 245–249.

HEYMAN, D. P., & MARSHALL, K. T. (1968) Bounds on the optimal operating policy for a class of single-server queues. *Operat. Res.* **16**, 1138–1146.

HILLIER, F. S. & YU, O. S. (with AVIS, D. M., FOSSET, L. D. & REIMAN, M. I.) (1981) *Queueing Tables and Graphs*. North Holland, New York.

HOCHSTÄDTER, D. (1969) *Stochastische Lagerhalterungsmodelle*. Lect. Notes Operat. Res. Math. Econ. **10**, Springer-Verlag, Berlin.

HOEFFDING, W. (1955) The extrema of the expected value of a function of independent random variables. *Ann. Math. Statist.* **26**, 268–275.

HOFFMANN-JØRGENSEN, J. (1977) Probability in Banach spaces. In: Lect. Notes Math. **598**, Springer-Verlag, Berlin, 1–186.

HOKSTAD, P. (1978) Approximations for the M/G/m queue. *Operat. Res.* **26**, 510–523.

HOLGATE, P., & LAKHANIE, K. H. (1967) Effect of offspring distribution on population survival. *Bull. Math. Biophys.* **29**, 831–839.

HOLTZMANN, J. M. (1973) On the accuracy of the equivalent random method with renewal inputs. *Bell System Tech. J.* **52**, 1673–1679.

HUBER, P. J. (1976) Kapazitäten statt Wahrscheinlichkeiten? Gedanken zur Grundlegung der Statistik. *Jber. Dtsch. Mathver.* **78**, 81–92.

HUMAK, K. M. S. (1977) *Statistische Methoden der Modellbildung, I*. Akademie-Verlag, Berlin.

HWANG, T.-Y., & WANG, N.-S. (1979) On best fractional linear generating function bounds. *J. Appl. Prob.* **16**, 449–453.

IOSIFESCU, M., & THEODURESCU, R. (1969) *Random Processes and Learning*. Springer-Verlag, Berlin.

JACOBS, D. R., & SCHACH, S. (1972) Stochastic order relationships between GI/G/k systems. *Ann. Math. Statist.* **43**, 1623–1633.

JAGERS, P. (1975) *Branching Processes with Biological Applications*. Wiley, London.

JANSEN, U., & KÖNIG, D. (1980) Insensitivity and steady-state probabilities in product form for queueing networks. *Elektron. Informationsverarb. Kybernet.* **16**, 385–397.

JANSSEN, J. (1976) Extension of a two-sided inequality of Daley and Moran to the completely semi-Markov queueing model. *Cahiers C.E.R.O.* **18**, 459–469.

JURKEVICH, O. M. (1973) Many-line queueing systems with arbitrary distribution of the length of servicing of impatient clients (in Russian). *Izv. Akad. Nauk. SSSR Tehn. Kibernet.* **1973**, 6: 80–89 (= *Engrg. Cybernetics* **11**, 954–963).

KALASHNIKOV, V. V. (1974) Uniform closeness of components of a Markov process with variation of its parameters (in Russian). *Izv. Akad. Nauk. SSSR Tehn. Kibernet.* **1974**, 4: 43–53 (= *Engrg. Cybernetics* **12**, 4: 32–41).

— (1975) Interaction between the stability of distributions and their limiting distributions. *Izv. Akad. Nauk. SSSR Tehn. Kibernet.* **1975**, 3: 103–108 (= *Engrg. Cybernetics* **13**, 3: 71–76).

— (1978) *Qualitative Analysis of the Behaviour of Complicated Systems Using the Method of Trial Functions* (in Russian). Nauka, Moskva.

— (1981) Estimates of convergence rate and stability for regenerative and renovative processes. In: *Proc. Coll. Point Processes and Queueing Theory, Kezsthely* 1978. North Holland, Amsterdam, 163–180.

KALASHNIKOV, V. V., & ANICHKIN, S. A. (1981) Continuity of random sequences and approximation of Markov chains. *Adv. Appl. Prob.* **13**, 402–414.

KALASHNIKOV, V. V., & TSITSIASHVILI, G. SH. (1972) On the stability of queueing systems with respect to disturbances of their distribution functions (in Russian). *Izv. Akad. Nauk. SSSR Tehn. Kibernet.* **1972**, 2: 41–48 (= *Engrg. Cybernetics* **10**, 211–217).

KALL, P., & STOYAN, D. (1981) Solving stochastic programming problems with recourse including error bounds. *Math. Operat.-forschung Statist., Ser. Optim.*, **12**, 431–447.

KALMYKOV, G. I. (1962) On the partial ordering of one-dimensional Markov processes. *Teor. Verojat. Primen.* **7**, 466–469 (= *Theory Prob. Appl.* **7**, 456–459).

— (1969) Semiordering of the probabilities of the first passage time of Markov processes. *Teor. Verojat. Primen.* **14**, 735–742 (= *Theory Prob. Appl.* **14**, 704–710).

KAMAE, T., & KRENGEL, U. (1978) Stochastic partial ordering. *Ann. Prob.* **6**, 1044 to 1049.

KAMAE, T., KRENGEL, U., & O'BRIEN, G. L. (1977) Stochastic inequalities on partially ordered spaces. *Ann. Prob.* **5**, 899–912.

KARLIN, S. (1960) Dynamic inventory policy with varying stochastic demands. *Management Sci.* **6**, 231–258.

— (1968) *Total Positivity, Vol. I.* Stanford University Press, Stanford, California.

KARLIN, S., & NOVIKOFF, A. (1963) Generalized convex inequalities. *Pacific J. Math.* **13**, 1251–1279.

KARLIN, S., & STUDDEN, W. J. (1966) *Tchebycheff Systems: With Applications in Analysis and Statistics.* Wiley, New York.

KARLIN, S., & RINOTT, Y. (1980) Classes of orderings of measures and related correlation inequalities: I. Multivariate total positive distributions. *J. Mult. Anal.* **10**, 467–498, II. *J. Mult. Anal.* **10**, 499–516.

KARPELEVICH, F. I., & KREININ, A. J. (1976) Estimates of the mean waiting time in a single-queue system (GI/G/1) (in Russian). *Izv. Akad. Nauk SSSR Tehn. Kibernet.* **1976**, 3: 105–107 (= *Engrg. Cybernetics* **14**, 3: 81–83).

— (1980) Some bounds for the mean waiting time in E_k/GI/1 queues. *Math. Operat.-forschung Statist., Ser. Statist.* **11**, 85–88.

KARR, A. F. (1975) Weak convergence of a sequence of Markov chains. *Zeit. Wahrscheinlichkeitsth.* **33**, 41–48.

KEILSON, J., & KESTER, A. (1977) Monotone matrices and monotone Markov processes. *Stoch. Processes Appl.* **5**, 231–241.

KEILSON, J., & SUMITA, U. (1982) Uniform stochastic ordering and related inequalities. *Canad. J. Statist.* **10**, 181–198.

KELLY, F. P. (1979) *Reversibility and Stochastic Networks.* Wiley, London.

KEMPERMANN, J. H. B. (1977) On the FKG-inequality for measures on a partially ordered space. *Indag. Math.* **39**, 313–331.

KENNEDY, D. P. (1972) The continuity of the single server queue. *J. Appl. Prob.* **9**, 370–381.

KERSTAN, J., & FRANKEN, P. (1968) Bedienungssysteme mit unendlich vielen Bedienungsapparaten. In: *Operationsforschung und Mathematische Statistik I* (Ed. O. BUNKE). Akademie-Verlag, Berlin, 67–76. [Math. Rev. 39 # 7693].

KIEFER, J., & WOLFOWITZ, J. (1955) On the theory of queues with many servers. *Trans. Amer. Math. Soc.* **78**, 1–18.

— (1956) On the characteristics of the general queuing process with application to random walks. *Ann. Math. Statist.* **27**, 147–161.

KINGMAN, J. F. C. (1962) Some inequalities for the queue G/G/1. *Biometrika* **49**, 315–324.

— (1964) A martingale inequality in the theory of queues. *Proc. Camb. Phil. Soc.* **59**, 359–361.

— (1965) The heavy traffic approximation in the theory of queues. In: *Proc. Symp. Congestion Theory.* (Ed. W. L. SMITH and W. E. WILKINSON), Univ. North Carolina Press, Chapel Hill, NC, 137–159.

— (1966) On the algebra of queues. *J. Appl. Prob.* **3**, 285–326.

— (1970) Inequalities in the theory of queues. *J. Roy. Statist. Soc.* B. **32**, 102 bis 110.

KIRSTEIN, B. M. (1976) Monotonicity and comparability of time-homogeneous Markov processes with discrete state space. *Math. Operat.-forschung Statist.* **7**, 151–168.

— (1978) Some monotonicity properties and bounds for GI/M/m systems without delay. *Elektron. Informationsverarb. Kybernet.* **14**, 203–216, Corr. **15**, 336.

— (1979) *Monotonie und Vergleichbarkeit Markowscher Prozesse mit Anwendungen in der Bedienungstheorie.* Dissertation, Humboldt-University, Berlin.

KLEINROCK, L. (1964) *Communication Nets — Stochastic Message Flow and Delays.* McGraw-Hill, New York.

— (1976) *Queueing Systems, Vol. 2: Computer Applications.* Wiley, New York.

KOBAYASHI, H. (1974) Application of the diffusion approximation to queueing networks I. *J. Ass. Comp. Mach.* **21**, 316–328.

KOFLER, E., & MENGES, G. (1976) *Entscheidungen bei unvollständiger Information.* Lect. Notes Econ. Math. Syst. **136**, Springer-Verlag, Berlin.

KOGAN, YA. A., & LITVIN, V. G. (1979) Piecewise diffusion approximations for queueing problems with heterogeneous arrivals and service. *Problems Control Inform. Theory/Problemy Upravlen. Teor. Inform.* **8**, 5–6: 433–443.

KOKOTUSHKIN, V. A. (1974) A generalization of the Palm-Khinchin theorem. *Teor. Verojat. Primen* **19**, 622–625 (= *Theory Prob. Appl.* **19**, 594–597).

KÖLLERSTRÖM, J. (1974) Heavy traffic theory for queues with several servers, I. *J. Appl. Prob.* **11**, 544–552.

— (1976) Stochastic bounds for the single-server queue. *Math. Proc. Camb. Phil. Soc.* **80**, 521–525.

— (1978) Stochastic bounds for the queue GI/G/1 in heavy traffic. *Math. Proc. Camb. Phil. Soc.* **84**, 361–375.

— (1979) Heavy traffic theory for queues with several servers, II. *J. Appl. Prob.* **16**, 393–401.

— (1981) A second-order heavy traffic approximation for the queue GI/G/1. *Adv. Appl. Prob.* **13**, 167–185.

König, D., Rolski, T., Schmidt, V., & Stoyan, D. (1978) Stochastic processes with imbedded marked point processes (PMP) and their application in queueing. *Math. Operat.-forschung Statist., Ser. Optim.* **9**, 125–141.

König, D., & Schmidt, V. (1980) Stochastic inequalities between customer stationary and time-stationary characteristics of queueing systems with point processes. *J. Appl. Prob.* **17**, 768–777.

König, D., Schmidt, V., & Stoyan, D. (1976) On some relations between stationary distributions of queue lengths and embedded queue lengths in G/G/s queueing systems. *Math. Operat.-forschung Statist.* **7**, 577–586.

Korzeniowski, A., & Opawski, A. (1976) Bounds for reliability in the NBU, NWU and NBUE, NWUE classes. *Zastosowania Mat.* **15**, 1–5.

Kosten, L. (1973) *Stochastic theory of service systems.* Pergamon Press, Oxford.

— (1978) Parameterization of overflow traffic in telecommunication and data handling. *Adv. Appl. Prob.* **10**, 337–338.

Kotzurek, M. (1979) Further bounds for waiting time distributions and their means in the case $\varrho < 1$. *Math. Operat.-forschung Statist., Ser. Optim.* **10**, 405–412.

Kotzurek, M., & Stoyan, D. (1976) A quantitative continuity theorem for the mean stationary waiting time GI/G/1. *Math. Operat.-forschung Statist.* **7**, 595–599.

— (1977) A remark on the stability of the stationary distribution of Markov chains. *Math. Nachr.* **78**, 7–12.

Kozlov, B. A., & Ushakov, I. A. (1970) *Reliability Handbook.* Holt, Rinehart and Winston, New York. (= Translation from 1966 Russian edition *Handbook for the Calculation of Reliability of Radioelectronic Devices,* Izdat. Sov. Radio, Moscow). [Maths. Rev. **43** # 5671].

— (1975) *Handbook for the Calculation of Reliability of Radioelectronic and Automatic Devices* (in Russian). Izdat. Sov. Radio, Moskva.

Kozlov, B. A., & Vasil'ev, V. A. (1969) The effect of the shape of the distribution on the reliability of two-unit systems (in Russian). In: *Reliability and Queueing Theory* (in Russian), Nauka, Moskva, 37–45.

Kozlov, B. A., & Solovyev, A. D. (1978a) Optimal servicing of renewable systems, I (in Russian). *Izv. Akad. Nauk. SSSR Tehn. Kibernet.* **1978**, 3: 75–80 (= *Engrg. Cybernetics* **16**, 3: 62–66).

— (1978b) Optimal servicing of renewable systems, II (In Russian). *Izv. Akad. Nauk SSSR Tehn. Kibernet.* **1978**, 4: 75–80 (= *Engrg. Cybernetics* **16**, 4: 53–58).

Krämer, W., & Langenbach-Belz, M. (1976) Approximate formulae for the delay in the queueing system GI/G/1. In: *Proc. 8th Int. Teletraffic Congr.* (ITC), Melbourne 1976, Congressblock, 235/1–8.

Krampe, H., Kubat, J., & Runge, W. (1974) *Bedienungsmodelle.* Verlag Die Wirtschaft, Berlin.

Kreinin, A. Ya. (1979) Bounds on the mean waiting time in single-server systems with warming-up (in Russian). *Izv. Akad. Nauk SSSR Tehn. Kibernet.* **1979**, 3: 207–210 (= *Engrg. Cybernetics* **17**, 3: 143–147).

— (1980) Bounds for mean characteristics of $E_k/GI/1/\infty$ queues. *Math. Operat.-forschung. Statist., Ser. Statist.* **12**, 515–519.

Kreinin, A. Ya. (1981) Bounds for mean characteristics of systems with Erlangian input and arbitrary service (in Russian). *Izv. Akad. Nauk SSSR Tehn. Kibernet.* **1981**, 3: 187–192. (= *Engrg. Cybernetics* **19**).

Krupin, B. G. (1974) Continuity of characteristics of many-line queueing systems with limited sojourn time (in Russian). *Izv. Akad. Nauk SSSR Tehn. Kibernet.* **1974**, 6: 132–138 (= *Engrg. Cybernetics* **12**, 6: 99–105).

— (1975) Continuity theorems for many-line queueing systems with bounded renewals (in Russian). In: *Trudy Mass. Obslush.*, Trudy Seminar Verojat. met. Tehn., Izdat MGU Moskva, 36–40.

LANGBERG, N. A., LEON, R. V., & PROSCHAN, F. (1978) Characterization of partially ordered classes of distributions. *Tech. Report* M 481, Statistics Dept., Florida State Univ., Tallahassee.

— (1980) Characterization of non-parametric classes of life distributions. *Ann. Prob.* **8**, 1163–1170.

LEE, A. M., & LONGTON, P. A. (1957) Queueing process associated with airline passenger check-in. *Operat. Res. Quart.* **10**, 56–71

LE GALL, P. (1962) *Les Systèmes avec ou sans Attente et les Processus Stochastiques.* Dunod, Paris.

LEHMANN, E. L. (1955) Ordered families of distributions. *Ann. Math. Statist.* **26**, 399–419.

— (1959) *Testing Statistical Hypotheses.* Wiley, New York.

LEVHARI, D., PARONSH, J., & PELEG, B. (1975) Efficiency analysis for multivariate distributions. *Rev. Econ. Stud.* **42**, 87–91.

LINDLEY, D. V. (1952) The theory of queues with a single server. *Proc. Camb. Phil. Soc.* **48**, 277–289.

LISEK, B. (1978) Comparability of special distributions. *Math. Operat.-forschung Statist., Ser. Statist.* **9**, 587–598.

— (1981) Stability theorems for queueing systems without delay, *Elektron. Informationsverarb. Kybernet.* **17**, 261–281.

LORDEN, G. (1970) On excess over the boundary. *Ann. Math. Statist.* **41**, 520–527.

LOULOU, R. (1973) Multi-channel queues in heavy traffic. *J. Appl. Prob.* **10**, 769–777

— (1974) On the extension of congestion theorems to multi-channel systems. In: *Mathematical Methods in Queueing Theory* (Ed. A. B. CLARKE), Lect. Notes Econ. Math. Syst. **98**, Springer-Verlag, Berlin, 185–197.

— (1978) An explicit upper bound for the mean busy period in a GI/G/1 queue. *J. Appl. Prob.* **15**, 452–455.

LOYNES, R. M. (1962) The stability of a queue with non-independent interarrival and service times. *Proc. Camb. Phil. Soc.* **58**, 494–520.

MAALØE, E. (1973) Approximation formulae for estimation of waiting-time in multiple-channel queueing systems. *Management Sci.* **19**, 703–710.

MACK, C. (1967) The pitfall in assuming exponential service times in queues. *New J. Statist. Operat. Res.* **3**, 17–21.

MAL'CEVA, N. I. (1970) Approximation of continuous distributions by a mixture of Erlang distributions (in Russian). *Problemy Pered. Inform.* **6**, 3: 89–92 (= *Problems Inform. Transmission* **6**, 266–268). [Math. Rev. **48** # 1285].

MANN, H. B., & WHITNEY, D. R. (1947) A test of whether one of two random variables is stochastically larger than the other. *Ann. Math. Statist.* **18**, 50 bis 60.

MARCHAL, W. G. (1976) An approximate formula for waiting time in single-serve queues. *AIIE Transactions* 8, 473–474.

MARCHAL, W. G., & HARRIS, C. M. (1976) A modified Erlang approach to approximating GI/G/1 queues. *J. Appl. Prob.* **13**, 118–126.

MARSHALL, A. W., & OLKIN, I. (1979) *Inequalities: Theory of Majorization and Applications.* Academic Press, New York.

MARSHALL, A. W., & PROSCHAN, F. (1970) Mean life of series and parallel systems. *J. Appl. Prob.* **7**, 165–174.

— (1972) Classes of distributions applicable in replacement, with renewal theory implications. *Proc. 6th Berkeley Symp. Math. Statist. Probab. 1970* Vol. 1, 495–515, Univ. California Press, Berkeley.

MARSHALL, A. W., & SHAKED, M. (1982) A class of multivariate new better than used distributions. *Ann. Prob.* **10**, 259–264.

MARSHALL, K. T. (1968a) Some inequalities in queueing. *Operat. Res.* **16**, 651–665.

— (1968b) Bounds for some generalisations of the GI/G/1 queue. *Operat. Res.* **16**, 841—848.

— (1973) Linear bounds on the renewal function. *SIAM J. Appl. Math.* **24** 245—250.

Marshall, K. T., & Wolff, R. W. (1971) Customer average and time average queue lengths and waiting time. *J. Appl. Prob.* **8**, 535—542.

Matthes, K., Kerstan, J., & Mecke, J. (1978) *Infinitely Divisible Point Processes.* Wiley, New York.

Meilijson, I., & Nadas, A. (1979) Convex majorization with an application to the length of critical paths. *J. Appl. Prob.* **16**, 671—677.

Miller, D. R. (1979) Almost sure comparisons of renewal processes and Poisson processes, with application to reliability theory. Math. *Operat. Res.* **4**, 406—413.

Miyazawa, M. (1976a) Stochastic order relations among GI/G/1 queues with a common traffic intensity. *J. Operat. Res. Soc. Japan.* **19**, 193—208.

— (1976b) On the role of exponential distributions in queueing models. Preprint, Dept. Inform. Sciences, Tokyo Institute of Technology.

— (1977) Time and customer processes in queues with stationary inputs. *J. Appl. Prob.* **14**, 349—357.

Moran, P. A. P. (1960) The survival of a mutant gene under selection, II. *J. Austral. Math. Soc.* **1**, 485—491.

Mori, M. (1975) Some bounds for queues. *J. Operat. Res. Soc. Japan* **18**, 152—181.

— (1976) Transient behaviour of the mean waiting time and its exact forms in M/M/1 and M/D/1. *J. Operat. Res. Soc. Japan* **19**, 14—31.

Narayan, P. (1981) On bounds for probability generating functions. *Austral. J. Statist.* **23**, 80—90.

Näther, W. (1972) Minimaxschätzungen im linearen Regressionsmodell. *Math. Operat.-forschung Statist.* **3**, 475—482.

— (1975a) Semi-orderings between distribution functions and their application to robustness of parameter estimators. *Math. Operat.-forschung Statist.* **6**, 179—188.

— (1975b) Optimalitätseigenschaften von Schätzungen nach der Methode der kleinsten Quadrate. *Freiberger Forschungsheft* D 92, Grundstoffverlag, Leipzig.

— (1979) Monotony relations between distribution functions and their application to experimental design. *Math. Operat.-forschung Statist., Ser. Statist.* **10**, 47—53.

Newell, G. F. (1973) *Approximate Stochastic Behaviour of n-Server Service Systems with Large n.* Lect. Notes in Econ. Math. Syst. **87**, Springer-Verlag, Berlin.

Nechepurenko, M. I. (1971) On bounding the mean waiting time (in Russian). In: *System Simulation* (in Russian), Vol. 2, Novosibirsk, 25—37.

Niu, S.-C. (1977) Bounds and comparisons for some queueing systems. *Operat. Res. Centre Report* 77—32, Univ. California, Berkeley.

— (1980a) Bounds for the expected delays in some tandem queues. *J. Appl. Prob.* **17**, 831—838.

— (1981) On the comparison of waiting times in tandem queues. *J. Appl. Prob.* **18**, 707—714.

Niu, S.-C. (1980b) A single server queueing loss model with heterogeneous arrival and service. *Operat. Res.* **28**, 584—593.

Nozaki, S. A., & Ross, S. M. (1975) Approximations in multi-server Poisson queues. *Tech. Report,* Dept. Industr. Engrg. Operat. Res., Univ. California, Berkeley.

— (1978) Approximations in finite capacity multi-server queues with Poisson arrivals. *J. Appl. Prob.* **15**, 826—834.

O'Brien, G. L. (1975a) The comparison method for stochastic processes. *Ann. Prob.* **3**, 80—88.

— (1975b) Inequalities for queues with dependent interarrival and service times. *J. Appl. Prob.* **12**, 653—656.

— (1980) A new comparison theorem for solutions of stochastic differential equations. *Stochastics* **3**, 245–249.

Page, E. (1972) *Queueing Theory in OR.* Butterworths, London.

Page, E. (1982) Tables of waiting times for M/M/n, M/D/n and D/M/n and their use for giving approximate waiting times in more general queues. *J. Operat. Res.* **33**, 453–473.

Pfanzagl, J. (1964) On the topological structure of some ordered families of distributions. *Ann. Math. Statist.* **35**, 1210–1228.

Pledger, G., & Proschan, F. (1973) Stochastic comparisons of random processes, with application in reliability. *J. Appl. Prob.* **10**, 572–585.

Polyak, D. G. (1968) Some problems in the theory of multichannel queueing systems with constant servicing time. *Izv. Akad. Nauk SSSR Tehn. Kibernet.* **1968**, 2: 94–102 (= *Engrg. Cybernetics* **1968**, 2: 86–95).

Potthoff, G. (1962) *Verkehrsströmungslehre.* Transpress, Berlin.

Prabhu, N. U. (1965) *Queues and Inventories.* Wiley, New York.

Prokhorov, Ju. V. (1956) Convergence of random processes and limit theorems in probability theory (in Russian). *Teor. Verojat. Primen.* **1**, 177–238 (= *Theory Prob. Appl.* **1**, 157–214).

Proschan, F. (1982) Coherent structure theory: a survey. In: *Encyclopedia of Statistical Sciences* (Ed: S. Kotz, N. L. Johnson), Wiley, New York.

Quine, M. P. (1976) Bounds for the extinction probability of a simple branching process. *J. Appl. Prob.* **13**, 9–16.

Rao, C. R. (1973a) *Linear Statistical Inference and Its Applications.* 2nd ed. Wiley, New York.

— (1973b) *Linear statistische Methoden,* Akademie-Verlag, Berlin (= German translation of first English edition (1965)).

Reuter, H., & Riedrich, T. (1981) On maximal sets of functions compatible with a partial ordering of distribution functions. *Math. Operat.-forschung Statist., Ser. Optimiz.* **12**, 597–606.

Robillard, B., & Trahen, M. (1976) Expected completion time in PERT networks. *Operat. Res.* **24**, 177–182.

Rogozin, B. A. (1966) Some extremal problems in the theory of mass service. *Teor. Verojat. Primen.* **11**, 161–169 (= *Theory Prob. Appl.* **11**, 144–151).

Rolski, T. (1972) On some inequalities for GI/M/n queues. *Zastosowania Mat.* **13**, 43–47.

— (1976a) Order relations in the set of probability distribution functions and their application in queueing theory. *Dissertationes Math.* **132**.

— (1976b) A note on orders in the set of distribution functions. *Bull. Acad. Polon. Sci. Ser. Math. Astr. Phys.* **24**, 513–516.

— (1977) On some classes of distribution functions determined by an order relation. In: *Proceedings of the Symposium to Honour Jerzy Neyman,* PWN — Polish Scientific Publishers, Warszawa, 293–302.

— (1978) A rate conservation principle for stationary piecewise Markov processes. *Adv. Appl. Prob.* **10**, 392–410.

— (1979) A note on queues with a common traffic intensity. *Math. Operat.-forschung Statist., Ser. Optim.* **10**, 413–419.

— (1980) Queues with non-stationary input stream: ROSS's conjecture. *Adv. Appl. Prob.* **13**, 603–618.

Rolski, T. (1983) Comparison theorems in queues with dependent interarrival times. Proc. Séminaire Int. Modélisation & Meth. d' Evaluation de Performance. Paris 1983, Vol. 3, 105–134.

Rolski, T., & Stoyan, D. (1974) Two classes of semi-orderings and their application in the queueing theory. *Zeit. angew. Math. Mech.* **54**, 127–128.

ROSS, S. M. (1974) Bounds on the delay distribution in GI/G/1 queues. *J. Appl. Prob.* **11**, 417–421.

— (1978) Average delay in queues with non-stationary Poisson input. *J. Appl. Prob.* **15**, 602–609.

— (1979) Multivalued state component reliability systems. *Ann. Prob.* **7**, 379–383.

ROSSBERG, H.-J. (1965) Über die Verteilung von Wartezeiten. *Math. Nachr.* **30**, 1–16.

— (1968) Optimale Eigenschaften einiger Wartesysteme bei regelmäßigem Eingang bzw. konstanten Bedienungszeiten. *Zeit. angew. Math. Mech.* **48**, 395–403.

— (1970a) Bemerkungen zu einer Arbeit von K. T. Marshall über das Wartemodell GI/G/1. *Zeit. angew. Math. Mech.* **50**, 423.

— (1970b) Verschärfung eines Satzes von B. A. Rogozin über extremale Eigenschaften der Wartemodelle D/G/1 und GI/D/1. In: *Operationsforschung und Mathematische Statistik II* (Ed. O. BUNKE), Akademie-Verlag, Berlin, 25–30.

ROSSBERG, H. J., & SIEGEL, G. (1974a) Die Bedeutung von Kingmans Integralungleichungen bei der Approximation der stationären Wartezeitverteilung im Modell GI/G/1 mit und ohne Verzögerung beim Beginn einer Beschäftigungsperiode. *Math. Operat.-forschung Statist.* **5**, 687–699.

— (1974b) The GI/G/1 model with warming-up time. *Zastosowania Mat.* **14**, 17–26.

ROTHCHILD, M., & STIGLITZ, J. E. (1970) Increasing risk, I: A definition. *J. Econ. Theory* **2**, 225–243.

— (1971) Increasing risk, II: Its economic consequences. *J. Econ. Theory* **3**, 66–84.

RUDOLPH, P. (1976) Bayesian estimation in linear models under different assumption about the covariance structure: Variance components models, equicorrelated models. *Math. Operat.-forschung Statist.* **7**, 649–664.

SAKASEGAWA, H. (1977) An approximation formula $L_q \approx \alpha\varrho\beta/(1-\varrho)$. *Ann. Inst. Statist. Math.* **29**, 67–75.

SAKASEGAWA, H., & YAMAZAKI, G. (1977) Inequalities and an approximation formula for the mean delay time in tandem queueing systems. *Ann. Inst. Statist. Math.* **29**, 445–466.

SCHASSBERGER, R. (1970) On the waiting time in the queueing system GI/G/1. *Ann. Math. Statist.* **41**, 182–187.

— (1973) *Warteschlangen.* Springer-Verlag, Wien/New York.

— (1978) The insensitivity of stationary probabilities in networks of queues. *Adv. Appl. Prob.* **10**, 906–912.

SCHMIDT, V. (1976) A remark on the non-equivalence of two criteria of the comparability of stationary point processes. *Zastosowania Mat.* **15**, 33–37.

— (1978) On some relations between stationary time and customer state probabilities for queueing systems G/GI/s/r. *Math. Operat.-forschung Statist., Ser. Optim.* **9**, 261–272.

— (1980) Equalities and inequalities between various queue characteristics. *Elektr. Informationsverarb. Kybernet.* **16**, 466–469.

SENETA, E. (1967) On the transient behaviour of a Poisson branching process. *J. Austral. Math. Soc.* **7**, 465–480.

— (1969) Functional equations and the Galton-Watson process. *Adv. Appl. Prob.* **1**, 1–42.

SHIMSHAK, D. G. (1979) A comparison of waiting time approximations in series queueing systems. *Naval Res. Logist. Quart.* **26**, 499–509.

SHNEPS, M. A. (1974) *Numerical Methods in Teletraffic Theory* (in Russian) Svyas, Moskva.

SIEGEL, A. F. (1978) Random space filling and moments of coverage in geometrical probability. *J. Appl. Prob.* **15**, 340–355.

SIEGEL, G. (1974) Abschätzungen für die Wartezeitverteilungen und ihrer Momente beim Wartemodell GI/G/1 mit und ohne „Erwärmung". *Zeit. angew. Math. Mech.* **54**, 609–619.

SIEGEL, G., & WÜNSCHE, S. (1979) Abschätzungen der Erneuerungsfunktion. *Math. Operat.-forschung Statist., Ser. Optim.* **10**, 265–275.

SIEGMUND, D. (1976) The equivalence of absorbing and reflecting barrier problems for stochastically monotone Markov processes. *Ann. Prob.* **4**, 914–924.

SIMONS, G. (1980) Extension of the stochastic ordering property of likelihood ratios. *Ann. Statist.* **8**, 833–839.

SMITH, D. R., & WHITT, W. (1981) Resource sharing for efficiency in traffic systems. *Bell System Tech. J.* **60**, 39–55.

SOBEL, M. J. (1980) Simple inequalities for multiserver queues. *Management Sci.* **26**, 951–956.

SONDERMAN, D. (1979a, b) Comparing multi-server queues with finite waiting rooms, I: Same number of servers; II: Different number of servers. *Adv. Appl. Prob.* **11**, 439–447 & 448–455.

— (1980) Comparing semi-Markov processes. *Math. Operat. Res.* **5**, 110 to 119.

SPATARU, A. (1974) Further results in stochastically monotone Markov chain theory. *Rev. Roum. Math.* **19**, 493–501.

STIDHAM, S. (1970) On the optimality of single-server queueing systems. *Operat. Res.* **18**, 708–732.

— (1974) A last word on $L = \lambda W$. *Operat. Res.* **22**, 417–421.

STONE, C. J. (1972) An upper bound for the renewal function. *Ann. Math. Statist.* **43**, 2050–2052.

STOYAN, D. (1972a) Über einige Eigenschaften monotoner stochastischer Prozesse. *Math. Nachr.* **52**, 21–34.

— (1972b) Halbordnungsrelationen für Verteilungsgesetze. *Math. Nachr.* **52**, 315–331.

— (1972c) Monotonieeigenschaften stochastischer Modelle. *Zeit. angew. Math. Mech.* **52**, 23–32.

— (1972d) Verlustsysteme aus der Bedienungs- und Dammtheorie bei „unvollständigen Informationen" über Verteilungsfunktionen. *Math. Operat.-forschung Statist.* **3**, 49–59.

— (1972e) Abschätzungen für die Bedienungs- und Zuverlässigkeitstheorie. *Zastosowania Mat.* **8**, 173–185.

— (1972f) Ein Stetigkeitssatz für einlinige Wartesysteme der Bedienungstheorie. *Math. Operat.-forschung Statist.* **3**, 103–111.

— (1973a) Queueing systems of the type M/G/n under conditions of small demand (in Russian). *Kibernetika* (Kiev) **1973**, 2: 108–110 (= *Cybernetics* **9**, 313–316 (1975)). [Math. Rev. **52** # 4458].

— (1973b) Bounds for the extrema of the expected value of a convex function of independent random variables. *Studia Sci. Math. Hungar.* 8, 153–159.

— (1973c) Monotonieeigenschaften einliniger Bedienungssysteme mit exponentiellen Bedienungszeiten. *Aplik. Mat.* **18**, 268–279.

— (1974a) Some bounds for many-server queues. *Math. Operat.-forschung Statist.* **5**, 117–129.

— (1974b) Some monotonicity properties of Galton-Watson processes. *Biom. Zeit.* **16**, 27–30.

— (1976) Approximations for M/G/s queues. *Math. Operat.-forschung Statist.* **7**, 587–594.

— (1977a) Inequalities for multi-server queues and a tandem queue. *Zastosowania Mat.* **15** 415–419.

— (1977b) Further stochastic order relations among GI/GI/1 queues with a common traffic intensity. *Math. Operat.-forschung Statist., Ser. Optim.* **8**, 541—548.
— (1978a) Queueing networks — insensitivity and a heuristic approximation. *Elektron. Informationsverarb. Kybernet.* **14**, 135—143.
— (1978b) A further bound for the mean stationary waiting time in M/G/s. *Elektron. Informationsverarb. Kybernet.* **14**, 393—395.
— (1980) On some qualitative properties of the Boolean model of stochastic geometry. *Zeit. angew. Math. Mech.* **59**, 447—454.
Stoyan, D., & Stoyan, H. (1969) Monotonieeigenschaften der Kundenwartezeiten im Modell GI/G/1. *Zeit. angew. Math. Mech.* **49**, 729—734.
— (1972) Bedienungstheoretische Anwendung von Halbordnungsrelationen für Verteilungsgesetze. *Wiss. Zeit. Techn. Univ. Dresden* **21**, 519—523.
— (1974) Inequalities for the mean waiting time in single-line queueing systems. *Izv. Akad. Nauk SSSR Tehn. Kibernet.* **1974**, 6: 104—106 (= *Engrg. Cybernetics* **12**, 6: 79—81).
— (1976) Some qualitative properties of single-server queues. *Math. Nachr.* **70**, 29—35.
— (1977) Bounds for the probability of wait in M/G/s queues. *Zeit. angew. Math. Mech.* **57**, 549—552.
— (1980) On some partial orderings of random closed sets. *Math. Operat.-forschung Statist., Ser. Optim.* **11**, 145—154.
Stoyan, H. (1973) Monotonie- und Stetigkeitseigenschaften mehrliniger Wartesysteme der Bedienungstheorie. *Math. Operat.-forschung Statist.* **4**, 155—163.
Strassen, V. (1965) The existence of probability measures with given marginals. *Ann. Math. Statist.* **36**, 423—439.
Sunaga, T., Biswas, S. K. & Nishida, N. (1982) An approximation method using continuous models for queueing problems II. *J. Operat. Res. Soc. Japan* **25**, 113—127.
Sunaga, T., Kondo, E., & Bisnos, S. K. (1978) Approximate methods for queueing problems. *J. Operat. Res. Soc. Japan* **21**, 29—42.
Suzuki, T., & Maruta, M. (1970) Inequalities for two bulk queues and a tandem queue. *Mem. Defense Acad.* **10**, 81—102.
Suzuki, T., & Yoshida, Y. (1970) Inequalities for many-server queue and other queues. *J. Operat. Res. Soc. Japan* **13**, 59—77.
Takacs, L. (1962) *Introduction to the Theory of Queues.* Oxford Univ. Press, Oxford.
Takahashi, Y. (1977) An approximation formula for the mean waiting time of a M/G/c queue. *J. Operat. Res. Soc. Japan* **20**, 150—163.
Tan, H. H. (1979) Another martingale bound on the waiting-time distribution in GI/G/1 queues. *J. Appl. Prob.* **16**, 454—457.
Taylor, J. M. (1983) Comparisons of certain distribution functions. *Math. Operat.-forschung Statist., Ser. Statist.* **13**.
Tchen, A. H. (1980) Inequalities for distributions with given marginals. *Ann. Prob.* **8**, 814—827.
Tijms, H. C., van Hoorn, M. H., & Federgruen, A. (1981) Approximations for the steady-state probabilities in the M/G/c queue, *Adv. Appl. Prob.* **13**, 186—206.
Tong, Y. L. (1967) Some theorems on stochastically larger random vectors. *Ann. Math. Statist.* **38**, 959.
— (1980a) *Probability Inequalities in Multivariate Distributions.* Academic Press, New York.
— (1980b) Some inequalities for the mixtures of distributions. *Tech. Report*, Univ. of Nebraska, Lincoln.
Tsitsiashvili, G. Sh. (1972a) Stability of many-line queueing systems relative to disturbances of their distribution functions (in Russian). *Izv. Akad. Nauk SSSR Tehn. Kibernet.* **1972**, 2: 49—54 (= *Engrg. Cybernetics* **10**, 218—223).

— (1972b) Stability of many-phase queueing systems with respect to underlying distribution functions (in Russian). *Izv. Akad. Nauk SSSR Tehn. Kibernet.* **1972**, 3: 97–105 (= *Engrg. Cybernetics* **10**, 451–459).

— (1973) Stability in a scheme of inactive redundancy with renewal (in Russian). *Izv. Akad. Nauk SSSR Tehn. Kibernet.* **1973**, 1: 62–66 (= *Engrg. Cybernetics* **11**, 55–59).

— (1975) Piecewise linear Markov chains and analysis of their stability (in Russian). *Teor. Verojat. Primen.* **20**, 345–358 (= *Theory Prob. Appl.* **20**, 338–350).

— (1979a) A survey of recent results in the stability of queueing systems (in Russian). Appendix to the Russian translation of the original German editio.

— (1979b) Construction of the Lyapunov function for the study of stability of piece-wise linear Markov chains. *Izv. Akad. Nauk SSSR Tehn. Kibernet.* **1979**, 2: 74–84 (= *Engrg. Cybernetics* **17**, 2: 45–55.

Turnbull, B. W. (1973) Inequalities for branching processes. *Ann. Prob.* **2**, 457–47.

Vainshtein, A. D., & Kreinin, A. Ya. (1981) On a bound for the mean queue length in single server queues (in Russian). *Avtomatika i Telemeh.* **1981**, 11, 60–65.

van Doorn, E. A. (1980) Stochastic monotonicity of birth-death processes. *Adv. Appl. Prob.* **12**, 59–80.

van Zwet, R. W. (1964) *Convex Transformations of Random Variables.* Math. Centrum, Amsterdam.

Vasicek, O. A. (1977) An inequality for the variance of waiting times under a general queue discipline. *Operat. Res.* **25**, 879–884.

Veinott, R. (1965) Optimal policy in a dynamic, single product, non-stationary, inventory model with several demand classes. *Operat. Res.* **13**, 761–778.

Veraverbeke, N., & Teugels, J. L. (1975) The exponential rate of convergence of the distribution of the maximum of a random walk. *J. Appl. Prob.* **12**, 279–288.

Vickson, R. G. (1977) Stochastic ordering from partially known utility functions. *Math. Operat. Res.* **2**, 244–252.

Volodin, S. V., & Kolin, K. K. (1972) On approximating the distribution and mean time in a single-server queue (in Russian). In: *Syst. Raspr. Inf.*, Nauka, Moskva, 36–41.

Waldmann, K. H. (1980) Bounds for the renewal function. *OR Spectrum* **2**, 75–78.

Walk, H. (1975) Inverspositive Operatoren und Taubersätze in der Erneuerungstheorie. *Monatsh. Math.* **79**, 333–346.

Whitt, W. (1971) Weak convergence of first passage time processes. *J. Appl. Prob.* **8**, 417–422.

— (1974a) The continuity of queues. *Adv. Appl. Prob.* **6**, 175–183.

— (1974b) Heavy traffic limit theorem for queues: a survey. In: *Mathematical Methods in Queueing Theory* (Ed. A. B. Clarke), Lect. Notes Econ. Math. Syst. **98**, Springer-Verlag, Berlin, 307–350.

— (1976) Bivariate distributions with given marginals. *Ann. Statist.* **4**, 1280–1289.

— (1979) A note on the influence of the sample on the posterior distribution, *J. Amer. Statist. Assoc.* **74**, 424–426.

— (1980a) Uniform conditional stochastic order. *J. Appl. Prob.* **17**, 112–123.

— (1980b) The effect of variability in the GI/G/s queue, *J. Appl. Prob.* **17**, 1062 to 1071.

— (1980c) Some useful functions for functional limit theorems. *Math. Operat. Res.* **5**, 67–85.

— (1980d) Contiunity of generalized semi-Markov processes. *Math. Operat. Res.* **5**, 494–501.

— (1981a) Comparing counting processes and queues. *Adv. Appl. Prob.* **13**, 207–221.

— (1981 b) The stationary distribution of a stochastic clearing process. *Operat. Res.* **29**, 294—308.

— (1981 c) On stochastic bounds for the delay distribution in the GI/G/s queue. *Operat. Res.* **29**, 604—608.

— 1981 d) The renewal-process stationary-excess operator. *Preprint Bell Laboratories.*

— (1982 a) On the heavy traffic Limit theorem for GI/G/∞ queues. *Adv. Appl. Prob.* **14**, 171—190.

— (1982 b) Multivariate monotone likelihood ratio and uniform conditional stochastic order. *J. Appl. Prob.* **19**, 695—701.

— (1982 c) Existence of limiting distributions in the GI/G/s queue. *Math. Operat.* Res. **6**, **7**, 88—94.

— (1983 a) Minimizing delays in the GI/G/1 queue. *Operat. Res.* **31**.

— (1983 b) The Marshall and Stoyan bounds for IMRL/G/1 queues are tight. To appear in *OR Letters.*

WHITTLE, P. (1955) The outcome of a stochastic epidemic — a note on Bailey's paper. *Biometrika* **42**, 116—122.

WIECZOREK, A. (1978) Ordering probabilities on an ordered measurable space. *Fund. Math.* **101**, 53—78.

WILKINSON, R. I. (1956) Theories for toll traffic engineering in the U.S.A. *Bell System Tech. J.* **35**, 421—514.

WITTING, H. (1966) *Mathematische Statistik.* Teubner-Verlag, Stuttgart.

WOLFF, R. W. (1977 a) An upper bound for multi-channel queues. *J. Appl. Prob.* **14**, 884—888.

— (1977 b) The effect of service time regularity on system performance. In: *Computer Performance* (Ed. K. M. CHANDY & M. REISER), North Holland, Amsterdam, 297—304.

YAMAZAKI, G. (1981) An ordering relation of the blocking two-stage tandem queue to the reduced single server queueing system. *Ann. Inst. Statist. Math.* **33**, 115—123.

YECHIALI, U. (1977) On the relative waiting time in the GI/M/s and GI/M/1 queueing systems. *Operat. Res. Quart.* **28**, 325—337.

YU, O. S. (1974) Stochastic bounds for heterogeneous-server queue with Erlang service times. *J. Appl. Prob.* **11**, 785—796.

ZOLOTAREV, V. M. (1975 a) Qualitative estimates in problems of continuity of queueing systems (in Russian). *Teor. Verojat. Primen.* **20**, 215—218 (= *Theory Prob. Appl.* **20**, 211—213).

— (1975 b) On the continuity of stochastic sequences generated by recurrent processes (in Russian). *Teor. Verojat. Primen.* **20**, 834—847. (= *Theory Prob. Appl.* **20**, 819—832).

— (1976 a) On stochastic continuity of queueing systems of type G/G/1 (in Russian). *Teor. Verojat. Primen.* **21**, 260—279 (= *Theory Prob. Appl.* **21**, 250—269).

— (1976 b) Metric distances in spaces of random variables and their distributions (in Russian). *Mat. Sb.* (*N.S.*) **101** (**143**), 3: 416—454, 456 (= *Math. USSR. Sbornik* **30**, 3: 373—401).

— (1977 a) Ideal metrics in the problem of approximating distributions of sums of independent random variables (in Russian). *Teor. Verojat. Primen.* **22**, 449—465. (= *Theory Prob. Appl.* **22**, 433—449).

— (1977 b) Quantitative estimates for the continuity property of queueing systems of type G/G/∞ (in Russian). *Teor. Verojat. Primen.* **22**, 700—711 (= *Theory Prob. Appl.* **22**, 679—691).

— (1977 c) General problems of the stability of mathematical models. *Proc. 41st Session Int. Statist. Inst. New Delhi* 382—401.

— (1979) Ideal metrics in problems of probability theory and mathematical statistics. *Austral. J. Statist.* **21**, 193—208; Corrigenda **22**, 97.

Appendix 1. Tables

Criteria for $F \prec G$ for certain gamma, Weibull, Gumbel, normal, exponential, log-normal, and beta distributions, $\prec$ denoting $\leqq_d$, $\leqq_c$, $\leqq_{cv}$, or $\leqq_L$. (Recall that $\leqq_d$ implies $\leqq_c$, $\leqq_{cv}$, and (for distributions on $(0, \infty)$) $\leqq_L$, and $\leqq_{cv}$ implies $\leqq_L$; write f, m_F for density and mean of F. Not all the constraints are mutually exclusive.)

Density: $f(x)$	Density: $g(x)$	Constraints on parameters	Ordering
Normal: $\frac{1}{\sigma\sqrt{2\pi}} \exp\left(-\frac{(x-\lambda)^2}{2\sigma^2}\right)$	Normal: $\frac{1}{\tau\sqrt{2\pi}} \exp\left(-\frac{(x-\mu)^2}{2\tau^2}\right)$	$\sigma = \tau,\ \lambda \leqq \mu$ $\sigma < \tau,\ \lambda \leqq \mu$ $\alpha < \tau,\ \lambda \geqq \mu$	$F \leqq_d G$ $F \leqq_c G$ $F \geqq_{cv} G$
Exponential: $\lambda\, e^{-\lambda(x-\alpha)}$ $(x \geqq \alpha)$	Exponential: $\mu\, e^{-\mu(x-\beta)}$ $(x \geqq \beta)$	$\alpha \leqq \beta,\ \lambda \geqq \mu$ $\lambda \geqq \mu,\ \alpha + \lambda^{-1} \leqq \beta + \mu^{-1}$ $\alpha \leqq \beta,\ \alpha + \lambda^{-1} \leqq \beta + \mu^{-1}$	$F \leqq_d G$ $F \leqq_c G$ $F \leqq_{cv} G$
Exponential: $\lambda\, e^{-\lambda(x-\alpha)}$ $(x \geqq \alpha)$	Normal: $\frac{1}{\sigma\sqrt{2\pi}} \exp\left(-\frac{(x-\mu)^2}{2\sigma^2}\right)$ or Gumbel for maxima: $\mu^{-1} \exp(-x' - e^{-x'})$ or Gumbel for minima: $\mu^{-1} \exp(x' - e^{x'})$, $x' \equiv (x-\beta)/\mu$	For arbitrary $-\infty < \xi < \infty$, set $\lambda(\xi) = g(\xi)/(1 - G(\xi))$, $\alpha(\xi) = \xi + [\ln(1 - G(\xi))]/\lambda(\xi)$; $\lambda \leqq \lambda(\xi),\ \alpha \geqq \alpha(\xi)$. $\lambda > \lambda(\xi)$ or $\alpha < \alpha(\xi)$, $F(x) = G(x)$ for $x = \xi_1, \xi_2$ with $\xi_1 < \xi_2$: $\lambda^{-1} e^{-\lambda(\xi_1-\alpha)} \geqq \int_{\xi_1}^{\infty} (1 - G(x))\, dx$ $\xi_2 + \lambda^{-1} e^{-\lambda(\xi_2-\alpha)} - (\alpha + \lambda^{-1}) \leqq \int_{-\infty}^{\xi_2} G(x)\, dx$	$F \geqq_d G$ $F \geqq_c G$ $F \geqq_{cv} G$

Continued

Density: $f(x)$	Density: $g(x)$	Constraints on parameters	Ordering
Gamma: $\dfrac{\lambda^\alpha x^{\alpha-1}\,e^{-\mu x\beta}}{\Gamma(\alpha)}$	Gamma: $\dfrac{\mu^\beta x^{\beta-1}\,e^{-\mu x}}{\Gamma(\beta)}$	$\alpha \leqq \beta, \quad \lambda \geqq \mu$ $\alpha > \beta, \; \alpha/\lambda \leqq \beta/\mu$ $\lambda > \mu, \; \alpha/\lambda \geqq \beta/\mu$	$F \leqq_d G$ $F \leqq_c G$ $F \geqq_{cv} G$
Weibull: $\lambda\alpha x^{\alpha-1}\,e^{-\lambda x^\alpha}$ $m_F = \lambda^{-1/\alpha}\Gamma(1 + 1/\alpha)$	Weibull: $\mu\beta x^{\beta-1}\,e^{-\mu x^\beta}$ $m_G = \mu^{-1/\beta}\Gamma(1 + 1/\beta)$	$\alpha = \beta, \; \lambda > \mu$ $\alpha > \beta, \; m_F \leqq m_G$ $\alpha > \beta, \; m_F \geqq m_G$	$F \leqq_d G$ $F \leqq_c G$ $F \geqq_{cv} G$
Gamma: $\dfrac{\lambda^\alpha x^{\alpha-1} e^{-\lambda x}}{\Gamma(\alpha)}$ $m_F = \alpha/\lambda$	Weibull: $\mu\beta x^{\beta-1}\,e^{-\mu x\beta}$ $m_G = \mu^{-1/\beta}\Gamma(1 + 1/\beta)$	$\alpha < \beta$, or $\alpha = \beta$ and $\lambda > \lambda_0$: $0 < \beta < 1$, or $\beta = 1$ and $\mu \leqq \lambda$ $\beta > 1$, or $\beta = 1$ and $\mu > \lambda$: $m_F \leqq m_G$ $m_F \geqq m_G$ $\alpha = \beta$ and $\lambda = \lambda_0$: $0 < \beta < 1$ $\beta = 1$ $\beta > 1$ $\alpha > \beta$, or $\alpha = \beta$ and $\lambda < \lambda_0$: $0 < \beta < 1$, or $\beta = 1$ and $\mu < \lambda$: $m_F \leqq m_G$ $m_F \geqq m_G$ $\beta > 1$, or $\beta = 1$ and $\mu \geqq \lambda$	$F \leqq_d G$ $F \leqq_{cv} G$ $F \geqq_c G$ $F \leqq_d G$ $F = G$ $F \geqq_d G$ $F \leqq_c G$ $F \geqq_{cv} G$ $F \geqq_d G$
$\lambda_0 \equiv (\mu\beta\Gamma(\alpha))^{1/\alpha}$			
Gumbel: $\lambda^{-1}\exp(-x' - e^{-x'})$ $x' \equiv (x - \alpha)/\lambda$ $m_F = \alpha + \lambda\gamma$	Gumbel: $\mu^{-1}\exp(-x' - e^{-x'})$ $x' \equiv (x - \beta)/\mu$ $m_G = \beta + \mu\gamma$	$\lambda = \mu, \; \alpha \leqq \beta$ $\lambda < \mu, \; \alpha + \lambda\gamma \leqq \beta + \mu\gamma$ $\lambda < \mu, \; \alpha + \lambda\gamma \geqq \beta + \mu\gamma$	$F \leqq_d G$ $F \leqq_c G$ $F \geqq_{cv} G$

$\gamma \equiv$ Eulers gamma constant

Log — Normal: $\frac{1}{x\sigma\sqrt{2\pi}} \exp\left(-\frac{(\ln x - \lambda)^2}{2\sigma^2}\right)$ $m_F = \exp(\lambda + \sigma^2/2)$	Log — Normal: $\frac{1}{x\tau\sqrt{2\pi}} \exp\left(-\frac{(\ln x - \mu)^2}{2\tau^2}\right)$ $m_F = \exp(\mu + \tau^2/2)$	$\sigma = \tau,\ \lambda \leqq \mu$ $\sigma \leqq \tau,\ m_F \leqq m_G$ $\sigma \leqq \tau,\ m_F \geqq m_G$	$F \leqq_d G$ $F \leqq_c G$ $F \geqq_{cv} G$
Gamma: $\frac{\lambda^\alpha x^{\alpha-1} e^{-\lambda x}}{\Gamma(\alpha)}$	Log — Normal: $\frac{1}{x\sigma\sqrt{2\pi}} \exp\left(-\frac{(\ln x - \mu)^2}{2\sigma^2}\right)$	$\mu + 1 + \ln(\lambda\sigma^2) \geqq \alpha\sigma^2$	$F \leqq_d G$
WEIBULL: $\lambda\alpha x^{\alpha-1} e^{-\lambda x^\alpha}$	Log — Normal: $\frac{1}{x\sigma\sqrt{2\pi}} \exp\left(-\frac{(\ln x - \mu)^2}{2\sigma^2}\right)$	$\alpha\mu + 1 + \ln(\lambda\alpha^2\sigma^2) \geqq \alpha^2\sigma^2$	$F \leqq_d G$
Beta: $\frac{x^{p_1}(1-x)^{q_1}}{\int_0^1 u^{p_1}(1-u)^{q_1}\,du}$, $(0 < x < 1)$; $m_F = (p_1+1)/(p_1+q_1+2)$.	Beta: $\frac{x^{p_2}(1-x)^{q_2}}{\int_0^1 u^{p_2}(1-u)^{q_2}\,du}$, $(0 < x < 1)$; $m_G = (p_2+1)/(p_2+q_2+2)$	$p_1 \leqq p_2,\ q_1 \geqq q_2$ $p_1 \geqq p_2,\ q_1 \geqq q_2,\ m_F \leqq m_G$ $p_1 \geqq p_2,\ q_1 \geqq q_2,\ m_F \geqq m_G$	$F \leqq_d G$ $F \leqq_c G$ $F \geqq_{cv} G$
Beta: $\frac{x^p(\gamma-x)^q}{\int_0^\gamma u^p(\gamma-u)^q\,du}$, $(0 < x < \gamma)$ $m_F = (p+1)\gamma/(p+q+2)$	Exponential: $\lambda e^{-\lambda x}$ $m_G = 1/\lambda$	$p < 0,\ q \geqq 0$ $p = 0,\ \lambda\gamma \leqq q + 1$ $\left.\begin{matrix} p > 0, \text{ or} \\ p = 0 \text{ and} \\ \lambda\gamma > q + 1 \end{matrix}\right\} \begin{cases} m_F \leqq m_G \\ m_F \geqq m_G \end{cases}$	$F \leqq_d G$ $F \leqq_d G$ $F \leqq_c G$ $F \geqq_{cv} G$

Appendix 2. Glossary of notation

A.2.1. Index of symbols (in alphabetic order)

A	Inter-arrival time d.f., 74
α_n	n^{th} inter-arrival time, 74
B	Service time d.f., 74
β_n	Service time of nth customer, 74
c_Σ	System quantity, 39
$\mathscr{D}$	Set of d.f.s., 2
$\mathrm{D}[0, a]$	Right-continuous real functions on $[0, a]$ with left limits, 152
$D(P, Q)$	Dudley's metric for distributions P and Q, 156
d	Skorokhod metric, 152
δ	Root of equation in GI/M/s queue, 77
E	State space, 26
$\mathrm{Erl}\,(\lambda, \alpha)$	Erlang d.f. with mean α/λ, 7
$\mathrm{Exp}\,(\lambda)$	Exponential d.f. with mean $1/\lambda$, 6
H, H_A	Renewal function (of d.f. A), 130
K	d.f. of $U_n = \beta_n - \alpha_n$, 74
$\mathfrak{R}_c(E)$, $\mathfrak{R}_{cv}(E)$	Space of monotonic and convex (concave) functions on E
$L_W, L_W{}^0$	Mean stationary queue size at arrival (arbitrary) epoch, 78
Λ	Set of increasing continuous functions with fixed points at 0, a, 152
λ	Intensity of arrival process ($\lambda = 1/m_A$), 74
$\lambda(t)$	Hazard rate, time dependent arrival rate, 7, 112
$\lambda(F, G)$	Levy metric for d.f.s F and G, 155
m_A	Mean inter-arrival time, 74
m_B	Mean service time, 75
m_W	Mean waiting time, 75
$\mathscr{M}$	Sigma-algebra of subsets of E, 26
μ	Service rate per server ($\mu = 1/m_B$), 77
$\mathbb{N}$	Set of natural numbers 1, 2, ...
ν_t	Number of customers in system at time t, 77
$\nu_n{}^*$	ν_t at $t_n{}^* + 0$, the departure epoch of the nth customer, 79
$\nu_n{}^0$	ν_t at $t_n - 0$, the arrival epoch of the nth customer, 80
p_W	Stationary probability that a customer waits ($p_W = 1 - W(0)$), 75
$P_{\mathscr{M}}$	Set of probability measures on $\mathscr{M}$, 26
$p(x, B)$	Markov chain transition function, 62
$p(x, y)$	Transition d.f. for Markov chain, 64
$\pi(P, Q)$	Levy-Prokhorov metric for distributions P and Q, 155
ϕ, ϕ_n	Mapping functions for recursively defined sample functions, 38
Φ	Mapping function for finitely defined sample function, 38
Q, Q_k	Stationary distributions for Markov chains, 167
$\mathbb{R}, \mathbb{R}^n$	Real-line, n-dimensional Euclidean space
$\mathbb{R}^\infty$ $(\mathbb{R}_+{}^\infty)$	Space of all infinite sequences of (non-negative) real numbers
$\mathscr{R}_<$	Class of $\prec$-monotone functions, 4
ϱ	Traffic intensity ($\varrho = \lambda m_B$), 74
$\varrho(F, G)$ or ϱ	Generic metric for d.f.s F and G
$\varrho_c(F, G)$	Metric on upper tails of d.f.s F and G, 156
$\varrho_n, \varrho_\infty$	Metric on $\mathbb{R}^n$, $\mathbb{R}^\infty$
s	Number of servers
Σ	Generic stochastic model (system), 39

σ_F^2	Variance of d.f. F (especially, σ_A^2, σ_B^2)
$\sigma(F, G)$	Variation metric for d.f.s F and G, 156
$\boldsymbol{T}$, $\boldsymbol{T}_k$	Transition operators, 62
$\Theta_c(.)$	D.f. degenerate at c ("Dirac" d.f.), 2
$u(F, G)$	Uniform metric on d.f.s F and G, 156
V	Test function, 171
W	Stationary waiting time d.f., 75
$\boldsymbol{W}$	D.f. of stationary waiting-time vector, 101
W_n, $\boldsymbol{W}_n$	D.f. of w_n and $\boldsymbol{w}_n$, 101
w_n	Waiting time of nth customer, 75
$\boldsymbol{w}_n$	Waiting time vector at n^{th} customer arrival epoch, 100
$w_{i,n}$	i^{th} component of $\boldsymbol{w}_n$, 100
Z, Z_n	R.v.s defined on a system Σ, 38

A.2.2. Partial order relations

$\prec$	Generic partial ordering, 2
$\leqq_c$	Convex ordering, 8
$<_c$	Convexity relation of d.f.s, 14
$\leqq_{cv}$	Concave ordering, 11
$\leqq_c^\alpha$, $\leqq_{cv}^\alpha$	α-moment analogues of $\leqq_c$, $\leqq_{cv}$, 22
$\leqq_d$	Stochastic ordering, 4
$\leqq_D$	Ordering of random vectors via d.f. ordering, 27
$\leqq_g$	Probability generating function ordering, 24
$\leqq_K$	Ordering of random vectors via tail d.f. ordering, 27
$\leqq_L$	Laplace-Stieltjes transform ordering, 22
$\leqq_{st}$	Same as $\leqq_d$
$<_*$	Star-shaped ordering, 14

A.2.3. Distribution function operations

For any distribution function F with mean $m_F \equiv m$:

$\overline{F}(x)$	$= 1 - F(x)$
$F_R(x)$	$= m^{-1} \int_0^x \overline{F}(u)\, du \quad (F(0-) = 0)$
F^*	Laplace-Stieltjes transform
F^{*n}	n^{th} convolution power
F^{-1}	Inverse function (see p. 5)

A.2.4. Kendall's queueing system terminology A/B/s/n

A	Arrival (or, input) process description
B	Server description
s	Number of servers (identical)
n	Number of places in waiting room
A/B/s = A/B/s/∞	
D	Constant (deterministic) inter-arrival or service time
GI	General independent inter-arrival or service times (thus, arrival process is a renewal process)
G	General inter-arrival or service times, not necessarily independent
E_k	Erlang inter-arrival or service times with k phases (i.e., Erl (., k))
M	Exponential inter-arrival or service times (M = E_1)

Name Index

Subject Index

DRAPER and SMITH • Applied Regression Analysis, *Second Edition*
DUNN • Basic Statistics: A Primer for the Biomedical Sciences, *Second Edition*
DUNN and CLARK • Applied Statistics: Analysis of Variance and Regression
ELANDT-JOHNSON • Probability Models and Statistical Methods in Genetics
ELANDT-JOHNSON and JOHNSON • Survival Models and Data Analysis
FLEISS • Statistical Methods for Rates and Proportions, *Second Edition*
FRANKEN, KÖNIG, ARNDT, and SCHMIDT • Queues and Point Processes
GALAMBOS • The Asymptotic Theory of Extreme Order Statistics
GIBBONS, OLKIN, and SOBEL • Selecting and Ordering Populations: A New Statistical Methodology
GNANADESIKAN • Methods for Statistical Data Analysis of Multivariate Observations
GOLDBERGER • Econometric Theory
GOLDSTEIN and DILLON • Discrete Discriminant Analysis
GREENBERG and WEBSTER • Advanced Econometrics: A Bridge to the Literature
GROSS and CLARK • Survival Distributions: Reliability Applications in the Biomedical Sciences
GROSS and HARRIS • Fundamentals of Queueing Theory
GUPTA and PANCHAPAKESAN • Multiple Decision Procedures: Theory and Methodology of Selecting and Ranking Populations
GUTTMAN, WILKS, and HUNTER • Introductory Engineering Statistics, *Third Edition*
HAHN and SHAPIRO • Statistical Models in Engineering
HALD • Statistical Tables and Formulas
HALD • Statistical Theory with Engineering Applications
HAND • Discrimination and Classification
HARTIGAN • Clustering Algorithms
HILDEBRAND, LAING, and ROSENTHAL • Prediction Analysis of Cross Classifications
HOAGLIN, MOSTELLER, and TUKEY • Understanding Robust and Exploratory Data Analysis
HOEL • Elementary Statistics, *Fourth Edition*
HOEL and JESSEN • Basic Statistics for Business and Economics, *Third Edition*
HOLLANDER and WOLFE • Nonparametric Statistical Methods
IMAN and CONOVER • Modern Business Statistics
JAGERS • Branching Processes with Biological Applications
JESSEN • Statistical Survey Techniques
JOHNSON and KOTZ • Distributions in Statistics
Discrete Distributions
Continuous Univariate Distributions—1
Continuous Univariate Distributions—2
Continuous Multivariate Distributions
JOHNSON and KOTZ • Urn Models and Their Application: An Approach to Modern Discrete Probability Theory
JOHNSON and LEONE • Statistics and Experimental Design in Engineering and the Physical Sciences, Volumes I and II, *Second Edition*
JUDGE, HILL, GRIFFITHS, LÜTKEPOHL and LEE • Introduction to the Theory and Practice of Econometrics
JUDGE, GRIFFITHS, HILL and LEE • The Theory and Practice of Econometrics
KALBFLEISCH and PRENTICE • The Statistical Analysis of Failure Time Data
KEENEY and RAIFFA • Decisions with Multiple Objectives